Fuels and Petrochemicals Division 2020

Held at the 2020 AIChE Annual Meeting

Online
16 – 20 November 2020

ISBN: 978-1-7138-2310-0

Printed from e-media with permission by:

Curran Associates, Inc.
57 Morehouse Lane
Red Hook, NY 12571

Some format issues inherent in the e-media version may also appear in this print version.

Printed with permission by Curran Associates, Inc. (2021)

For permission requests, please contact AIChE
at the address below.

AIChE
120 Wall Street, FL 23
New York, NY 10005-4020

Phone: (800) 242-4363
Fax: (203) 775-5177

www.aiche.org

Additional copies of this publication are available from:

Curran Associates, Inc.
57 Morehouse Lane
Red Hook, NY 12571 USA
Phone: 845-758-0400
Fax: 845-758-2633
Email: curran@proceedings.com
Web: www.proceedings.com

TABLE OF CONTENTS

601066 Process Design of Light Naphtha Purification By Type III Deep Eutectic Solvents

Abstract:

Problem Statement: Sulfolane, though effective at removing benzene, toluene, and xylene (BTX) from naphtha, poses a serious risk to the environment. Type III deep eutectic solvents (DES) made from phosphonium or ammonium salts (hydrogen bond acceptors or HBA) paired with a hydrogen bond donor (HBD) may be able to replace sulfolane in the naphtha purification process through using similar absorption mechanisms. The experimental evaluation of Type III DES is problematic in that slight changes in concentration or functionality (e.g. changing the HBA from methyltriphenylphosphonium bromide to ethyltriphenylphosphonium) can have drastic effects on the properties of the DES. For the efficient screening of DES candidates for naphtha purification, a method that is based on knowledge of sulfolane absorption is needed. Sulfolane is an effective solvent for BTX removal from naphtha due to its high capacity for BTX (due to similar electron density distribution and favorable Van Der Waals interactions with BTX) and its high selectivity (due to its polarity differences with naphtha).

Proposed Solution: In this study, conductor like screening model (COSMO) was adopted to screen suitable HBA (quaternary ammonium and polyphenol phosphonium salts) and HBD (glycol, glycerol, and carboxylic acids) for the removal of BTX from light naphtha (modeled as pentane) based on selectivity and maximum capacity. The best performing DES where then evaluated in a technoeconomic analysis model and compared to sulfolane. A model process consisting of an absorber, desorption unit, solvent regeneration and heat recovery system was used with each of the selected DES's thermophysical properties to obtain mass and energy balances for a naphtha purification process. Processing parameters such as maximum temperature and pressure were kept similar to sulfolane, with temperatures below 100°C and pressures 2 atm. This process was designed partially in ASPEN. A sensitivity analysis evaluating the effect of key parameters such as capacity, selectivity, heat recovery, etc., was used to determine which variables should be prioritized in guiding future endeavors to modify physical absorption units for use with DES.

612551 Hinging Hyperplanes Crude Oil Mixing Model for Production Planning Optimization

Abstract:

Refinery feed, usually a mixture of several crude oils, is separated via crude distillation unit (CDU). Crude true boiling point (TBP) distillation curve determines the amount of products which can be obtained via CDU separation. TBP curve of the mixed crude feed is calculated by blending pseudocomponents from each individual crude; the blending calculation introduces a lot of nonlinear terms and makes it difficult to optimize the crude selection and the downstream refinery operation. This work proposes a continuous piecewise linear (CPWL) model to approximate the crude blending, thereby eliminating the nonlinear blending terms. A hinging hyperplanes (HH) model is employed to formulate the CPWL model. Additionally, its parameters are computed by a two-layer perceptron, which is built to simulate the HH model. This leads to a mixed-integer linear model suitable for crude oil selection and cut point optimization. The proposed model enables accurate computation of the mixed crude feed TBP curve, while eliminating nonlinear blending terms. Accuracy of the proposed approximation model has been tested on mixtures of two crudes and five crudes. It is shown that a high approximation accuracy is attained by having a sufficient number of hinges which correspond to the nodes. At the same time, the model still retains an easy to solve (small size) MILP problem in the feedstock selection optimization. Since pseudocomponent based mixing model leads to an MINLP formulation, its execution times are significantly higher than the hinging hyperplanes based mixing model. At the same time, the high accuracy of the hinging hyperplanes model ensures that its use in the planning models will lead to the same results as the pseudocomponent based crude mixing model. Hence, it is suitable for inclusion in refinery planning models dealing with optimal crude selection.

Selection and Comparison of thermic fluids in a fired heater

Avinashkumar V. Karre, Worley Group Inc, Baton Rouge, LA 70809-3433, United States

Abstract:

Hot fluid properties such as density, thermal conductivity, heat capacity, vapor pressure, and viscosity of hot oils such as 100N, ULSD, Dowtherm A, Dowtherm G, and Therminol are used to estimate their thermal and hydraulic performance parameters. Thermal performance parameters include the Reynolds number, Prandtl number, Nusselt number, and heat transfer coefficient. Hydraulic performance parameters include line pressure drop, friction factor, hydraulic horse power, velocity, and flow rate. Cost benefit evaluation based on the hydraulic performance parameters was performed for different hot oil. Even though the cost of 100N and ULSD per barrel is 23 times inexpensive than Therminol and Dowtherm, these cannot be used when operating temperatures are above 600°F due to thermal degradation. Also, though Therminol-71 has higher heat transfer coefficients, Dowtherm G is the optimal hot oil due to its lower operating costs and lower vapor pressures. This study shows a systematic approach for selection of hot oil fluid. This study can be utilized in optimization and debottlenecking studies of existing or new heater projects for wide varieties of temperature range.

Introduction:

Hot oil, also called thermic fluid, is a heat transfer fluid (HTF) used in oil refineries and chemical plants. A HTF transfers heat to the process fluid using sensible heat. There are many limitations to using steam as a heat source. The steam pressure needs to be high enough to get the high temperatures required for the process. A steam system also requires water treatment and condensate collection systems. Steam systems are also a common leak point within plants. On the other hand, steam is non-toxic and non-flammable, so there is zero negative impact to the environment. Hot oil can also leak but it does not require a high-pressure system and by selecting an appropriate HTF, high temperatures can easily be reached. The hot oil process is a closed loop process. HTF's undergo thermal degradation overtime due to the high temperature process and many times will need to be refilled by an external make-up. The hot oil system consists of a fired heater, circulation pumps, expansion tanks, filters, a hot oil cooler, process users, and a slop oil system. The respective process equipment is located close to its process user to minimize both the piping and the required hot oil volume. Efficient design of hot oil equipment and the proper selection of the HTF is crucial to the performance of the hot oil system [1]. The circulating hot oil systems can limit the capital expenditure depending on the chosen HTF. They are most advantageous to areas of the plant in which there are multiple users that require process temperatures above what can be achieved with high pressure steam. In some select cases, a circulating hot oil system can be utilized in which only one fired heater is required vs installing independent fired heaters for each of the process users.

Many HTFs are available from manufacturers based on various properties such as temperature range, density, thermal conductivity, heat capacity, vapor pressure, viscosity, corrosivity, flammability, flash point, fire point, thermal expansion, thermal stability, freezing point, flash point, and autoignition temperature. A few of the HTF manufacturers are Dow,

Therminol, and Duratherm. The main parameters such as heat transfer coefficient and line pressure drop are used to determine the suitability of the hot oil.

The refinery clent of Worley currently uses Ergon 100-N oil as the hot oil medium. 100-N degrades rapidly at an operating temperature between 575-600°F, causing fouling issues in the pumps and exchangers. A Worley client refinery condensate fractionation project required a reboiler duty of 48.1 MMBtu/hr. The additional duty for the project needed a comparative study of different HTF system for the service in the naphtha splitter reboiler since the plant experienced a degradation problems with the 100N hot oil. The project team evaluated different hot oils which remain stable and are expected to have a 20-year life span for operating temperatures at or above 680°F. This study evaluates the hot oils by comparing their heat transfer characteristics, hydraulic benefits, cost benefits, and other (vapor pressure, thermal stability, etc.) very important process parameters. The aims of the study are :

(1) Investigate thermal and hydraulic performance and suitability of 100N and ULSD hot oils at an operating temperature above 575-600°F.

(2) Compare different HTF's and recommend an optimal hot oil for operating temperature above 680°F.

It is important to note that several operating refineries and chemical plants are still using the same hot oil that was selected 40-50 years ago. These old plants could be running their hot oil systems very inefficiently and could be experiencing degradation problems as mentioned above. Any plant or refinery can use this approach to optimize solve a degradation problem and appropraitely select an optimal HTF suitable for their system.

Method parameters used for comparison

Ultra-low sulfur diesel (ULSD), Dowtherm A, Dowtherm G, and Therminol 72 are selected for comparison in this study because they exhibit the desired properties for the temperature range of 550-680°F. Note that ULSD also has been observed to break down fairly rapidly above 600°F similar to 100N. The following parameters are estimated for each hot oil fluid. The description and significance of each parameter is also mentioned below.

1.1 Vapor pressure: The vapor pressure is the pressure exerted by a pure component at equilibrium, at any temperature, when both liquid and vapor phases exist [2]. Vapor pressure increases with an increase in temperature. Low vapor pressures are preferred in hot services to avoid any vaporization within the system. A two-phase hot oil system has higher frictional losses and requires more pumping energy, which means there will be a higher operating cost. Also, a two-phase system requires additional piping supports which adds to the capital costs for the project. Vapor pressure values for Dowtherm [3,4] and Therminol are obtained from their websites [5]. Typical refinery diesel stream data and an ASPEN HYSYS simulation are used for estimating ULSD properties.

1.2 Fluid properties: Basic fluid properties such as heat capacity (Cp, Btu/lb-F), thermal conductivity (k, Btu/hr-ft-F), viscosity (μ, cP), and density (ρ,lb/ft3) are used to estimate the thermal and hydraulic process parameters. Fluid properties for Dowtherm [3,4] and Therminol are from their websites [5]. Typically, the refinery diesel stream data and an ASPEN HYSYS simulation are used for estimating USLD properties. All of these fluid properties play an important role in determining the thermal and hydraulic performance of a hot oil.

1.3 Absorbed heat duty and mass flow rate: The heat duty (Q) is the heat absorbed by the hot oil heater, which is 48.1 MMBtu/hr. Based on the process requirements, the supply temperature from the hot oil heater is 680°F and the return temperature is 551°F. The differential temperature is the difference between the supply and return temperatures (i.e. $\Delta T = 680°F - 550°F = 129°F$). The duty and differential temperature are used to estimate the mass flowrate of a hot oil through the hot oil system. The mass flow rate can then be used to estimate volumetric flow rate of a hot oil.

The heat equation (1) is,

$$Q = m \times Cp \times \Delta T \qquad ...(1)$$

Where m is the mass flow rate of the hot oil in lb/hr, Q is the heat absorbed, and Cp is the heat capacity of hot oil in Btu/lb-F, and ΔT is the differential temperature in °F.

1.4 Volumetric flow rates: The hydraulic equation (2) is used to convert the mass flow rate into the volumetric flow rate (VF). The volumetric flow rate can then be used to estimate the velocity and the Reynolds number.

$$VF(gpm) = \frac{m\,(\frac{lb}{hr})}{500.5 \times SG} \qquad ...(2)$$

Where SG is specific gravity $= \frac{\text{density of hot oil (lb/ft3)}}{\text{density of water (62.4 lb/ft3)}}$

1.5 Velocity, Reynolds number, and Prandtl number: The velocity through an 8" nominal diameter pipe (7.981" inner diameter) is estimated for each hot oil. This diameter is selected based on an approximate 10 ft/s velocity for liquids. The corresponding Reynolds (Re) and Prandtl (Pr) numbers are also calculated [6]. Re and Pr numbers are then used to estimate the heat transfer coefficients.

$$V\left(\frac{ft}{\text{sec}}\right) = \frac{VF\ (gpm)}{Area\ of\ tube} = \frac{VF\ (gpm) \times 1\ USG \times 1\ minute}{7.48\ ft^3 \times 60\ sec \times \frac{3.142}{4} \times (d/12)^2} \qquad ...(3)$$

Where d is the diameter of the hot oil tube in inches.

$$Re = \frac{\rho\left(\frac{lb}{ft^3}\right) \times V\left(\frac{ft}{sec}\right) \times d\left(\frac{inch \times 1\,ft}{12\,inch}\right)}{\mu\left(\frac{cP \times 0.000672 \frac{lb}{ft-sec}}{1\,cP}\right)}$$

...(4)

$$\text{Pr} = \frac{Cp\left(\frac{Btu}{lb-F}\right) \times \mu\left(\frac{cP \times 0.000672 \frac{lb}{ft-}}{1\,cP}\right)}{k\left(\frac{Btu}{hr-ft-F}\right)}$$

...(5)

1.6 Nusselt number and heat transfer coefficient: The Nusselt number (Nu) and the heat transfer coefficients (h) are based on the Dittus and Boelter equation [6]. Since d and 0.023 are constant for all of the hot oils, equation (6) is rearranged to equation (7) and is defined as the modified heat transfer coefficient, H. The net change in modified heat transfer coefficient (ΔH) is also defined in order to compare the heat transfer performance of the new hot oils with 100N. The required heat exchange area decreases as H and ΔH increase, because the heat transfer coefficient is inversely proportional to the surface area.

$$Nu = \frac{h \times d}{k} = 0.023 \times Re^{0.8} \times Pr^{0.4} \qquad ...(6)$$

$$\frac{h \times d}{0.023} = k \times Re^{0.8} \times Pr^{0.4} = H \qquad ...\ (7)$$

$$\Delta H = \left(\frac{heat\ transfer\ coefficient\ of\ hot\ oil - heat\ transfer\ cofficient\ of\ 100N}{heat\ transfer\ cofficient\ of\ 100N}\right) \times 100 \qquad ...\ (8)$$

1.7 Frictional factor and pressure drop: The friction factor (f) and line pressure drop (ΔP) are estimated per approximations of the Colebrook equation [7]. These parameters are used for estimating the hydraulic horsepower requirement and associated cost.

$$f = \frac{0.25}{\left[log\left(\frac{\varepsilon}{3.7 \times d\frac{inch \times 1ft}{12\ inches}} + \frac{5.74}{Re^{0.9}}\right)^2\right]} \qquad ...(9)$$

Where $\mathcal{E}$ is the pipe roughness factor of 0.0005 ft.

$$\Delta P\ (Psi) = \frac{f \times L\ (ft) \times V^2\ (\frac{ft}{sec})^2 \times \rho\left(\frac{lb}{ft^3}\right)}{d\left(\frac{inch \times 1\ ft}{12\ inc}\right) \times 2 \times g\ (\frac{ft}{sec^2}) \times 144} \quad \ldots (10)$$

Where g is acceleration due to gravity (32.17 ft/sec^2). ΔP per 100 ft of pipe ($\frac{\Delta P}{100\ FT}$) is estimated by substituting 100 ft for pipe length (L) in equation (10). ΔP per 2,000 ft of pipe is also estimated by using 2,000 ft for L.

1.8 Horsepower [7]: Equivalent hydraulic horsepower (HH) and the actual horsepower (AH) for the calculated line pressure loss was estimated using a pump equation (11a and 11b)

$$HH(hp) = \frac{VF\ (gpm) \times \Delta P(Psi)}{1714} \quad \ldots(11a)$$

Assuming a pump efficiency of 60%.

$$Actual\ horsepwer\ (AH, hp) = \frac{HH\ (hp)}{Pump\ efficiency} = \frac{HH\ (hp)}{0.6} \quad \ldots (11b)$$

1.9 Cost of horsepower: The cost of horsepower is estimated based on the cost of industrial electricity in client's location, and with the assumption of 8,000 hours of operation. The estimated costs are compared for the different hot oils.

1 hp = 0.7457 kw

Cost of industrial electricity in client's location, USA = 6.17 cents /kwh [8]

Industrial hours in a year = 8,000 hours

$$Cost = 0.7457 \times 6.17 \times AH \times 8{,}000 \times 0.01 = USD \quad \ldots (12)$$

1.10 Cost of thermic fluid:

The cost of all of the Dowtherm and Therminol products are very high ($1,600/bbl). 100N wax free oil and ULSD are normally less than $100/bbl – more like $70 /bbl on average.

Results and discussions

Table 1. Fluid properties of hot oils at 680°F and vapor pressures at 680°F and 551°F.

Thermic fluids	Density, lb/ft3	Viscosity, cP	Cp, Btu/lb-F	k, Btu/hr-ft-F	Vapor pressure, psia at 680°F	Vapor pressure, psia at 551°F
100N	40.19	0.20	0.746	0.02756	37.8	15.16
ULSD	35.13	0.10	0.762	0.02358	133.9	50.85
Dowtherm A	45.94	0.15	0.608	0.04870	90.4	27.51
Dowtherm G	49.10	0.21	0.651	0.05404	**48.8**	**14.35**
Therminol 72	48.10	0.16	0.591	0.05680	68.0	20.00

From Table 1, 100N has the lowest vapor pressure at 680°F, but is not suitable at the operating temperature. 100N degrades rapidly above 600°F, which causes fouling issues within the pumps and exchangers. ULSD has the highest vapor pressure which means the system pressures need to be higher in order to avoid two-phase conditions within the system. Dowtherm G has the lowest vapor pressure at both 680°F and 551°F, second only to 100N. As shown in Table 1 below, the viscosities for all of the hot oils are relatively low when compared to water (1 cP). A lower viscosity leads to lower energy requirements for circulation within the system.

Table 2. Estimated mass and volumetric flow rates of different hot oils

Thermic fluids	Absorbed heater duty required, MMBtu/hr	m, lb/hr	VF, gpm
100N	48.1	499,823	1,551
ULSD	48.1	**489,328**	1,737
Dowtherm A	48.1	613,270	1,664
Dowtherm G	48.1	572,762	**1,454**
Therminol 72	48.1	631,911	1,635

Table 3. Hydraulic performance of different hot oils

Thermic fluids	F	ΔP/100ft	ΔP for 2,000 ft	HH, hp	AH, hp	USD/year
100N	0.0186	**1.2**	**24.01**	**21.7**	**36.2**	**13,312**
ULSD	0.0185	1.31	26.2	26.5	44.2	16,257
Dowtherm A	0.0185	1.57	31.46	30.6	51	18,772
Dowtherm G	0.0186	**1.29**	**25.78**	**21.9**	**36.5**	**13,435**
Therminol 72	0.0185	1.59	31.81	30.4	50.7	18,649

Tables 2 and 3 show the hydraulic performance of different hot oils. The frictional factors for all of the hot oils are close to each other. ULSD requires the lowest mass flow rate. 100N requires the lowest volumetric flow rate, but is not selected for use due to known fouling issues. Dowtherm G requires the lowest volumetric rate. Dowtherm G also has lower in ΔP/100FT, line pressure losses, and hydraulic horse power requirement compared to the other hot oils. The lowest cost found for Dowtherm G shows that approximately $13,435 is needed per year, which is about 16% lower than ULSD and about 39% lower than both Dowtherm A and Therminol 72. Lowest pumping cost for the Dowtherm G is due to the lowest volumetric flow rate compared with other hot oils.

Table 4. Thermal performance of different hot oils

Thermic fluids	V, ft/s	Re	Pr	H	ΔH%
100N	9.9	1.95E+06	0.003703	316	-
ULSD	11.1	4.06E+06	0.002075	386	22
Dowtherm A	10.7	3.24E+06	0.001258	545	73
Dowtherm G	9.3	2.16E+06	0.0017	493	56
Therminol 72	10.5	3.13E+06	0.00112	**589**	**87**

As shown in Table 4, Dowtherm G has a higher heat transfer coefficient than both ULSD and 100N, but is still slightly lower than Dowtherm A. Therminol 72 has a slightly higher heat transfer coefficient than both Dowtherm A and Dowtherm G. The higher the heat transfer coefficient, the lower the required area. The capital cost related to equipment could be lower due to the reduced area requirement, but the operating cost (pumping) would still be much higher for Therminol 72, as shown in Table 3. Dowtherm A and Therminol 72 have higher ΔH values than Dowtherm G, but Dowtherm G has a lower operating cost and lower volumetric rates.

Every hot oil system needs a certain amount of hot oil volume in the system and frequent refill is required to account for losses and thermal degradation. Therminol and Dowtherm are about 23 times expensive compared with ULSD and 100N but they provide higher operating temperature required for the process.

Summary and conclusion

Fluid properties and cost data of hot oils can be used to determine their suitability. Properties such as temperature range, density, thermal conductivity, heat capacity, vapor pressure, and viscosity are used to estimate the hydraulic and thermal performance of a hot oil. The pressure drop through a line and the heat transfer coefficient are essential for the selection of a hot oil. The original hot oil, 100N, was not selected due to its poor fluid properties and degradation at a temperature above 600°F. The low vapor pressure of Dowtherm G is needed at a high operating temperature of 680°F, to keep the hot oil in the liquid phase. Other hot oils with a higher vapor pressure like Dowtherm A, ULSD, and Therminol-72, require a higher system operating pressure which increases the operating costs of pumping. Dowtherm G has the lowest line pressure drop and lowest cost per year, which minimizes the operating costs of pumping. Even though

Therminol-72 has the highest heat transfer coefficient, it is not preferred due to the higher pumping related operating cost. The cost of Dowtherm and Therminol per barrel is very high compared with 100N and ULSD. Dowtherm G was chosen for the project because of stability at very high temperature and lowest pumping cost compared with Therminol-72 and Dowtherm A. 100N and ULSD are not suitable for an operating temperature above 600°F.

Over a 10 – 20 year cycle, a hot oil system has exchanger leaks, flange leaks, solids building up in the system, etc. All of these can amount to substantial volume loss. Also, deinventorying the system for 10 year vessel inspections typically result in some fairly high system losses as some will ultimately get pushed to the flare, regardless of how well the plant minimize the losses. Large systems are harder to manage in a refinery system. Due to the sheer volume of the hot oil system, and the fact that the refinery frequently contaminate and purge high volumes of hot oil material (especially during an outage), the price rendered specialized heat transfer fluids (Dowtherm and Therminol) as infeasible. Refinery went with a superheated 450# steam on some other new project for this reason. If a hot oil system does lose an appreciable volumes of material, or simply need a fresh load, it is going to take at least a month for Dowtherm or Therminol to generate and supply the necessary volumes. The refinery still uses hot oil systems such as 100N and ULSD for operating temperature below 600°F. Through trial and error in the refinery, the plant noticed that if the existing system is operated near 575°F, and virtually no thermal degradation occurs. Making 100N and ULSD thermally stable below 575°F ideally. Although, 100N and ULSD are economically very attractive, these can only be used when the system operating temperatures are below 575°F.

This research can be utilized in any revamp, new grass-root refinery, or a chemical project where hot oil fluids need to be compared. It is very important to look for temperature range of process and vapor pressure is very critical parameter in selecting an appropriate hot oil fluid for a system. With the presented calculation approach, a defining heat transfer and hydraulics process parameters can be estimated to properly chose a hot oil type. Also, it is very important to note that, single parameter cannot conclude but rather analysis of more than one parameter can give a bigger picture of the system. Also, a refinery experience with operation of a particular hot oil fluid and together with a cost of hot oil plays a ctitical role in chosing a hot oil for the new projects.

References

[1] Mitra, S., A technical report on design of hot oil system, 2015, 10.13140/RG.2.1.4159.0882.

[2] James D.R., Speight G., Environmental Inorganic Chemistry for Engineers, 1st Edition, ISBN 9780128498910, 2017

[3] Dowtherm G

http://msdssearch.dow.com/PublishedLiteratureDOWCOM/dh_098b/0901b8038098b395.pdf?filepath=/heattrans/pdfs/noreg/176-01466.pdf&fromPage=GetDoc

[4] Dowtherm A

http://msdssearch.dow.com/PublishedLiteratureDOWCOM/dh_098b/0901b8038098b391.pdf?filepath=heattrans/pdfs/noreg/176-01463.pdf&fromPage=GetDoc

[5] Therminol 72

https://www.therminol.com/sites/therminol/files/documents/TF-13A_Therminol_72.pdf

[6] J.P. Holman, Heat Transfer, Tata McGraw-Hill publishing company limited, ISBN 0-07-049547-5

[7] Crane Company, Flow of fluids, Technical Paper Number 410, 2010, ISBN 1-40052-712-0

[8] Cost of electricity in Kentucky https://www.electricitylocal.com/states/kentucky/louisville/

612552 Product Tri-Section Based Crude Distillation Unit Model for Refinery Production Planning and Refinery Optimization

Abstract:

Crude distillation unit (CDU) is the start of the production process in a refinery. Accuracy of a CDU model has a very significant impact on the refinery production plan. High model accuracy is typically accomplished via nonlinear models, which leads to convergence difficulties when the entire refinery model is optimized. CDU model presented in this work is an MILP model with a modest number of binary variables, and it has accuracy on par with rigorous tray to tray CDU models. The model relies on the observation (Fu et al., 2016) that a line through the middle of the TBP curve of a CDU product (which is easily modeled via linear partial least squares) depends on the crude feed properties and the yields of the adjacent products. Novelty of the model presented in this work (product tri-section model) is that it does not require models of individual distillation towers comprising CDU, thereby leading to a much simpler model structure that is amenable to representation as a MILP. The model predicts the product true boiling point (TBP) temperatures and the bulk properties (e.g., specific gravity, sulfur content). High accuracy and mixed-integer linear structure make this model eminently suitable for the plant-wide refinery production planning optimization. Significant reduction in the computational effort required for the optimization of nonlinear refinery models is illustrated by comparison with previous work, while its high accuracy is illustrated by comparison with a rigorous tray to tray model in AspenPlus.

603506 Development of an Equation Oriented Model Applied to a Hydrodesulfurization (HDS) Process

Abstract:

The presence of impurities in refining units compromises the quality of its final products. For example, sulfur, nitrogen, oxygen and some metals when linked to organic compounds constitute the so-called oleolytic impurities. Among these, sulfur proves to be harmful because it is corrosive, increases the polarity of oils, changes the color and odor of final products, and is a contaminant to various catalysts in the refining process. Hydrotreating (HDT) remains one of the most effective processes in the removal of oleolytic impurities[1]. The hydrodesulfurization (HDS) process consists of treating the contaminated oil charge with sulfur through a hydrogenation reaction in the presence of catalysts under controlled operating conditions. In general, this process occurs in trickle-bed reactors operating in one or more stages under high pressure. The concentration of sulfur in the oil and the quality of the associated desired final product will define the treatment intensity and influence the overall performance of the process[2]. Therefore, a detailed phenomenological model in a process simulator is a useful tool to describe the hydrodesulfurization reactions and their particularities.

In this work, the implementation of an equation-oriented model of an HDS reactor will be presented. The first step corresponds to the development of a reaction kinetic block model to describe the hydrodesulfurization (HDS) process in the AVEVA SimCentral Simulation Platform. In its simulation structure, SimCentral allows total autonomy in the evaluation and development of new models, without a deep knowledge about computational programming. The HDS block is based on an already developed and validated phenomenological model[3,4] and has been implemented in SimCentral using the model writing functionality. The developed HDS block model will then be used as the base submodel for the second step that corresponds to a more detailed version of the HDT (Hydrotreating) trickle-bed reactor, where a set of typical reactions will be defined. The HDS developed model and the simulator block are validated using data from literature and from a large Latin America Refining Company.

References

[1] Swain, E. J. (2018) "US Refining Crude Slates Continue Towards Heavier Feeds, Higher Sulfur Contents", *Oil Gas Journal*, 96, pp. 43-48;

[2] Ancheyta, J.; Alvarez, A.; Leyva, C. (2016) "Hydrotreating of Oil Fractions" in Zeynep Ilsen Önsan and Ahmet Kerim Avci (ed.). *Multiphase Catalytic Reactors: Theory, Design, Manufacturing, and Applications*, pp. 295-327;

[3] Korsten, H.; Hoffmann, U. (1996) "Three Phase Reactor Model for Hydrotreating in Pilot Trickle Bed Reactors", *AIChE Journal*, 42, pp. 1350-1360;

[4] Tavernard, A.; Fernandes, T.; Bispo, H.; Brandão, A. Nicácio, J. (2020) "Three-Phase Trickle-Bed Reactor Model for Industrial Hydrotreating Processes" *submitted to Fuel Processing Technology.*

598502 Novel Representation of Saturated Acyclic Hydrocarbon Chains and Its Application in Stochastic Reconstruction Model

Abstract:

The stochastic reconstruction method(SR) is currently one of the most popular methods in the area of molecular reconstruction of petroleum fractions[1]. To generate pseudo mixtures, the SR method uses a building diagram to determine the homologous series and the carbon number of each molecule in the generation loop. Then the predefined molecules are selected based on their homologous series and carbon number. However, it is expensive and even not accessible to build the predefined molecules library for certain petroleum samples. Meanwhile, commonly used property estimation methods in molecular reconstruction process, for example, group contribution methods, cannot recognize full structural difference between molecular components. To solve these problems, a novel structural vector including eleven structural descriptors is proposed to represent the structure of saturated acyclic hydrocarbon chains. The novel structural vector can represent the uniqueness of molecular components at the level of property estimation[2]. A building diagram is designed to couple the novel representation method with stochastic reconstruction method[3]. The novel stochastic reconstruction model starts with predefined rings and eleven structural descriptors and can generate mixtures containing pseudo components, which are unique at the level of property estimation. With enough analytical knowledge, exact chemical structures of generated pseudo molecules can be transformed from the representation method in this work.

Eleven structural descriptors(SDs) are proposed as building increments of hydrocarbon chains, as shown in Table 1. $SD_{I\_to\_IX}$ are consistent with functional groups in group contribution method[2]. Therefore, contribution values on properties can be directly used for $SD_{I\_to\_IX}$. SD_X and SD_{XI} are created to count the number of partial overlapping sections between $SD_{I\_to\_V}$. Contribution values of SD_X and SD_{XI} can be derived from $SD_{I\_to\_IX}$. The number of eleven structural descriptors form the structural vector to represent the exact structure of saturated acyclic hydrocarbon chains at the level of property estimation.

We designed a building diagram to apply the novel structural vector into stochastic reconstruction method. Fig. 1 shows the difference between traditional SR method and the SR method in this work. In traditional SR method, building diagram just focuses on configuring structural attributes of molecules, including the number of rings and side chains and the length of each side chain. Then according to the determined structural attributes, predefined molecules in corresponding categories are selected. While in this work, we present different building strategies. In the step of configuring rings, we also determine the number of rings, including aromatic rings and aliphatic rings. In configuration of structures of paraffin and side chains, if they are isomeric, we determine the number of eleven structural descriptors, instead of the length of chains. For ring-containing components, it is necessary to determine the structure of binding sites of side chains. The hierarchy strategy can present the uniqueness of molecular components at the level of property estimation and simultaneously reduce the complexity of building procedure in molecular reconstruction.

A structural library of rings is predefined in this work. Due to the limits of boiling point temperature and carbon number of components, the library includes cyclopentane, cyclohexane and benzene.

There are six types of binding sites. For naphthenes, they are CH_{cyc}-CH_2, CH_{cyc}-CH and CH_{cyc}-C. For aromatics, they are aC-CH_2, aC-CH and aC-C. CHcyc-CH_3 and aC-CH_3 are not included because they are regarded as normal side chains. It is necessary to determine the structure of binding sites to calculate the properties of full molecules. Heteroatom-containing hydrocarbons are not considered in this work because the naphtha samples used in this paper only contain few content of sulfur. In this work, it is assumed that each generated mixture contains 5000 molecules.

Histogram and gamma PDFs are used in sampling process. For determining the type of molecules and whether the chain is normal, histogram is reasonable. Gamma PDFs are used in determining the number of rings and number of eleven structural descriptors. Since the chemical analysis suggests that weight fractions follow a gamma distribution against boil point temperature in homologous series, we impose the gamma PDFs on weight fractions in homologous series against boiling point temperature in this work.

An objective function is proposed by minimizing the relative error of the bulk properties of naphtha samples between experiment and mixing rules. Properties used in this work includes basic properties, distillation curve and detailed weight fractions of homologous series. The average properties of the generated mixture are calculated by mixing rules.

The new SR model were tested with 50 naphtha samples. Properties of the generated mixture in each run are very close to experimental values. The average relative errors of basic properties and distillation curve are all under 3.8%. The relative errors of basic properties and distillation curve for 50 tests are all under 4.95%. The model also shows good performance to predict detailed weight fractions of homologous series. This work decoupled the indispensable relationship between models and predefined molecules and can contribute to the study of molecular reconstruction under the shortage of analytical information. With enough prior knowledge, analytic information can also be coupled with this model.

1. Ren, Y., et al., *Molecular reconstruction: Recent progress toward composition modeling of petroleum fractions.* 2019. **357**: p. 761-775.
2. Hukkerikar, A.S., et al., *Group-contribution + (GC +) based estimation of properties of pure components: Improved property estimation and uncertainty analysis.* Fluid Phase Equilibria, 2012. **321**: p. 25-43.
3. Neurock, M., et al., *Monte carlo simulation of complex reaction systems: molecular structure and reactivity in modelling heavy oils.* Chemical Engineering Science, 1990. **45**(8): p. 2083-2088.

Table. 1 The eleven structural descriptors

Index	SD_I	SD_{II}	SD_{III}	SD_{IV}	SD_V	SD_{VI}
Structure	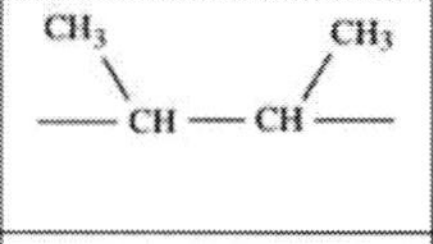	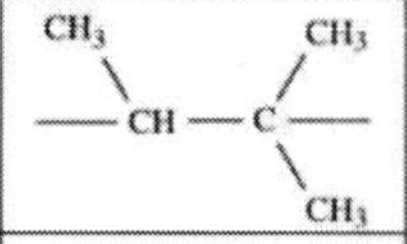	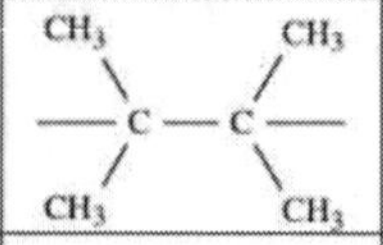	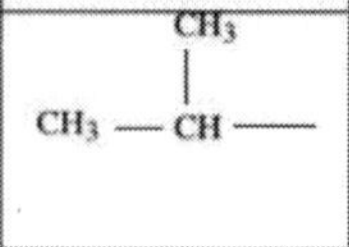	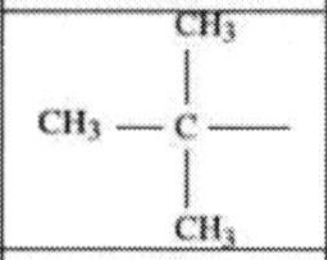	
Index	SD_{VII}	SD_{VIII}	SD_{IX}	SD_X	SD_{XI}	-
Structure	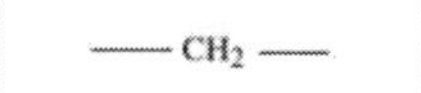	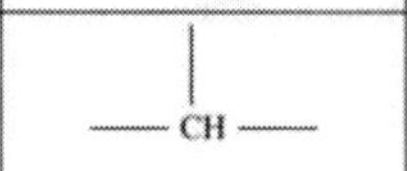	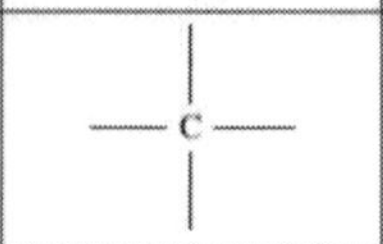	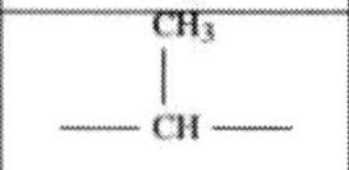	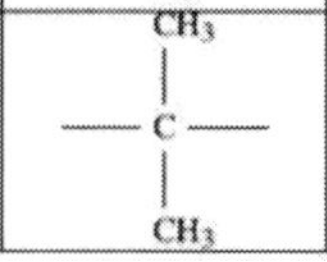	-

Fig. 1 Comparison of strategies to determine the exact structure between traditional SR models and this work

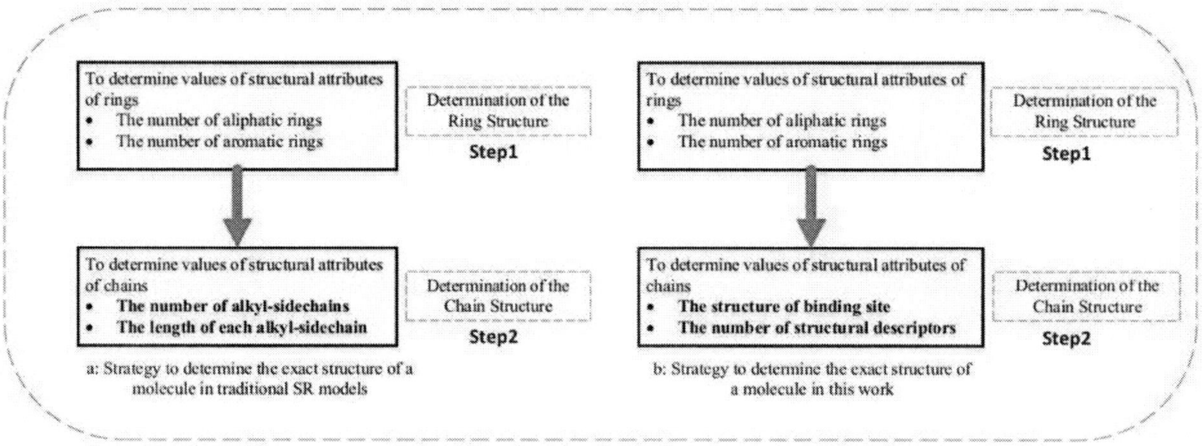

595980 Conceptual Design of the Fluid Catalytic Cracking Process for Maximizing the Yields of Olefins and Aromatics (FMOA)

Abstract:

Fluid Catalytic Cracking (FCC) is one of the core processes in petrochemical refineries. It plays a significant role in converting heavy oil to valuable products such as transport fuels and light olefins [1,2]. For the efficient utilization of oil resources, the high conversion of gas oil is commonly required, while the yields of the dry gas and coke must be restricted at low levels during FCC processes. However, the yields of the dry gas and coke always go up as the conversion of the gas oil increases. Especially when its conversion surpasses 80%, the yields of the dry gas and coke grow rapidly in an exponential manner[3]. In order to getting rid of high yields of the dry gas and coke, the conversion of gas oil must be maintained below 80% which can be operated under a moderate operating condition. Therefore, some researchers call the process as the moderate catalytic cracking (MCC) process [3]. The process is shown as the Case A process in the attached figure. It is obvious that in the MCC process, low yields of the dry gas and coke are obtained at the cost of the relatively low conversion of feedstock. For the purpose of promoting the conversion of the feedstock, the FCC gas oil (FGO), which is made up of the diesel and slurry, can be recycled back to the FCC riser reactor after being hydrogenated. It is believed that with the hydrogenation, the polycyclic aromatic hydrocarbons (PAHs) in the FGO can be converted into the mononuclear aromatics, olefins, or alkanes which can further be cracked readily into light hydrocarbons at moderate operation conditions [4]. This is the Case B process shown in the attached figure. Furthermore, as we all know, the ethylene and propylene are produced from the cracking reactions of the gas oil, gasoline and butylene during FCC processes. Compared to the gas oil and gasoline, relatively high temperature for the butylene cracking is commonly required, which inevitably gives rise to high yields of the dry gas and coke [5]. However, if the butylene is oligomerized firstly to be C8 and C12 olefins, the very components of the gasoline, then cracked into the propylene and ethylene in another reactor together with the gasoline olefins, which comes from the gasoline mixtures through solvent extractions, the yields of the dry gas and coke will decrease dramatically as the temperature for the cracking of the gasoline olefins is much lower than that for the butylene. At the meantime, the benzene, toluene and xylene (BTXs) can also be produced with the physical solvent extraction operation. This is the Case C process shown in the attached figure.

Table 1 in the attached figure compares the product yield distributions of the above three FCC processes. Since the operating temperature is a bit lower in the Case A process compared to conventional FCC processes, the conversion of the gas oil is only 78.13%. Meanwhile, the yields of the dry gas and coke are also much lower than that in conventional FCC processes. As for the Case B process in which the hydrogenated FGO is recycled to the FCC riser reactor, while the total yields of the dry gas and coke increase slightly from 8.58% to 9.78%, the yields of gasoline increase dramatically from 33.32% to 47.45%, and the yield of propylene increases from 18.56% to 19.87%. Regarding the Case C process, the yield of BTXs is predicted to be 30%. The estimated yields of the propylene and ethylene are 46% and 4.6%, respectively. Since the gasoline is not a target product for the Case C process, its yield is none, which indicates that FCC units can be designed and constructed to produce only petrochemicals (i.e., propylene, ethene, and BTXs) other than transport fuels. We call the Case C process as the FCC for Maximizing yields of the Olefins and Aromatics (FMOA) process.

The FMOA process integrates technologies of the MCC, FGO hydrogenation, solvent extraction and butylene oligomerization followed by re-cracking. The significance of the FMOA technology is that the flexibility of production plans has changed fundamentally. The FCC process can be designed to produce only basic chemicals (i.e., propylene, ethene, and BTXs).

References

1. Yupeng Du, Hui Zhao, An Ma, Chaohe Yang (2015), "Equivalent reactor network model for the modeling of fluid catalytic cracking riser reactor," Industrial & Engineering Chemistry Research, 54(35), pp. 8732-8742.
2. Yupeng Du, Lejing Sun, Abdallah Berrouk, et al. (2019), "Novel integrated reactor-regenerator model for the fluidized catalytic cracking unit based on an equivalent reactor network," Energy & Fuels, 33(8), pp. 7265-7275.
3. Youhao Xu and Shouye Cui (2018), "A novel fluid catalytic cracking process for maximizing iso-paraffins: From fundamentals to commercialization," Frontiers of Chemical Science and Engineering, 12(1), pp. 9-23.
4. Xin Zhou, Hui Zhao, Xiang Feng, et al. (2019), "Hydrogenation and TMP coupling process: novel process design, techno-economic analysis, environmental assessment and thermo-economic pptimization," Industrial & Engineering Chemistry Research, 58(24), pp. 10482-10494.
5. Jun Long, Youhao Xu, Jian Zhang, et al. (2011), "Consider new processes for clean gasoline and olefins production advanced technologies promote propylene yield while reducing olefins in gasoline," Hydrocarbon Processing, 97(9), pp. 85-91.

Acknowledgements

Financial supports from the National Natural Science Foundation of China (21908186), Shandong Provincial Natural Science Foundation (ZR2017LB022), and Shandong Province Higher Educational Science and Technology Program (J17KB075) are acknowledged.

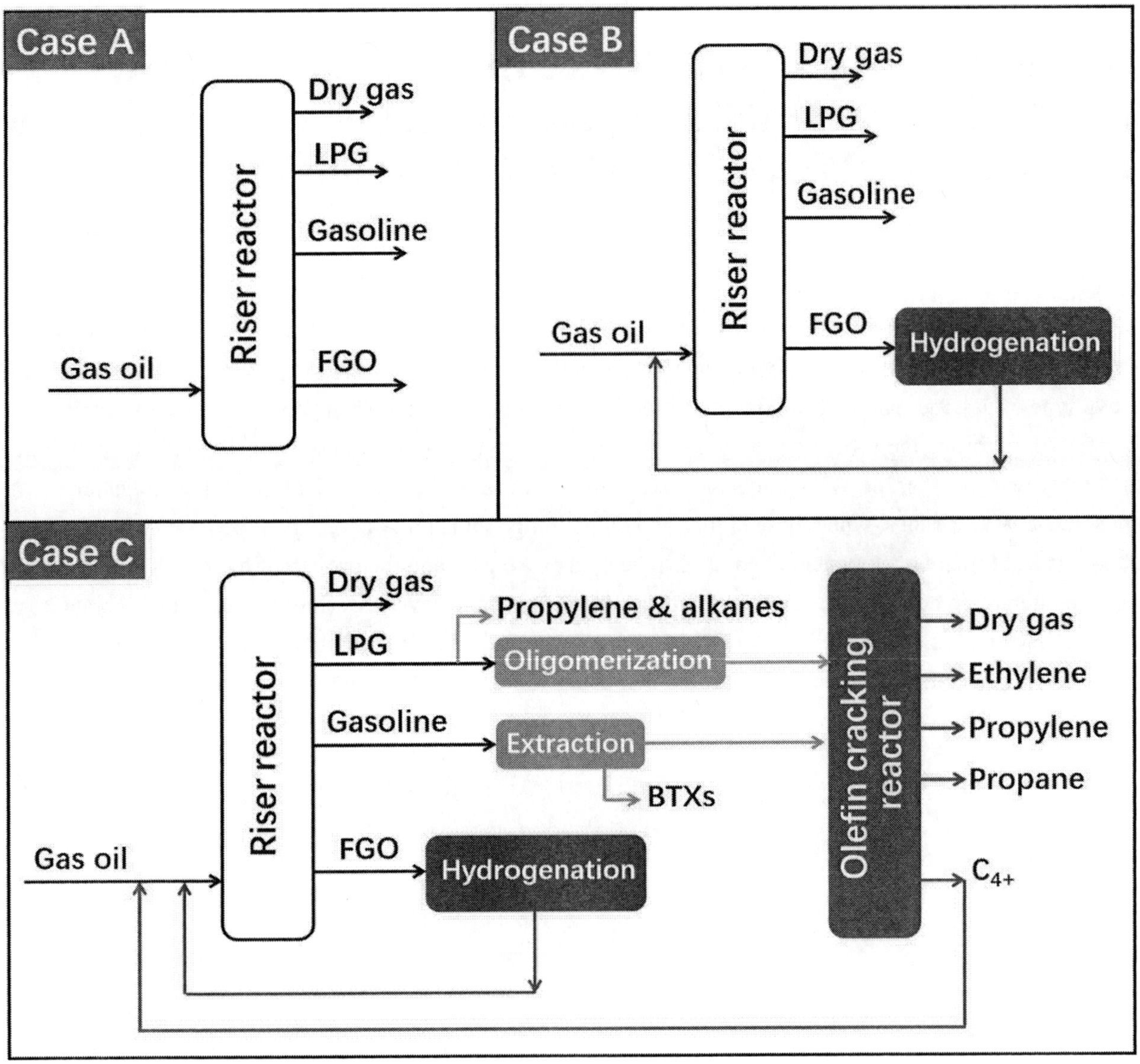

Table 1. Product yield distributions for the above three processes.

Products	Case A	Case B	Case C #
Dry gas	3.65	4.24	7.00
Ethylene	*1.80*	*2.66*	*4.60*
LPG	36.23	42.77	57.00
Propylene	*18.56*	*19.87*	*46.00*
Gasoline	33.32	47.45	***
BTX	*N/A*	*N/A*	*30.00*
FGO	21.87	***	***
Coke	4.93	5.54	6.00

Product yields for the Case C process is predicted by simulation tools.

606581 The Synergic Effect of Wax Inhibitor and Kinetic Hydrate Inhibitor on Inhibiting the Formation of Cyclopentane Hydrate in Highly Wax-Containing W/O Emulsion

Abstract:

In this work, a kind of co-polymer wax inhibitor EVA (Ethylene-vinyl acetate) and a type of kinetic hydrate inhibitor (KHI) PVP are doped into W/O emulsion system that containing high content of wax, aiming at testing the effect of two type of polymer on affecting the formation of cyclopentane hydrate respectively, as well as their synergic effect of inhibiting the formation of hydrate with wax in the W/O emulsion system. The results have shown that in the highly wax –containing W/O emulsion system, the wax can extend the critical time of cyclopentane hydrate to some extent. Besides, when these highly wax-containing emulsion systems are doped with EVA co-polymer, the critical time of cyclopentane hydrate is extended obviously, and the growth of hydrate is also inhibited because of the interaction between the EVA copolymer and the wax in the system, which can affect the crystallizing behavior of wax. Moreover, when the KHI is added in the aqueous phase of W/O emulsion, the inhibition of hydrate can be further strengthened, which has demonstrated that using wax inhibitor and the kinetic hydrate inhibitor simultaneously can produce synergic effect on inhibiting the formation and growth of cyclopentane hydrate.

606173 An Improved Method for Predicting Autoignition Temperatures Based on First Principles

Abstract:

Methods for predicting autoignition temperatures (AIT) have been historically inaccurate and are rarely based on the underlying physical phenomena leading to observed AIT. Previous attempts at predicting AIT have been developed based on Quantitative Structure-Property Relationships (QSPR) and additive group contribution methods. This work presents an improved method for predicting AIT based on the method originally developed by the late Dr. William H. Seaton. The method of Seaton is described in detail including its use of first principles to predict AIT. Improvements on the method of Seaton and underlying principles are presented and discussed.

611821 Water Washing of Startup Crude Oil

Abstract:

During startup of cold sub-sea wells, methanol is injected at the wellhead to prevent the formation of hydrates in the pipeline/riser system. Methanol injection continues until the fluid arrival temperature exceeds hydrate-formation temperature at the operating pressure, and the typical duration is a few days. Methanol contamination is problematic, and hence the methanol specification of the crude oil is usually less than 20 ppmw. Water or brine wash is often used to reduce the methanol content of the startup crude oil. This paper studies the design of these systems using process simulation. The enabling need is for a reliable and accurate thermodynamic model. The model must describe the three-phase system containing the aqueous and oil liquid phases and the vapor phase, and the key needed feature is description of the distribution coefficient of methanol between the oil and aqueous phases. A model has been developed based upon the ElectrolyteNRTL model in Aspen Plus, and the presentation describes the key data sources and the parameter development. The model enables design of an optimum process flowsheet, and a two-stage countercurrent has been determined to be the optimum option.

607019 Universal Correlation for the Binary Interaction Parameter of Hydrogen with Petroleum Fractions in Refinery Hydroprocessing Applications

Abstract:

Hydrotreating and hydrocracking applications in oil refineries use large quantities of hydrogen. Hydrotreating removes hetero-atoms of sulfur, nitrogen and oxygen and can saturate aromatic ring structures, producing more desirable transportation fuels, such as ultra-low sulfur diesel. Hydrocracking is a more severe reaction, which breaks carbon-carbon bonds and improves the ratio of hydrogen to carbon in the fluid, again, making more suitable transportation fuels. Hydrogen is expensive and is either made on-site, perhaps through steam reforming of methane, or is bought from outside as a utility.

To design new hydroprocessing facilities and to improve the performance of existing plant, it is essential to be able to predict the solubility of hydrogen in typical hydrocarbon fluids. The hydrogen has to be separated from the product streams in high and low pressure separators and is then recycled to the reactor feed, where it is combined with hydrogen make-up.

Although Detailed Hydrocarbon Analysis (DHA) enables us to define the composition of reactor feedstocks in terms of a larger slate of defined chemical components, it is still common practice to represent the heavier portion of the feed in terms of undefined components, or petro fractions, characterized by narrow cut boiling points and densities from a petroleum assay.

Cubic equations of state (EoS), such as Soave-Redlich-Kwong and Peng-Robinson are the workhorse thermodynamic methods for calculating vapor liquid equilibrium (VLE) in refinery processes. From the study of the VLE of binary systems of hydrogen as a solute in a defined hydrocarbon solvent, it is evident that a significant correction to the basic Van der Waals mixing rule in the EoS is required if we are to predict accurately the composition of hydrogen in the liquid phase. Moreover, we can show that this correction, through a binary interaction parameter (k_{ij}) needs to be temperature-dependent. While we can correlate k_{ij} for hydrogen binaries with defined components through regression of experimentally measured VLE data, the problem is how we can predict the interaction between hydrogen and any petro fraction? We observe that increasingly large values of k_{ij} are required as the molecular weight of the hydrocarbon increases and any petro fractions are likely to be in the higher molecular weight range. Ignoring the interaction parameter between hydrogen and a petro fraction will result in significant inaccuracy in any prediction of solubility.

In this study, we will show how the wealth of binary VLE data for hydrogen with defined hydrocarbon solvents can be reduced through regression and conversion to Henry's Law constants that correlate with the reduced temperature and critical pressure of the solvent [1,2]. We propose a new universal correlation for these Henry's Law constants that will allow the subsequent prediction for k_{ij} between hydrogen and a petro fraction, based only on knowledge of the critical temperature, critical pressure and acentric factor of the petro fraction that can be calculated from well-established estimation methods. The validity of this approach will be shown by comparing the predicted solubility of hydrogen in a well-analyzed diesel stream with a base case using methods currently described in the Technical Data Book of the American Petroleum Institute [3].

[1] Trinh TKH, de Hemptinne JC, Lugo R, Ferrando N, Passarello JP, "Hydrogen Solubility in Hydrocarbon and Oxygenated Organic Compounds." J. Chem. Eng. Data 2016,61,119-34

[2] Fernández-Prini R, Alvarez JL, Harvey AH, "Henry's Constants and Vapor–Liquid Distribution Constants for Gaseous Solutes in and at High Temperatures "J. Phys. Chem. Ref. Data, 3012, 32, 903-912

[3] API Tech Data Book 10, 2019

600821 Distillation Curve and Vapor Pressure Predictions of Alcohol-Gasoline Mixtures Using an Advanced Equation of State

Abstract:

Due to federal environmental regulations, such as the Energy Policy Act of 2005 in the United States, ethanol is frequently blended into gasoline in an effort to reduce tailpipe emissions. Moreover, there is a shift towards increasing the use of more sustainable energy resources such as biofuels, which includes ethanol, as transportation fuels. A fundamental understanding of how blending alcohol into a gasoline stream affects the properties of said stream is of critical importance. Specifically, the change in the distillation curve and vapor pressure is essential from a manufacturing and engine performance perspective. In the plant setting, there exists empirical correlations to predict these changes. However, these are typically very limited in scope as a correlation is only applicable to a specific plant gasoline mixed with a specific alcohol in a certain blend ratio.

Process simulators of petroleum process rely on equation of state (EoS) predictions of the phase behavior of petroleum pseudo-components defined by normal boiling point T_b, specific gravity *SG*, and *MW*. For gasoline blending these pseudo-components are typically generated from D86 curves. Hence, any method to be widely used in the prediction of alcohol / gasoline blending in refining processes should be accurate for petroleum pseudo-components.

In this work, we present a fundamental methodology for alcohol / gasoline blend property prediction using the polar PC-SAFT EoS[1] combined with the recently developed petroleum parameterization methodology, TAPPS (formerly referred to as EMPETRO)[2-6]. TAPPS is a highly accurate methodology used to determine hydrocarbon PC-SAFT parameters for petroleum species defined solely by T_b, *SG*, and *MW*. In addition to highly accurate predictions for pure component vapor pressures and densities, TAPPS accounts for polar features of petroleum such as the fact that aromatic species receive hydrogen bonds. This makes it well suited to describe alcohol / gasoline blending.

The approach presented here employs the base gasoline D86, average gasoline SG, and dry vapor pressure equivalent (DVPE) to generate gasoline pseudo-components. Using these pseudo-components, it is demonstrated the PC-SAFT / TAPPS approach allows for the quantitative prediction of the D86 and DVPE of alcohol / gasoline mixtures. We validate the approach by reproducing both literature[7,8] and ExxonMobil refinery data[9] for various blends. This work demonstrates the strength of PC-SAFT / TAPPS in predicting the physical properties of non-traditional blends of petroleum streams and lays the foundation for the development of a predictive blending tool that will be valuable when considering emerging biofuel markets.

References

[1] J. Gross; G. Sadowski, *Ind. Eng. Chem. Res.*, **40** (2001) *1244*.

[2] B. D. Marshall; C. P. Bokis, *Fluid Ph. Equilibria*, **489** (2019) *83*.

[3] B. D. Marshall; C. P. Bokis, *Fluid Ph. Equilibria*, **478** (2018) *34*.

[4] B. D. Marshall, *Fluid Ph. Equilibria*, **493** (2019) *153.*

[5] B. D. Marshall, *Fluid Ph. Equilibria*, **497** (2019) *79.*

[6] B. D. Marshall, *Fluid Ph. Equilibria*, **507** (2020) *112420.*

[7] V. F. Andersen; J. E. Andersen; T. J. Wallington; S. A. Mueller; O. J. Nielsen, *Energy Fuels,* **24** (2010) *2683.*

[8] V. F. Andersen; J. E. Andersen; T. J. Wallington; S. A. Mueller; O. J. Nielsen, *Energy Fuels,* **24** (2010) *3647.*

[9] J. R. Vella; B. D. Marshall, *Ind. Eng. Chem. Res.* (2020) *accepted.*

606540 Development of Mean Ionic Activity Coefficients for Modification of the Krichevsky – Ilinskeya Equation for Methylimidazolim – Type Ionic Liquids

Abstract:

Removal of H_2S from oil and gas streams has been a priority, oftentimes a necessity, for upstream crude oil/gas facilities for decades. However, the advances in production, namely horizontal drilling and hydraulic fracking, have produced oils and natural gases that contain higher concentrations of H_2S than previously seen. Crude oils containing > 0.5 % sulfur are considered sour and must undergo separation and/or conversion processes to lower the H_2S content prior to entering the midstream pipelines and railcars. H_2S scavengers can be categorized into two types: non-regenerative and regenerative. Non-regenerative scavengers, such as triazines, use an irreversible reaction to convert H_2S forming sulfur-containing hydrocarbons. With regenerative scavengers, such as alkanolamines, absorb H_2S using reversible absorption with stripping recovery processes.

Ionic Liquids (ILs) are a new family of absorbent materials that are being sought to replace amine-type reversible absorbents due to lower production of volatile organic carbons (VOCs) and reduced costs of the regenerative stripping phase. The ILs that have been chosen for this work were 1-butyl-3-methylimidazolium hexafluorophosphate, 1-butyl-3-methylimidazolium chloride, and 1-butyl-3-methylimida-zolium tetrafluoroborate. In this work, the mean ionic activity coefficient and the adjustable parameters for the Pitzer Ion Interaction Model have been determined. An infinite dilution procedure has been employed where a concentrated mixture of H_2S in IL was stripped using helium gas. The decrease in H_2S concentration (measured through gas chromatography) over time was used to determine the pertinent thermodynamic parameters.

In addition, the Henry's constant and partial molar volume of H_2S in IL were determined. The Krichevsky – Kasarnovsky equation has been shown to be applicable over a wide range of temperatures, pressures, and solute/solvent systems. However, in some cases (i.e. where T is near T_c) the molar volume ( ) is not constant, and there are strong deviations between the measured solubility and the calculated Henry's constants. From preliminary estimates, the Krichevsky – Ilinskeya equation (a modification of the Krichevsky – Kasarnovsky equation) does not model the phenomenon well. Ultimately, this work seeks to establish a new thermodynamic model that replaces the 2-suffix Margules component of the Krichevsky – Ilinskeya equation with the mean ionic activity coefficient found in Pitzer Ion Interaction model. The experimental results and model development will be presented.

612425 Catalytic Methane Pyrolysis in Polyvalent Molten Salt Mixtures

Abstract:

Methane pyrolysis (MP) for hydrogen production is one potentially cost-effective approach for the utilization of low-cost, natural gas without generation of CO_2. A major roadblock to industrial implementation is management of the stoichiometric solid carbon product; solid carbon quickly deactivates solid catalysts and there is currently no effective method for CO_2-free regeneration. Molten metal and molten salt environments do not suffer this deactivation by providing a continuously-renewed gas-liquid reaction interface and a means for continuous removal of the carbon via solid-liquid separations [1-3]. Molten salts (e.g., NaCl and KCl) are more attractive to industry; they are low-cost, non-toxic, and a trace residue in the solid carbon product may be acceptable for certain applications or disposal strategies. However, these monovalent molten alkali-halide salts (NaCl, KCl, NaBr, KBr), were found to have minimal activity for MP. Mixtures of these simple salts with higher-valency salts such as $MnCl_2$ and $FeCl_3$ resulted in considerable catalytic activity, which we attribute to the formation of polyvalent molecular ions, such as $[MnCl4]^{2-}$ and $[FeCl_4]^-$, which are present at the gas-liquid interface and catalyze methane activation and conversion.

In this talk we will report on the activity of salt mixtures such as $MnCl_2$-KCl, $FeCl_3$-NaCl-KCl, and others. We demonstrate with CH_4-D_2 exchange experiments that the catalytic transformation pathways on these complex salt surfaces is unique from established routes for either the gas-phase or on solid catalysts. We confirm the formation of polyvalent ions using molar volume (density) measurements as a function of bulk concentration, which reveal an excess molar volume when a polyvalent ion is formed in salt mixtures. More importantly, we confirm their presence at the gas-liquid interface in molten salt bubble column reactors using surface tension measurements. Anionic speciation is carried out using spectroscopy such as Raman, UV-vis, and XPS. Continued work is focused on developing an overall understanding of which molten salt polyatomic ions are active for MP based on their valency (acidity) and surface concentration.

References

1. Upham, D. C., Agarwal, V., Khechfe, A., Snodgrass, Z. R., Gordon, M. J., Metiu, H., McFarland, E. W. *Science* 358, 6365 (2017).
2. Palmer, C.; Tarazkar, M.; Kristoffersen, H. H.; Gelinas, J.; Gordon, M. J.; McFarland, E. W.; Metiu, H. *ACS Catal.* 9, 8337 (2019).
3. Kang, D.; Rahimi, N.; Gordon, M. J.; Mectiu, H.; McFarland, E. W. *Catal. B-Environ.* 254, 659 (2019).c

597525 Diesel Fuel Flashpoint Reduction and Microbial Induced Corrosion in Underground Fuel Tanks Due to Gasoline Contamination

Abstract:

Due to the large number of diesel flash point failures, the Georgia Department of Agriculture has shut down more than 200 retailer diesel service stations every year since 2014. Tremendous efforts have been devoted to find the root causes of the diesel flash point depression.

Automobile industries design their combustion engines based on the properties of the fuels they are intended to consume. Fuels are formulated with additives to meet the requirements of the engines. Flash point has been a key criterion in these properties. If the ignition is too slow, the engine may not combust the fuel appropriately, resulting in difficult startup or incomplete burning. On the other side, using a fuel with an ignition point that is too fast might result in premature ignition. Both conditions could cause severe engine damage. In addition, the flashpoint of a volatile material is the lowest temperature at which it can vaporize to form an ignitable mixture in air. Ignition may continue and lead to explosion. The lower the flashpoint, the more of a fire danger the material is. Flashpoint is one of the number of properties which must be measured in assessing the overall flammability hazard of a material. To ensure that a fuel meeting the performance and safety standards for the intended application, ASTM D975 specification for diesel fuels have been developed and published. Flash point minimum of 125°F is one of the key parameters in this specification.

Based on Georgia Department of Agriculture investigation, gasoline is likely the main source of contamination of diesel fuel to depress the flash point at the local retailers. The contamination is most likely due to the transport of diesel to the retailers through a gasoline contaminated truck compartment, which is known as switch loading (hauling gasoline prior to hauling diesel fuel). Gasoline contaminated diesel not only causes the flash point depression, but also may induce the microbial corrosion of underground fuel storage tanks.

In this presentation, we provide the technical information where the contamination has been introduced and the recommendations to resolve the issue.

612579 Process Analytical Technologies in Biofuel Research

Abstract:

Biofuels represent a promising solution to replace fossil fuels and especially second generation biofuels which derived from lignocellulosic feedstock such as forestry and/or agricultural residues widely available in countries such as Canada. However, the development of the second-generation biofuels by integrated biorefineries is facing several challenges. For example, the lignocellulosic biomass is highly heterogenous and standard laboratory characterization protocols (ASTM, TAPPI, NREL) require 8 to 10 days to be completed in order to determine the biomass chemical composition such as ash and moisture contents, extractives, cellulose, hemicelluloses and lignin. Unfortunately, this time frame is incompatible with the development of integrated biorefineries because the economic value of a shipment and the selected conversion process depend on the biomass chemical composition and a long storage time could degrade the biomass.

Recently, Process Analytical Technologies (PAT) have gained a strong interest with the development of new instrumentations. PAT can be defined as a tool box including multivariate analysis, data acquisition and analysis tools capable of analyzing, controlling and improving processes.

In the field of pharmaceuticals, PAT using spectroscopies such as Near-Infrared (NIR), Mid-Infrared (MIR) and Raman have been successfully applied to characterize industrial processes. These techniques are now being deployed in others fields such as biofuels. PAT are very promising because they could potentially allow characterizing reactants, intermediates and products which could facilitate the development of actual and new conversion processes and could reduce the time required for characterization and especially for biomass feedstock. In addition, recent works have also demonstrated the benefits of PAT in terms of waste reduction, energy efficiency and real-time analysis.

596830 Classical Density Functional Theory for Surface Charge Regulation of Oil-Brine-Rock Systems

Abstract:

Low salinity water-flooding (LSW) can enhance oil recovery by modifying the interfacial properties of reservoir systems, but the underlying mechanism and potential adverse effect are not well understood. Existing models can explain wettability changes in LSW processes with only limited success. Herein we report a coarse-grained model for oil-brine-rock systems within the classical density functional theory (cDFT) framework. In addition to the phase behavor of reservoir fluids, the coarse-grained model allows us to capture ionic size effect and electrostatic correlations essential for predicting the interfacial properties of oil-brine-rock systems. We demonstrate that our model is able to make quantitative predictions of surface charge density and zeta potential at rock-brine interface as a function of important parameters such as temperature, pH and brine composition.

597933 Properties and Phase Behavior of Some Heavy Oils from Kuwaiti Fields

Abstract:

Some heavy oils produced in Kuwait such as those from Marrat and Minagish Oil fields exhibit asphaltene precipitation issues which cause difficulties during production processes. Understanding the behavior and quantity of asphaltene deposition during the production processes such as EOR process depend on properties and phase behavior of the oil under the operating and reservoir conditions. In this paper some general data measured in laboratory for a live crude oil produced from these fields are presented. These properties include density, zero flash data, GOR, SARA analysis, bubble point which are used to determine the conditions that asphaltene precipitation in the field may occur. Parameters that influence stability of asphaltene in the crude are also discussed in this presentation.

610558 Absolute Adsorption of Methane in Shale Nanoporous Media: Challenges Arising from Dual-Heterogeneity

Abstract:

Accurate estimation of methane adsorption amount is essential for reliable gas-in-place (GIP) evaluation and well productivity prediction in shale media. While only excess adsorption isotherm can be obtained by experimental measurements, the absolute adsorption which represents the actual adsorption amount needs to be converted from the excess adsorption. In previous works, it has been shown that methane adsorbed phase density is dependent on pressure and temperature, indicating that the constant pre-determined adsorbed phase density may not be applicable for absolute adsorption calculation. The widespread pore size distribution (PSD) in shale reservoirs leads to varying adsorption mechanisms in micropores and mesopores. Our recent studies have shown that it is necessary to explicitly consider PSD in the estimation of absolute adsorption in kerogen from excess adsorption. On the other hand, shale consists of organic (kerogen) and inorganic matters (clay and quartz, etc), while both can contain an extensive number of nano-scale pores with drastically different pore surface properties. The pore size heterogeneity and rock heterogeneity, the so-called dual-heterogeneity, imposes a grand challenge in methane absolute adsorption estimation in shale media.

In this work, we use grand canonical Monte Carlo (GCMC) simulations to describe methane adsorption behavior in various sizes of organic and inorganic nanopores. For simplicity, we use carbon and illite slit nanopores to represent organic and inorganic nanopores. The methane adsorption in nanopores is divided into six and five adsorption types for carbon and illite slit pores, respectively, based on density profiles from GCMC simulation. The characterized adsorption models with lumping in terms of the corresponding pore size are applied in the Ono-Kondo (OK) model. The validity of our proposed OK model is examined by 1000 sets of artificially generated PSDs ranging from 0.7 to 30 nm. We find that by fitting the excess adsorption isotherm, OK model coupled with new adsorption models shows a good agreement with GCMC simulation in terms of absolute adsorption. The proposed model can faithfully capture the methane adsorption in organic-rich, clay-rich and organic-clay mixed (each contribution is comparable) shale nanoporous media. On the other hand, we find that without considering the dual-heterogeneity, the popularly used models, such as Langmuir and SDR, show noticeable deficiencies. Our work shows the applicability of OK model for accurate calculation of methane absolute adsorption in shale nanoporous media and the necessity of explicitly considering dual heterogeneity in shale gas-in-place estimation.

IGPL EXPERIENCE OF O4-66PLUS CATALYST REPLACEMENT WITH NEXT GENERATION CATALYST O4-88 FOR PRODUCTION OF PHTHALIC ANHYDRIDE

Rama Raghava Kumar Kotti

Panchi P M

IG Petrochemicals Limited (IGPL), Navi Mumbai, India

Biography:

Rama Raghava Kumar Kotti, worked as Head Plant Operations at IGPL, B.Tech in Chemical Engineering from Andhra University, having 30 years of experience in Fertilizers including 2 years in Petrochemicals sector. Commissioned Urea and Carbon dioxe recovery Plants and revamped Ammonia, Urea fertilizer plants. Written and published more than 12 papers on plant performance improvement, commissioning etc. Senior member in AIChE.

Panchi, President Production & Technology at IGPL, B.Tech from National Institute of Technology, Surathkal, Karnataka, India & having 35 years of experience in PA plants operation, project execution etc.

Introduction

IGPL has replaced the earlier O4-66 Plus catalyst with O4-88 in Phthalic Anhydride plant III for better results. Higher yield, more life and long term stable operation were few of the expectations before installing 04-88, compared to O4-66 plus. Based on detailed study conducted for nearly 300 days of Operation, this paper articulates anticipated benefits after replacement, actual results achieved so far and future recommendations.

M/s. IGPL is one of the largest manufacturers of Phthalic Anhydride (PA) product in India. The plant is located at Navi Mumbai, western coast of Maharashtra. IGPL has 3 Phthalic Anhydride plants, 3 Maleic Anhydride plants and one benzoic acid plant supported by all offsites and utilities. IGPl is right now commissioning one more additional streams of PA and MA plants. First PA plant is commissioned in 1992, second plant in 1996 and third plant in year 2013. PAIII plant has only oxidation section and crude PA produced by PAIII plant is processed by PA I and PA II distillation sections. The technology is supplied by M/s. PA consultants for PAIII plant and the process is based on using Ortho xylene as main raw material. IGPL is administering ISO 9002 and ISO 14001 systems.

I. Brief Process

PAIII reactor has around 15026 tubes with a tube length of around 3.7 meters. ID is 25 mm with a thickness of 5 mm. The plant designed for a loading of 100 g o-xylene / Nm^3 @ 4 Nm^3/h of air per reactor tube per hour. The plant is designed to produce crude Phthalic Anhydride out of o-xylene with a capacity of 6,672 kg/h equivalent to a name plate capacity of 50,000 t/a. First time catalyst was replaced in 2016 with 04-66 Plus.

The catalyst has given a normal life of around three years and after completion of three years in 2019, the catalyst was replaced with improved generation of catalyst i.e O4-88. Catalyst vendor details kept confidential for necessary protection.

Reactor tubes are filled with a multilayer catalyst. The reaction heat is absorbed by molten salt (mixture of sodium nitrite and potassium nitrate) circulated around the tubes (shell side of Reactor) from bottom to top. Two axial pumps are attached to the ring channels opposite to each other for circulating the salt melt through the reactor and ancillary equipment like, salt bath cooler (Heats Boiler feed water), steam super heater and electric start-up heater.

Both catalysts contain vanadium pentoxide and titanium dioxide as the main ingredients. The other components are patented by vendor. Main reaction is the formation of phthalic anhydride and water is as follows:

C8H10 + 3 02 ⇒ C8H403 + 3 H2O

O-xylene is heated and vaporised and Air is also heated and introduced in the reactor. To some extent there is a total combustion to carbon dioxide, carbon monoxide (referred as combustion products) and water. Other byproducts are maleic anhydride, benzoic acid, citraconic anhydride which are over oxidation products. Similarly, unconverted Ortho-Xylene, O-Toluyl Aldehyde and phthalide are under oxidation products.

In first few centimetres of reactor tubes the gas is heated up to reaction temperature. Then the exothermic reaction starts and the gas is further heated up by the reaction heat until a maximum is reached, the so-called hot spot temperature (HST). This same phenomenon is repeated in every layer.

Most of the conversion takes place in the top layer, where most of the reaction heat is released. This layer is decisive for the overall selectivity (PA conversion is desired reaction) above all others. The duty of the following layers is mainly to convert under oxidation by-products as well as unreacted ortho-xylene into PA.

II. Catalyst configuration

O4-66 Plus was 5 layer catalyst, where first two layers were selective layers (CL1 and CL2) and the last two layers were active layers (CL3 and CL4). In between inert layer was there. The purpose of inert layer is to reduce temperature of reaction gas by around 10 to 15 Deg C as formation total combustion products formation will be less in active layers. Higher yield and stable operation for longer period are few of the anticipated benefits from replacing O4-66 with next generation catalyst. Whereas O4-88 is a 5 layer catalyst, with 3 selective layers (CL0, CL1, CL2) and remaining are 2 active layers (CL3, CL4), which will give better product quality.

Refer the diagram below for pattern of different catalysts filled:

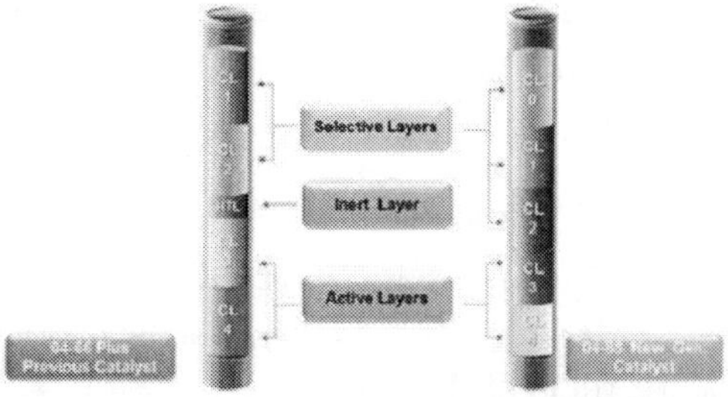

Figure.1 : Configuration of catalyst layers for both catalysts

III. Temperatures monitoring inside the Reactor

In order to effectively control the reactions in the reactor, thermocouples are provided inside the tubes. There are two types of temperature indications in the reactor. Mutli point (MP) temperature indications and single point temperature indications. The tubes which are having Multi point and single point thermocouples, are also filled with catalysts, which will represent the temperature conditions inside the reactor of other tubes.

Multi Point temperature indication: Several thermocouples which are very thin (special design) are provided in the catalyst tubes. These thermocouples are uniformly placed at different heights in a tube, which will enable operating people to know, at what height hot spot is prevailing and how it is moving down over a period of time. This movement is very carefully observed and trends are monitored. There are 8 multi point tubes which are filled with thermocouples in the reactor.

Single Point temperature indication: A single thermocouple is provided inside the tube at a particular height, which is specified by catalyst vendor. Every multi point is associated two single point tubes on either side.

IV. Methodology of performance comparison

Having the PA III plant run nearly 300 days with O4-88, realization of benefits attained so far and catalyst behaviour were examined in this paper. Hot spot profile pattern, Salt bath temperature, Yield, Plant load, under oxidation products, over oxidation products and combustion products were compared for knowing the differences in real behaviour of catalysts and possible prediction of stability. The reactor outlet gases are analysed on a continuous basis. Variations are monitored and parameters are adjusted based on requirement. This analysis has been done by comparing with O4_88 with O4-66 for the initial period of ~300 days in three stages.

1. **After completing 100 days of operation**
2. **Between 101 to 200 days of operation**
3. **Between 201 to 300 days of operation**

Also, the other plants experience was taken into consideration to the possible extent for comparing benefits with O4-88.

1. Comparison after 100 days of Operation

In the first year of operation, performance of O4-88 is similar to O4-66plus & about 1% higher yield is expected more from second year of operation. In both cases, catalyst was filled with care and differential pressures (DP) were within acceptable ranges. For O4-66 Plus, DP's for all the tubes were ranging from 56 mbar to 62 mbar and whereas for O4_88, DP's were ranging from 53 mbar to 58 mbar.

O4_88 and O4_66 Plus catalyst operating parameter data was compared for initial 100 days of operation. With O4-88, multi point (MP) temperature was 16 Deg c lesser than O4-66 Plus. Where as, single point (SP) temperature O4-88 was more by 16 Deg c. For the almost same load, SBT was kept low with O4_88 case.

Table.1 :

S.no	Description	O4-88	O4-66 Plus	Difference
1	Multi Point Temp, Deg C	424	440	-16
2	Single Point Temp, Deg C	431	415	16
3	Air flow per tube, Nm3/Hr	4.0	4.0	0
4	O-Xylene Load, g/Nm3	87	86.7	0.3
5	Salt Bath Temp, Deg C	350.3	354	-3.7
6	Diff.Pressure in Bar	0.339	0.346	-0.007

As per the general hot spot temperature(HST) guide lines provided by vendor for a load more than 80 g/nm3, after 3 to 6 months of Operation with O4_88, CL1 temperature shall be less than 440 Deg C and CL 0 temp was supposed to be less than 410 Deg C. In the first year of operation, CL0 temperatures shall be lower than CL1. CL1 temp was maintaining 16 Deg c less, but CL0 temp was more by around 15 Deg C indicating higher activity of CL0 due to unexpected plant interruptions. Refer the trends of temperature profile and yield below:

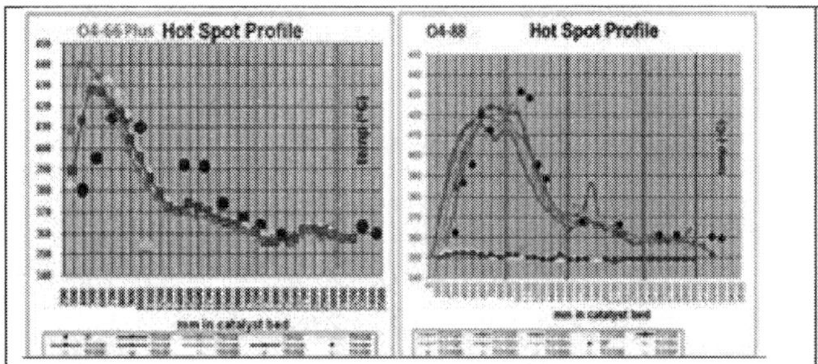

Figure.2: O4-66 Plus (previous catalyst) Vs. O4-88 (New Gen.Catalyst) Hot spot Profile comparison after 100 Days of Operation.

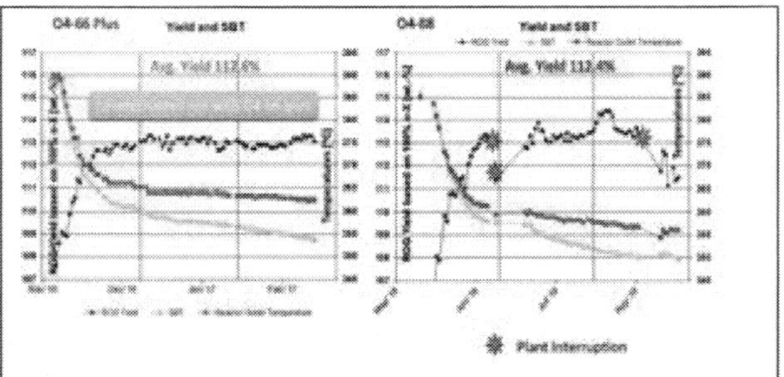

Figure.3: O4-66 Plus (previous catalyst) Vs. O4-88 (New Gen.Catalyst) Yield comparison after 100 Days of Operation.

In spite of low Salt Bath Temperature (SBT) and also plant outages caused by other than catalyst reasons, the yield remained almost same for both. O4_88 (yield was less by 0.2%.) relatively.

Yield was less for O4_88 due to obvious reasons of unexpected shut downs. Under oxidation products with O4-88 were consistently less. During this period, on an average 0.2% weight less compared to O4_66 Plus, especially O-xylene slippage was less by around 0.14%wt. with O4_88. But over oxidation products were almost same around 4.7% wt. for both types of catalysts. Combustion products were also relatively same at around 1.15%, figures were corrected to 40 gms load, which will provide more accurate results.

Major Observations during initial 100 days of Operation:

1. No major concerns with new catalyst, in fact HST's maintained relatively less with O4_88, compared to O4_66 Plus.
2. After restarting with O4_66 Plus in 2016, plant III did not trip for 196 days. The plant had started on 29.11.2016 and run up to 13.06.2017. Whereas after restarting the plant with O4-88, plant tripped III for 4 times due to reasons not were not related to catalyst. This had caused less Yield during this period. Especially, after the plant planned shut down on 26.08.2019 under inevitable circumstances, yield effected significantly.
3. For almost same plant load, reactor DP remained slightly low with O4_88.
4. Under oxidation products were slightly less with O4_88.

2. Comparison between 101 to 200 days of Operation

O4_88, Hot spot Multi point temperature is maintaining 16 Deg C less compared to O4_66 Plus. Overall, Hot spot temperatures were maintaining less with O4_88 catalyst. Further, PA III Plant also loaded 3.9 g/nm3 additionally with O4_88. Salt bath temperature is less by 5.7 Deg C.

Table.2 :

S.no	Description	O4-88	O4-66 Plus	Difference
1	Multi Point Temp, Deg C	409	425	-16
2	Single Point Temp, Deg C	409	420	-11
3	Air flow per tube, Nm3/Hr	4.06	4.03	0.03
4	O-Xylene Load, g/Nm3	93	89.1	3.9
5	Salt Bath Temp, Deg C	346.6	352.3	-5.7
6	Diff.Pressure in Bar	0.343	0.369	-0.026

With O4-88, CL1 temperature was less than 440 Deg C as per vendor recommendations. CL 0 temp was supposed to be less than 410 Deg C and it was 409 Deg C only. As mentioned by vendor in the first year of operation, CL0 temperatures were maintaining lower than CL1.

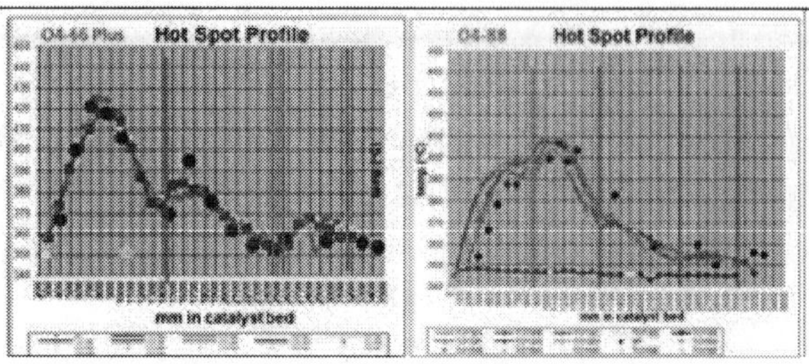	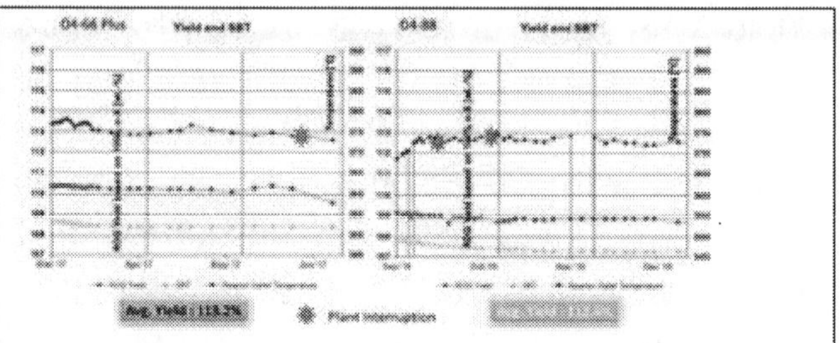
Figure.4: 04-66 Plus (previous catalyst) Vs. 04-88 (New Gen.Catalyst) Hot spot Profile comparison after 200 Days of Operation	**Figure.5:** 04-66 Plus (previous catalyst) Vs. 04-88 (New Gen.Catalyst) Yield comparison between 101 to 200 Days of Operation

During this period, Yield was maintaining low with O4_88 compared to O4_66 Plus. However, advantage gained due to running the plant at high load.

Between 101 to 200 days of operation also, under oxidation products with O4-88 were consistently less. During this period, on an average, under oxidation products were less by 0.23% weight when compared to O4_66 Plus, especially O-xylene slippage was less by around 0.13% with O4_88. But, with O4-88, over oxidation products were around 4.62% and this was 0.27% wt. more compared to O4_66 Plus. Out of which, Maleic Anhydride (MA) and Citroconic Anhydride (CA) contents were produced more marginally with O4_88. Combustion products, when corrected to 40 gms load, were around 1.17% with O4_88 and this was 0.04% wt. higher when compared to O4_66 Plus.

Major Observations during between 101 to 200 days of operation:

1. Yield was low with O4_88 and average yield was less by 0.6% during this period. With O4-66 Plus PA III plant has tripped once on 13.06.2019 due to other issues. After plant restart, yield was slightly affected. Where as in case of O4_88, yield came down to around >112% previously (after 26th August – Refer Figure.3 – After 100 days of Operation) but did not recover, but maintained at almost same level although plant tripped 2 more times.
2. In general, HST were less with O4_88 compared to O4_66 plus.
3. Overall, Yield was slightly less and over oxidation products slightly increasing were the main observations during this period.

3. Comparison between 201 to 300 days of operation

O4_88, Overall, hot spot multi point temperature were maintaining 9 Deg C less compared to O4_66 Plus. Further, PA III Plant also loaded up to 95.1 g/nm3 with O4_88. Salt bath temperature is less by 2.5 Deg C.

Table.3 :

S.no	Description	O4-88	O4-66 Plus	Difference
1	Multi Point Temp, Deg C	419	410	9
2	Single Point Temp, Deg C	408	406	2
3	Air flow per tube, Nm3/Hr	4.05	4.12	-0.07
4	O-Xylene Load, g/Nm3	95.1	88	7.1
5	Salt Bath Temp, Deg C	346.1	348.6	-2.5
6	Diff.Pressure in Bar	0.351	0.388	-0.037

In case of O4-66 plus, hot spots are maintaining at around 1700 mm location, whereas for O4_88 hot spots are maintaining at 1200 mm of length of tube, which was an indication of more catalyst life with O4-88.

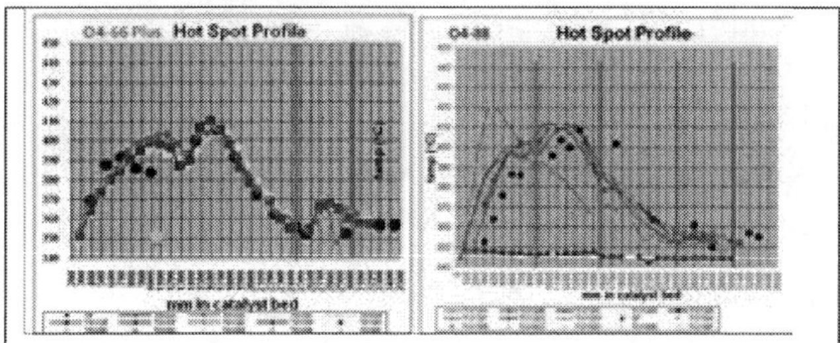

Figure.6: 04-66 Plus (previous catalyst) Vs. 04-88 (New Gen.Catalyst) Hot spot Profile comparison after 300 Days of Operation

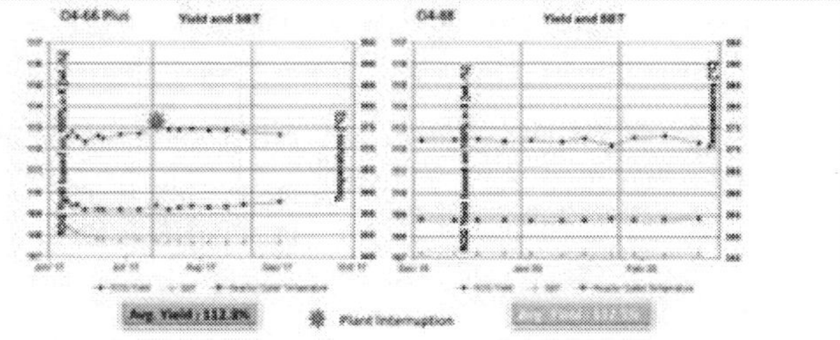

Figure.7: 04-66 Plus (previous catalyst) Vs. 04-88 (New Gen.Catalyst) Yield comparison between 201 to 300 Days of Operation

It can be clearly observed that, with more O-xylene load and less salt bath temperature, the yield is comparable. Between 201th day to 300th day, O4_88 has produced under oxidation products less by around 0.07% wt. compared to O4-66 Plus. O-xylene slippage is more by around 0.04% wt. with O4_88. But, during the same period, under oxidation products were showing increasing tendency with O4_66 Plus, whereas O4_88 under oxidation was steady in nature. With O4_88, over oxidation products also increased to around 5.0% and this was 0.53% more compared to O4_66 Plus. Again this was due to MA and CA contents were more marginally with O4_88. Combustion products, when corrected to 40 gms load, were around 1.27% with O4_88 and when compared to O4_66 Plus, it was more by 0.11%.

Major Observations during between 201 to 300 days of operation:

1. During this period, the average Yield was less by 0.3% with O4_88.
2. With O4-66 Plus, the plant had tripped once on 26.07.2017, but it had not affected the yield much. However, it can be seen that difference in yield difference gap between O4_88 and O4_66 Plus was coming down and O4_88 yield trend is steady.
3. Plant has been loaded further by 7.1 gms/nm3, during the same period with O4-88, when compared to O4-66 Plus. This has resulted in additional production of 775 T due to usage of O4_88. It was considered that plant could be loaded additional due to usage of higher version of catalyst.

4. Both under oxidation and over oxidation products were maintaining more with O4_88 when compared to O4_66 Plus.

V. Overall Observations for 300 days

1. Until 300 days of operation, both O4-66 Plus and O4-88 catalysts have given same average yield i.e around 112.7%.
2. However, average MA concentration was higher by 0.14% with O4-88 catalyst.
3. By maintaining less SBT temperature with O4_88, catalyst life is getting extended, but this is more useful if yield is more.
4. Lower SBT is supposed to result reduction in over oxidation products, but actually it is more. The reason can be attributed to multiple shut downs occurred in case of O4-88.
5. In order to get maximum benefit, the present charge of O4-88 shall be run minimum 4 years, provided yield is maintaining consistently. PA plant with a capacity of 50,000 MTPA at Ludwigshafen has given a life of 5 years, which is based on O-xylene with a minimum yield of 113.8 %. But in Ludwigshafen case, the initial yield started at high rate i.e, more than ~114% and maintained for 1.4 years. After that only, yield started came down. But in IGPL case, yield with O4_88 is 112.5% with less than one year of operation. In order to get life up to 4 years maintaining the same load as present, sustaining yield at this rate is important. The same shall be possible with cooperation of Catalyst vendor in IGPL case for getting the maximum benefit.
6. After completion of around 300 days of operation, in case of O4-66 the expected reduction in yield was about 1%, where as in case on O4-88, drop in yield is expected about 0.5% only.
7. As per the vendor, compared to O4-66 plus, O4-88 will perform better in the next year and yield will be 1% more in case of O4-88.
8. O4-88 is designed in such a way that, in second year also reaction will occur in CL-1 and CL-2 unlike in case of O4-66 plus.

VI. Conclusions

1. Over all yield is marginally less, but production was more with O4_88. The additional production achieved due to replacement with new type of catalyst i.e O4_88 was 1039 T, so far.
2. O4-66 life is maximum 3 years only, whereas O4-88 life is 4 years. Hence, I year additional life available with O4-88.
3. So far, benefits obtained are relatively low when compared to other plants which are operating with O4_88. Reason could be number of outages. However, second year performance of 04-88 is stable now.
4. In O4-88, CL-0 is also a selective layer. Hence, total 3 selective layers are available i.e Cl-0, Cl-1 and Cl-2. Hence, more life and better performance are expected in future.

   In conclusion, it is lucrative and more benefits are expected with new generation catalyst in future.

598845 Molecular Weight Distribution Control of Autoclave Ldpe/EVA Product Using CFD Modeling

Abstract:

In low density polyethylene (LDPE) and ethylene vinyl acetate (EVA) production through high pressure autoclave reactor, mixing dynamics has a great effect on product properties like molecular weight distribution (MWD) and melt index (MI). In commercial reactor, the inside temperature distribution is not uniform due to the difference between the reaction rate and the dispersion rate of the raw materials including the initiator. There is a slight difference between the prediction on ideal and real mixing case in physical properties due to non-uniform temperature distribution [1]. Thus, in increasingly competitive polyolefin market, accurate MWD prediction and control are very important to achieve high productivity with product quality maintained and to make quick decisions to develop highly valued products.

In previous studies, MWD prediction was performed by assuming that each of the autoclave zones is fully mixed like CSTR [2, 3]. For the consideration of the imperfect mixing effect on reactor performance, the simulation by dividing it into modules that calculate networks between flow compartments derived from CFD and modules that calculate reactions in each compartment was performed [4, 5]. These methods are mainly to predict the more detailed MWD because the flow calculation is simple with assumed to be fully mixing or small scale network of CSTRs. Another method is to calculate flow and reaction simultaneously in the CFD code. However, this method could only express a simple property average MW using the method of moment [6]. No study of full scale CFD reactor model has been applied to complex commercial reactors for detailed MWD prediction.

In this study, commercial scale of CFD model was developed to accurately predict the mixing behavior, temperature distribution and the trend of detailed MWD of autoclave LDPE/EVA products. The model performance was validated with commercial reactor products. The characteristic of autoclave LDPE MWD is the formation of a shoulder in the high molecular weight. The analysis of detailed MWD with variation of operating conditions such as pressure and temperature was conducted. Furthermore, the developed model can predict the high molecular weight shoulder, which is a key feature of autoclave LDPE products. To predict this, the method of moment for LCB-class was applied [7]. On the other hand, since EVA does not have an asymmetric shape like LDPE, the method of general moment can be applied.

[1] Wells GJ and Ray WH. Prediction of Polymer Properties in LDPE Reactors. Macromolecular Materials and Engineering, 1996;290: 319-346.

[2] Ham JY, Rhee H-K. Modeling and control of an LDPE autoclave reactor. Journal of Process Control. 1996;6(4):241-246.

[3] Chien IL, Kan TW, Chen B-S. Dynamic simulation and operation of a high pressure ethylene-vinyl acetate (EVA) copolymerization autoclave reactor. Computers & Chemical Engineering. 2007;31(3):233-245.

[4] Wells GJ, Ray WH. Methodology for modeling detailed imperfect mixing effects in complex reactors. AIChE Journal. 2005;51(5):1508-1520.

[5] Zheng H, Huang Z, Liao Z, Wang J, Yang Y, Wang Y. Computational Fluid Dynamics

Simulations and Experimental Validation of Macromixing and Flow Characteristics in Low-Density Polyethylene Autoclave Reactors. Industrial & Engineering Chemistry Research. 2014;53(38):14865-14875.

[6] Zhou W, Marshall E, Oshinowo L. Modeling LDPE Tubular and Autoclave Reactors. Industrial & Engineering Chemistry Research. 2001;40(23):5533-5542.

[7] Pladis P, Kiparissides C. A comprehensive model for the calculation of molecular weight–long-chain branching distribution in free-radical polymerizations, Chemical Engineering Science, 1998; 53(18):3315-3333,

600055 Novel pH-Switchable Viscosity Modifiers to Improve Rheology and Cost Efficiency of Hydraulic Fracturing Fluids

Abstract:

The composition of the fracturing fluid used during hydraulic fracturing plays an eminent role in determining the efficacy of the process. Being able to design fluids that carry proppants with enough driving force into fissures while keeping costs low has become an interesting research problem.

Herein, we report a simple formulation based on the dynamic binary complexation of a long chain acidic compound and triamine. At low weight percentages (<2.5%) of the mixture in water, we report large viscosities of the solution in certain pH ranges, at high shear rates. Changes to the pH reduces the viscosity once again, creating an easily switchable system. This in turn will further enable change of the settling velocity of sand and other proppant particles by a few orders of magnitude by simply controlling the pH. Particle settling tests were deployed to investigate this. Several other tests were also conducted to gain a better understanding of the interfacial phenomena involved in the system, as well as to facilitate rheological predictions. Furthermore, the effects of temperature and salinity on the properties of the mixture were investigated to compare it to real world conditions, such as those observed in major shale basins across the USA.

In summary, there is tremendous potential for the use of such supramolecular formulations in the hydraulic fracturing space. They may go a long way in producing a cost-effective solution to some of the key issues that currently plague this domain such as cost, sustainability, recyclability, etc.

606885 NH3 As the H2 Carrier and Breakthrough Method of NH3 Decomposition

Abstract:

Here I suppose to offer the method of the hydrogen extraction from ammonia by its oxidation partly, at moderate temperatures. Such hydrogen can be served as a clean and cheap and non- poisoning fuel for H2-air Fuel Cells for high energy effective powering without any harmful exhaust being the water vapor and nitrogen only.

At present, decomposition of ammonia needs noble metal catalysts, being very expensive and rare ones, so such hydrogen is very expensive, and it is a big obstacle for large-scale introducing it as a common technology in the world practice. .

The problem of substitution the noble metal catalyst by something more cheap and common is a hot engineering problem being now not solved.

(see Bell, T.E., Torrente-Murciano, L. H_2 Production via Ammonia Decomposition Using Non-Noble Metal Catalysts: A Review. *Top Catal* **59,** 1438–1457 (2016). https://doi.org/10.1007/s11244-016-0653-4)

Here just below I suppose a following chemistry offered by me:

NH3 + 0.2 O2 = 0.5 N2 + 0.4 H2O + 1.1 H2.

So effectiveness of the chemistry by hydrogen production would be

K = 1.1 * 2 / 3 = 73%, and supposing Fuel Cell energy efficiency as 67 %,

the system's total energy efficiency would be 73% * 67% ~= 50 % by NH3 as a fuel

Its Gibbs free energy is –

dG = dH – T * dS, hence

dG = 0.4 * 242,000 – 46,000 - T * (0.5 * 200 + 0.4 * 189 + 1.1 * 131 – 193 – 0.2 * 205) =

dG = 51,000 - T * (320 – 234) = 51,000 – 86 * T,

Supposing dG =< - 35,000, so

T >= 1000 K ~= 700 C

The higher temperature T, the higher a productivity our chemistry offered.

Note: The problems of very cheap ammonia and very cheap & abundant & inexhaustible renewable energy of any designed supply schedule can be used for the said cheap ammonia production, it was and is my especial efforts and numerous results..

610396 Vegetable Oils Extracted from Carnauba Seed with Potential for Bio-Jet Fuel Production

Abstract:

Due to the growing demand for energy, it is essential to develop products and technologies based on a diversified and clean matrix of energy and fuel production. Nowadays has increased the development with a sustainable and renewable cycle in order to reduce greenhouse gas (GHG) emissions. Biofuels are emerging as a viable alternative to fossil energy dependence and contributing to reducing GHG and other substances harmful to human health and the environment. Bio-jet fuel, a drop-in type of biofuel for aviation, based on carbon and hydrogen are generally derived from a broad spectrum of feedstock and thus has a very different composition. One of the most common bio-jet fuels pathway production l is the HEFA (hydrogenated esters and fatty acids) due to its conversion and yield. HEFA is a process to hydrotreat the triglycerides, saturated or unsaturated fatty acids in the vegetable oils, used cooking oils and animal fats to produce bio-jet fuels. Vegetable oils are mainly constituted of triglycerides (95-98%) and a mixture of smaller components (2-5%) from a wide range of chemical compounds. However, in the same feedstock, the type of organic solvent used in the extraction and refining procedures can modify the final oil composition. Nonpolar or weakly polar compounds are soluble in nonpolar or low polar solvents, while compounds with high polarity are soluble in polar solvents, due to the organic solubility be directly related to the molecular structure. This work proposed the oil extraction from Carnauba seed, sustainable biomass, using ethanol and hexane as solvents, to be used as feedstock for bio-jet fuel. The oil was extracted using a Soxhlet extractor at 75 ° C for 3 h for n-hexane and 8 h for ethanol. The products obtained were cooled and stocked in a desiccator until constant weight. The Soxhlet extractions using two solvents were carried out to evaluate the solvent influence on the carnauba oil solubility. After the extraction, the carnauba oil produced was purified by a silica column. In the case of hexane as a solvent, the material recovered was apolar, and ethanol was polar, followed by distillation to separate the oil from the solvent. For polar fraction, using ethanol as solvent, a reddish colored substance was obtained with the oil, and for apolar fraction, greenish color with oil was achieved. This difference in color becomes evident in the composition of the oils obtained. These purified oils will be used as raw material to be tested in the bio-jet fuel production.

611257 Removal of Carbon Monoxide in a Post-WGS Gas Stream By Methanation Using Ni-Ru Catalysts

Abstract:

Recently, hydrogen energy has received attraction as one of the future renewable energy sources to reduce greenhouse gases. Hydrogen can be produced from fossil fuels, such as natural gas, by hydrocarbon reforming process. Production of high-purity hydrogen is crucial to enhance the long-term stability of PEMFCs. CO is one of the main impurities that need to be removed from the reforming gas. The remaining CO from the hydrocarbon reforming process can be removed with PROX and methanation. The PROX system needs additional units for oxygen supply, cooling system, and a mixing device for the reformate gas and air. These characteristics lead to system expansion and increased costs. Therefore, methanation has been considered as an alternative because it is a low-cost, compact process that is oxygen-free and reuses the produced methane. This study aims to suggest optimal design guidelines of methanation. Ni-Ru/CGO catalysts were used because they tend to promote methanation at low temperatures. Experiments were performed by changing the nickel and ruthenium contents of the Ni-Ru catalysts. The reaction feed gas comprised 50% H_2, 20% CO_2, 1% CO, 14% N_2, and 15% H_2O (volume percent). The temperature and GHSV were 200~300°C and 1,000/2,500/5,000/10,000 h^{-1}, respectively. We analyzed the catalysts' CO selectivity, which indicated whether the catalysts converted more CO than CO_2. Low GHSVs clearly led to higher CO conversion and CO selectivity. The Ni-Ru/CGO catalysts achieved a CO concentration less than 100 ppm. The catalyst samples were analyzed to investigate the effects of the metal loading. Through our work, we can produce optimal catalysts for CO methanation and gain insight into designing a methanation reactor on a larger scale.

594919 Compact Hydrogen Fueling Hazars Free at Mild Temperature and Pressure

Abstract:

Compact hydrogen direct storage hazard free, in operative quantity at mild conditions for personal cars, had been practically unsolvable problem, so hydrogen economics and hydrogen energy there are not in practice for personal cars till now, except of scientific dreams.

Really, direct storage in a car for 650 km mileage, even if with very energy effective H2-air FC and an electric mover per a wheel, needs about 6.0 kg hydrogen, and even if under enormous big pressure (let be 700 atm.), it would be enormous big tank of 150 liter (42 kg/m3), and it is very hazard deal, and for refueling especially. A little more adequate science research is a porous containers of gaseous hydrogen, being compact (70.0 kg/m3, 85 liter + adsorbent), but it needs cryogenic 87 K, a little bit higher, than liquefied nitrogen (75 K), being very unsuitable for personal cars.

Hazard free in my interpretation is one of much more practical solutions, was offered by me. It is compact storage - 120 kg/m3, 34 kg = 6 * 17 /3 kg = 3 * 16 liter = 48 liter totally for 650 km mileage, and at normal temperature and 20 atm. Moreover, it allows our hydrogen to store and use at wide range of T and P conditions can be simply adjusted for different kinds of fuel calls and their variable operational needs.

My above mentioned technology is a very interdisciplinary, considering wide environment of problems of hydrogen economics and hydrogen energy. Non hazard hydrogen storage and its usage were a small part of them, but a key part.

Detailed disclosure.

Here I consider a cars' design allows 1000 km mileage by H2-air fuel cell fed from small bottle with gaseous hydrogen at normal conditions outside the bottle, and at T and P inside the bottle, designed for FC used, in quantity of some seconds of the car's running, when in its turn, the bottle intake is coming from another tiny bottle (TB), served for its ammonia electro-thermal decomposition and placed being protected on outer wall of the car and fed through a check valve opened electrically in pulse regime with time controlled delay, by liquid ammonia stored on-board of the car at not extreme conditions, e.g. it is liquid at about 10.0 atm., if room temperature.

And all those can be ordinary approach to safety problem, because it well known, that hydrogen and ammonia in tiny quantity, even if occasionally ignited in large air environment are practically hazard free, and any small leakage of ammonia stored in operative quantity in a car, it can be immediately detected by its strong pungent, but not lethal odor, and by sensor and alarm, and they allow immediately to retire to a safe distance.

My own particular innovation is including above mentioned, as its key part, but additionally I suppose the following chain of ammonia technologies and devices for them.

They are:

(1) Green ammonia production from air and water very cheaply by extremely cheap energy of a special wind power plant deployed near a refueling station. It is a long story of inventions, but was

earlier disclosed and evaluated by me.

(2) A safe refueling station, and method and devices for safe refueling, near but outside of settlements.

(3) Related details of car's design (see below),

They totally can be breakthrough innovation opening the door for practical realization of hydrogen economics and hydrogen energy for towns and cars aren't concentrated in small area, and not for large city. .

Behind the saloon's wall, there is a locked section, and there are three tilted rectangular cylinders (e.g. 80 cm long) one above another with 16 kg liquid ammonia each. It can be enough for 1000 km mileage for ordinary car with H2-air FC of 65% energy efficiency, and an electric mover per each wheel. The cylinders' ends can be colored differently. The front end serves for refueling through a check valve. The back end provides the TB by liquid ammonia, as was described above..

Note: I can suppose any personal cars to have nothing perspective for not so far future, especially cars fueled in any configuration by hydrogen, in favor to city-cars as electric grid element and to flying vehicles for suburb and country. So hazard free hydrogen fueling can be theoretical consideration only.

608694 Amorphous Cobalt-Cerium Binary Metal Oxides As High Performance Electrocatalyst for Oxygen Evolution Reaction

Abstract:

The amorphous Co-Ce binary metal oxides $Co_{1-y}Ce_yO_x$ prepared by the photochemical metal-organic deposition (PMOD) method are developed as high performance electrocatalysts for oxygen evolution reaction (OER). The influence of Ce content on the OER activity is investigated in terms of geometric and electronic factors. Ce can remarkably enhance the OER activity of CoO_x due to the synergistic effect of surface area, Co^{3+} content, and metal-OH bond strength when the content of Ce is less than 60%. $Co_{0.9}Ce_{0.1}O_x$ supported on fluorine-doped tin oxide (FTO) coated glass substrate shows overpotential of 320 (2) mV at 10 mA cm^{-2} in 1 M KOH solution. The OER mechanism exploration reveals that the rate determining step changes with the Ce content due to the variation of metal-OH bond strength. This work sheds light on the design of high-performance yet cost-effective OER catalysts.

Automatic Generation of Liquid-Phase Kinetic Models for Fuel Autoxidation

Minh Duy Le, Valérie Warth, Romain Privat, Pierre-Alexandre Glaude, Roda Bounaceur, René Fournet, Baptiste Sirjean

LRGP, CNRS-Université de Lorraine, 1, rue Grandville, 54000 Nancy, France

Introduction

The European regulations impose an increased share of biofuels in the transport sector, which will be blended with conventional transportation fuels. One of the main obstacles to the widespread use of biofuels is their tendency to downgrade the oxidation stability (resistance to aging) of conventional fuels. During storage, transport and injection phases in engine operations, liquid fuels are in contact with air, and undergo significant variations in temperature and pressure, which can favor the fuel autoxidation. This phenomenon leads to changes in the physical and chemical fuel characteristics through the formation of oxidation products, which are gum and deposits precursors. For example, the viscosity of *n*-alkanes could be raised by a factor of up to 65 [1] due to autoxidation, which can affect negatively engine components as injector, pump, and filter system. The oxidation products can also alter the combustion quality [2] and enhance the particle emission [3]. In the literature, the chemical process governing fuel autoxidation remains poorly understood [4-7]. The chemical mechanism underlying the liquid–phase oxidation of fuels is characterized by a mechanism of free-radical chain reaction. The rigorous modeling of this phenomenon requires the use of detailed chemical kinetic models composed of hundreds to thousands elementary reactions. Automatic generation of detailed kinetic models is a powerful tool to create and handle such large mechanisms. However, if their development is well-established for the generation of gas-phase kinetic models (combustion), this is not the case for liquid-phase oxidation. Currently, RMG software from MIT [8] is the only tool able to generate kinetic models adapted to the liquid phase. The methodological approach used by these authors relies on automated corrections applied to the generated gas-phase oxidation models. In this work, we present new approaches to capture solvent effects on thermodynamic, diffusion and intrinsic rate constants, and their implementation in our gas-phase automatic generator of detailed kinetic models named EXGAS [9]. Validation tests are performed against experimental data of the literature for the liquid-phase oxidation of *n*-butane.

Numerical method

The automatic generation of a liquid-phase oxidation kinetic model for *n*-butane starts with the generation of a gas-phase mechanism in EXGAS. This software has been extensively validated for combustion kinetics [10-12]. The generated kinetic model contains a list of several thousands of elementary reactions as well as the associated thermodynamic and kinetic data (ideal gas). New procedures have been coded in EXGAS to adapt these data to the liquid phase.

Thermodynamic data. Gas-phase thermodynamic data are adapted to the liquid phase using free energies of solvation (ΔG_{solv}) corrections. This approach is also used in RMG, where ΔG_{solv} is calculated using an empirical linear free energy relationship (LFER) from the literature, which can be applied to 25 solvents. In our work, ΔG_{solv} is automatically computed for all the species of the mechanism using the UMR-PRU (Universal Mixing Rule Peng-Robinson UNIFAC) equation of state [13,14]. The computed ΔG_{solv} values were shown to accurately predict thousands of experimental values, with a mean absolute deviation of 0.36 kcal/mol, for a large variety of solute/solvents mixtures at different temperatures [15]. This new approach therefore considerably expands the number of fuels (solvents) and the temperature range that can be treated with our code, compared to LFER approaches defined at ambient temperatures.

Diffusion corrections. Bimolecular gas-phase rate constants are corrected with a diffusion rate constant calculated using a Stokes-Einstein approach for the diffusion of spherical particles through liquids. Molecular radii are calculated based on the UNIFAC molecular size parameter included in UMR-PRU equation of state and temperature dependant viscosities that are read in database.

New rate rules for kinetic parameters. A set of new reaction rate rules, adapted to the liquid-phase oxidation of *n*-alkanes, is established for the decomposition of hydroperoxides, the self-termination reactions of peroxy radicals, and the H-atom abstractions by a peroxy and an alkoxy radical. These reaction rate rules are defined based on literature data. When there was a lack of experimental data, theoretical chemistry calculations have been used. Gas-phase rate constants (k_{gas}) are calculated based on potential energy surfaces computed at the CBS-QB3 level of theory and transition state theory with rigid rotor harmonic oscillator approach (1-DHR-U treatment of internal rotations) and Eckart tunnelling [16]. The computed k_{gas} were then corrected with ΔG_{solv} of activation computed at the M06-2X/6-31+G(d) level of theory with the SMD solvation model [17].

Results and conclusions

Based on the numerical method described above, the liquid-phase kinetic model of *n*-butane has been generated with EXGAS. The validation of this model is performed against the experimental data of Mill et al. [18] who conducted experiments on the liquid-phase oxidation of *n*-butane in a batch reactor. A homogeneous liquid batch reactor (constant T, P and V) model has been developed based on Chemkin II library [19]. The developed model reasonably simulates the conversion of n-butane as shown in Figure 1. Table 1 summarizes the initial conditions of the experiments presented in Figure 1.

Table 1: Experimental conditions of liquid-phase oxidation of n-butane [18]

Experiment #	T (K)	V (ml)	[*n*-Butane] (mol/cm^3)	[Di-tert-butyl peroxide] (mol/cm^3)	Reaction time (min)
1	373	17	8.12 x 10^{-3}	4.18 x 10^{-5}	1439
2	373	8.9	8.12 x 10^{-3}	6.32 x 10^{-5}	1462
3	373	4.3	8.12 x 10^{-3}	6.59 x 10^{-5}	4339
4	398	7.4	7.04 x 10^{-3}	3.32 x 10^{-5}	178

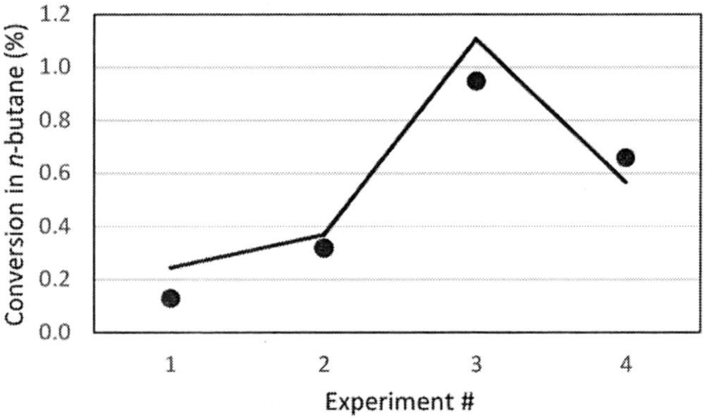

Figure 1: Simulation of the conversion of n-butane during liquid-phase oxidation [18]. Experiments conditions presented in table 1. Symbols: experiment. Line: simulation.

The good simulation results obtained in the case of *n*-butane confirm the robustness of the method proposed in this study to simulate liquid-phase oxidation. On-going works are performed to expend and validate our generator of liquid-phase oxidation mechanisms to other hydrocarbons and oxygenated fuels. Once validated, this tool will be used to simulated and understand the fundamental phenomena that affect the oxidation stability of conventional fuels when biofuels are added.

References

[1] Blaine, S. et al., Ind. Eng. Chem. Res. 30 (1991) 2185–2191.
[2] Kuboyama, T. et al., *SAE Int. J. Engines* 8 (2015) 529–537.
[3] Pourkhesalian, A. M. et al., Atmos. Chem. Phys., 15 (2015) 9099–9108.
[4] Chatelain K. et al., Energy Fuels 30 (2016) 1294-1303.
[5] Mielczarek D.C. et al., Energy Fuels 31 (2017) 12893-12913.
[6] Chatelain K. et al., Energy Fuels 32 (2018) 9415-9426.
[7] Alves-Fortunato M. et al., Fuel 268 (2020) 117074.
[8] Green et al., RMG−Reaction Mechanism Generator; Department of Chemical Engineering, MIT, 2013.
[9] Warth V. et al., Computers & chemistry 24 (2000) 541-560.
[10] Schönborn M. et al, Combustion and Flame 205 (2019) 466-483.
[11] Schönborn M. et al, Energy Fuels 33.12 (2019) 12704 – 12713.
[12] Gornay J. et al., Journal of Analytical and Applied Pyrolysis 146 (2020), 104767
[13] Voutsas E. et al., Fluid Phase Equilib. 241, 216 (2006).
[14] Voutsas E. et al., Ind. Eng. Chem. Res. 43, 6238 (2004).
[15] Moine E. et al., Pure and Applied Chemistry 91 (2019) 1295-1307
[16] Lizardo-Huerta J. C. et al., Phys. Chem. Chem. Phys., 2016, 18, 12231– 12251
[17] Marenich A. V. et al., J Phys Chem B 113.18 (2009) 6378-6396
[18] Mill, T et al., Journal of the America Chemical Society 94.19 (1972) 6802-6811
[19] Kee R. J. et al., CHEMKIN-II: a Fortran chemical kinetics package for the analysis of gas-phase chemical kinetics. Sandia Report SAND89-8009, Sandia National Laboratory, Livermore, CA, 1989

An Experimental and Modelling Study of *n*-Decane Autoxidation

Minh Duy Le, Zaki El Sayah, Salma Khalloufi, Valérie Warth, Pierre-Alexandre Glaude, Romain Privat, René Fournet, Baptiste Sirjean

LRGP, CNRS-Université de Lorraine, 1, rue Grandville, 54000 Nancy, France

Introduction

Fuel reactivity is a key element of fuel/engine adequacy governing engine efficiency and emission. While gas-phase reactivity has been thoroughly studied, literature studies on liquid-phase reactivity are scarce. From storage to injection in the combustion chamber, significant variation of temperature and pressure occurs that leads to fuel autoxidation (aging). This unwanted phenomenon leads to changes in physical and chemical characteristics of the fuel because of the formation of oxidation products and deposits. Autoxidation can negatively affect engine components such as injectors, pumps, and filter system, and significantly deteriorates the combustion quality [2] and enhances particle emissions [3]. Currently, the rigorous simulation of fuel autoxidation processes remains a challenge in the literature [5-8]. A better understanding of liquid-phase oxidation is required to have a better control on this natural phenomenon. This study aims at experimentally and numerically investigating the reactivity of *n*-decane in the liquid phase. The liquid-phase oxidation mechanism of *n*-decane developed in this work will be a building block for the development of a detailed chemical kinetic model of a surrogate fuel autoxidation model.

Experimental method

In this study, the liquid-phase oxidation of *n*-decane was conducted in a PetrOxy apparatus, which is part of the ASTM D7545 standard test. Initial oxygen pressure was set at 700 kPa. The temperature of PetrOxy was held constant. Fuel oxidation stability is quantified by an induction period (IP). The latter parameter corresponds to the time required to observe a decrease from the gas-phase maximum pressure by 10%. IP measurements were conducted at two temperatures, 408 and 413 K. During a PetrOxy experiment, liquid phase samplings were also conducted. These samples were then analyzed by iodometric titration. This method allows to determine the total concentration of hydroperoxides formed during the autoxidation process. IP and hydroperoxide speciation as a function of time are valuable data used to validate the liquid-phase kinetic model of *n*-decane.

Numerical method

Numerical efforts have been performed in the literature to investigate the autoxidation of *n*-decane. Chatelain et al. [5] developed a detailed liquid-phase chemical mechanism of *n*-decane using an automated mechanism generator, RMG [9]. For the mechanism generation, RMG adapts an initial gas-phase kinetic model to the liquid-phase by correcting the gas-phase thermodynamic data with free energies of solvation. The latter correction is estimated based on a linear solvation energy relationship (LSER) [10]. The model of Chatelain et al. overestimated IP of *n*-decane with in a factor of 3 [5].

In this study, the development of the liquid-phase oxidation kinetic model of *n*-decane is based on the gas-phase one. The modelling work in this study consists of three main steps. In the first step, the gas-phase mechanism of *n*-decane is generated by the EXGAS software [11], which is an automatic generator of detailed kinetic model of combustion developed in our group. This software has been successfully employed to generate robust gas-phase kinetic models [12-14]. The second step allows adapting the gas-phase model to a liquid-phase model using corrections of thermochemistry data and rate constants. Thermodynamic data are corrected with free energies of solvation computed with the UMR-PRU equation of state [15]. Diffusion rate constants are calculated to correct bimolecular rate constants using the Stoke-Einstein equation, with molecular radii calculated with the UNIFAC molecular parameters and temperature dependent viscosities taken from experimental databases. Rate constants for the most sensitive reaction for *n*-decane conversion and hydroperoxide formation were determined based on literature review and theoretical chemistry calculations to propose new reaction rate rules. The final step is dedicated to validating the liquid-phase kinetic model using PetrOxy data generated during the experimental work and other experimental data obtained from well-defined reactors in the literature. PetrOxy experiments can be modeled as a homogenous batch reactor [5,6].

Results and conclusions

The measured IP of *n*-decane at 413 K is 189 min, which is in good agreement with the one measured by Chatelain et al. (184 mins) [5]. Figure 1 illustrates the simulation of the formation of total hydroperoxides as a function of reaction time during the PetrOxy experiment. By iodometric titration, it was found that initial liquid phase contained a non-negligible amount of hydroperoxides (0.016 mol/L). The concentration of hydroperoxides increases during the IP measure and is correctly captured by the developed kinetic model.

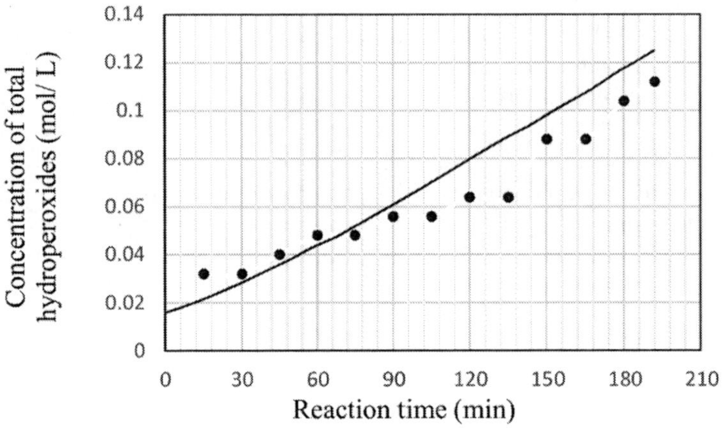

Figure 1: Simulation of the formation of total hydroperoxides during a PetrOxy experiment of n-decane at 700 kPa, 413 K.

Further validations of the developed model of *n*-decane were conducted with literature data [5, 16]. Figure 2 presents the prediction of *n*-decane IP according to various temperatures (393 – 433 K). The measured IP were conducted by Chatelain et al. [5]. IP of *n*-decane decreased with temperatures. The simulated IP are in good agreement with measured ones. The most significant discrepancy is observed at the lowest temperature (393 K) where the model overestimated the IP by a factor of 1.5.

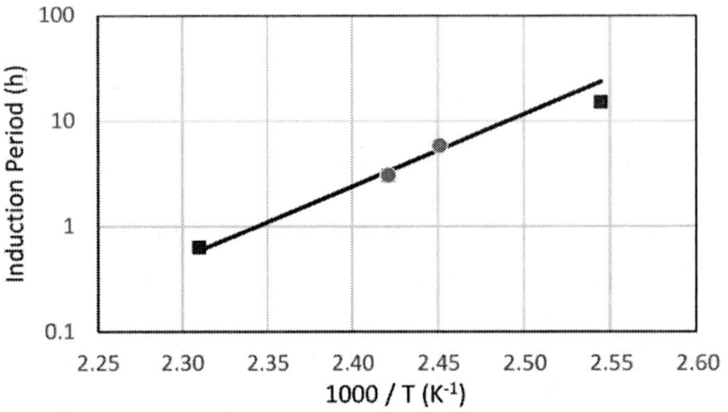

Figure 2: Simulation of IP of n-decane at 700 kPa. Red symbols: experiments conducted in this study. Black symbols: experiments conducted by Chatelain et al. [5]. Line: simulation with the developed model.

Hydroperoxide formation during the liquid-phase oxidation of *n*-decane was quantified in the work of Cullis et al. [16]. These authors conducted autoxidation experiments of *n*-decane in a reflux reactor at 423 K. Figure 3 presents the simulation of *n*-decane monohydroperoxides formation by the developed model.

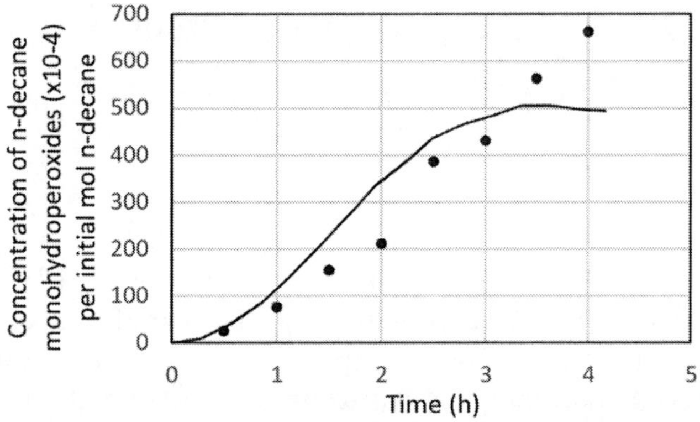

Figure 3: Formation of n-decane monohydroperoxides during the liquid-phase oxidation of n-decane at 423 K. Symbols: experiments conducted by Cullis et al. [16]. Line: simulation with the developed model.

The developed kinetic model of *n*-decane is validated using the experimental data in this study and the literature. In simulations, several important reactions governing liquid-phase reactivity of *n*-decane are determined including (1) H-atom abstractions from *n*-decane by alkoxy and peroxy radicals; (2) unimolecular decomposition of hydroperoxides; and (3) termination reactions of peroxy radicals.

This study proposes the most extensively validated kinetic model of *n*-decane autoxidation of the literature. This study provides insight into the comprehension of the autoxidation phenomenon of *n*-decane and liquid fuels in general. This facilitates the control of fuel reactivity in both gas and liquid phases satisfying the fuel/engine adequacy.

References

[1] Blaine, S. et al., Ind. Eng. Chem. Res. 30 (1991) 2185–2191.
[2] Kuboyama, T. et al., SAE Int. J. Engines 8 (2015) 529–537.
[3] Pourkhesalian, A. M. et al., Atmos. Chem. Phys., 15 (2015) 9099–9108.
[4] Yaakob Z. et al., Renewable and Substainable Energy Reviews 35 (2014) 136-153.
[5] Chatelain K. et al., Energy Fuels 30 (2016) 1294-1303.
[6] Mielczarek D.C. et al., Energy Fuels 31 (2017) 12893-12913.
[7] Chatelain K. et al., Energy Fuels 32 (2018) 9415-9426.
[8] Alves-Fortunato M. et al., Fuel 268 (2020) 117074.
[9] Green et al., RMG–Reaction Mechanism Generator; Department of Chemical Engineering, MIT, 2013.
[10] Abraham et al., Journal of Chromatography A 842 (1999) 79-114.
[11] Warth V. et al., Computers & chemistry 24 (2000) 541-560.
[12] Schönborn M. et al, Combustion and Flame 205 (2019) 466-483.
[13] Schönborn M. et al, Energy Fuels 33.12 (2019) 12704-12713.
[14] Gornay J. et al., Journal of Analytical and Applied Pyrolysis 146 (2020), 104767.
[15] Moine E. et al., Pure and Applied Chemistry 91 (2019) 1295-1307.
[16] Cullis C. F. et al., Proc. R. Soc. Lond. A 375 (1981) 543-563.

602668 Experimental Comparison of Metal Oxide Materials for Solar Thermal Water Splitting

Abstract:

Renewable H_2 is produced from water using concentrated sunlight to drive the endothermic water splitting reaction via a solar thermochemical hydrogen (STCH) redox cycle. These redox cycles use metal oxide intermediates (fluorites, perovskites, spinels) at high temperatures. Although much effort has been directed towards computational analysis to discover new materials, less effort has been given to directly compare these materials experimentally to date. Here, we compare different solar thermal water splitting (STWS) materials and evaluate their performance under various operating conditions for H_2 production capacity, kinetics, water consumed to hydrogen produced (H_2O:H_2) ratio, and stability (robustness) using a stagnation flow reactor. A direct comparison of materials performance provides researchers with a path forward regarding process efficiency and cost.

606059 Optimal Water Management in a Macroscopic System That Includes a Shale Gas Exploitation System

Abstract:

According to the US Energy Information Administration (EIA), about 200 shale gas resource opportunities have been identified in five geologic provinces in eastern Mexico. Prospective regions include 1) Paleozoic shale gas in Chihuahua region; 2) Cretaceous shale gas in the Sabinas Burro-Picachos region; 3) Cretaceous shale gas in the Burgos Basin; 4) Jurassic shale gas in Tampico-Misantla; and 5) unspecified shale gas potential in Veracruz. Technically recoverable shale resources have been estimated at 545.2 trillion cubic feet of natural gas and 13.1billion barrels of oil and condensate. However, the exploitation of shale gas in Mexico is still under investigation because of its potential impacts on technical, environmental (water footprint) and social aspects.

This work presents a mixed-integer nonlinear program (MINLP) to assess the water impact in any of the macroscopic geologic regions where a shale gas exploitation system is technically feasible. The model allows finding the optimal distribution of water in a hydrological region that involves different users: domestic, industrial (including shale gas exploitation) and agricultural; conventional and unconventional water sources (rainwater storage system) are considered. In particular, water withdrawals and wastewater discharges arising from a shale gas distribution system are included. The formulation has been applied to a case study for a shale gas exploitation system in the Burgos basin, located in Tamaulipas, Mexico. The water impact is estimated by using the AWARE methodology, which evaluates the demand for water with respect to the total area of the region analyzed. The model parameters are taken from national geographic databases, which have been analyzed using the ESRI ArcGIS geographic information system. The model is solved by using the GAMS® modeling environment. Results provide the selection of a distributed treatment system that minimizes the levels of pollution in the water bodies and reduces the consumption of fresh water.

604317 Functional and Chemical Properties of Bio-Oil Produced from Biomass Residues

Abstract:

In the present work the pyrolysis of different types of walnut shells (WS) has been studied at different temperatures (450 oC and 550 oC) and particle sizes (2mm and 0.1mm) at a constant heating (20 oC/min) and sweep gas flow (100 cm3/min) rates. The different walnut shells considered are paper (PSW), thin (TSW), medium (MSW) and hard (HSW) walnut shell. Experiment has The yield for bio-oil obtained from PSW, TSW, MSW and HSW are 48.2% (550 oC) and 32% (450 oC), 41.7% (550 oC) and 36.15% (450 oC), 39% (550 OC) and 32% (450 oC), and 49.7% (550 oC) and 36% (450 oC), respectively. Maximum oil (49.7 wt%) and gas (37%) yields have been obtained for HSW at a pyrolysis temperature of 550 oC, while minimum yield is obtained for PSW and MSW at an operating temperature of 450 OC. Calorific values of bio-oil obtained from different walnut shells are 26.8, 26, 26.5 and 27 MJ/kg for PSW, TSW, MSW and HSW, respectively. Chemical composition analysis of both fractions of bio-oils (organic and aqueous) has been investigated using GC-MS, FTIR, 1H-NMR and 13C-NMR. It is found that the pyrolytic bio-oil composition is predominated by oxygenated compounds. The oxygen compound is imitated by the presence of oxygenated fractions such as carbonyl and carboxyl groups formed by pyrolysis of the cellulose and phenol and methoxy groups formed by pyrolysis of lignin. From the structural analysis of the bio-oil, it can be concluded that the bio-oil obtained from WS may be used as a feedstock for furnaces, as a fuel, and also used as a lubricant after small modifications.

608300 A Rigorous Refinery Dynamic Model for Advanced Process Control and Operational Optimization

Abstract:

Today's environmental regulations, stringent competition, and oil market uncertainty call for a tighter integration of the decision making process in oil refining operations. This, in turn, requires strengthening the communication between production planning, short-term production scheduling and advanced process control layers of the decision-making hierarchy. To this end, the development of accurate and computationally tractable dynamic mathematical models of refining facilities becomes critical. Taking advantage of modern simulation, control and optimization tools requires equation-oriented modeling and open model architectures, that allow for flexibility in modifying the representation of unit operations, adding new units and investigating new control structures. To the knowledge of the authors, most of the models available in the open literature are structurally incomplete (i.e., they do not cover all the key units of a refinery) and/or may be outdated implementation-wise [1,2,3,4,5,6].

In this paper, we address the aforementioned issues by developing an open-source model of a key section of an oil refinery, the fluidized-bed catalytic cracker (FCC). Our model represents a significant extension of the FCC model originally introduced by McFarlane et al [1]. It includes the cracker unit, as well as the fractionator column and wet gas compressor, and by considers a large number of (pseudo) components such that the streams ("cuts") drawn from the fractionation column are representative of those obtained in industry [7].

Against this backdrop, we develop a model predictive control structure based on the dynamic matrix control paradigm. Our model is freely available for download and can be used for simulation as well as a basis for further studies on, e.g., big data analysis, fault detection.

References

1. R.C. McFarlane et al., Dynamic simulator for a Model IV fluid catalytic cracking unit, Computers and Chemical Engineering 17 (1993) 275-300
2. H. Huang, Simulation and control of complex distillation columns, PhD dissertation, Texas Tech University, Lubbock, Texas (2000)
3. L.F.L Moro, D. Odloak, Constrained multivariable control of fluid catalytic cracking converters, Journal of Process Control, 5 (1995) 29-39
4. C.B. Chung, J.B. Riggs, Dynamic simulation and nonlinear-model-based product quality control of a crude tower, AIChE 41 (1995) 122-134
5. J. M. Pinto et al., Planning and scheduling models for refinery operations, Computers and Chemical Engineering 24 (2000) 2259-2276
6. O. Santander, M. Baldea, On the interaction and integration of production planning and (advanced) process control , Computers and Chemical Engineering 133 (2020) 1-17
7. M. Heydari et al., Study of seven-lump kinetic model in the fluid catalytic cracking unit, American Journal of Applied Sciences 7 (2010) 71-76

PREDICTION OF THE EQUILIBRIUM CONDITIONS OF CLATHRATE HYDRATES USING UPDATED PRESSURE DEPENDENCE OF THE LANGMUIR ADSORPTION CONSTANT IN THE VAN DER WAALS-PLATTEEUW MODEL

Min-Kang Hsieh, Department of Chemical Engineering, Columbia University, New York City, NY, USA*
Shiang-Tai Lin, Department of Chemical Engineering, National Taiwan University, Taipei, Taiwan
Li-Jen Chen, Department of Chemical Engineering, National Taiwan University, Taipei, Taiwan

**Corresponding authors: Min-Kang Hsieh, E-mail: mh3254@columbia.edu*

Introduction

Gas hydrates, also known as clathrates or clathrate hydrates, are solid mixtures of gas molecules and water, in which water molecules form large enough cavities to encapsulate small gas molecules and hydrocarbons. There are three main structures of hydrates, structure I (sI), structure II (sII), and structure H (sH)[1], depending on the encapsulated molecules or the mixtures of them. The molecular size as small as methane and carbon dioxide can be encaged in the small (5^{12}) and the large ($5^{12}6^2$) water formed cavities in sI hydrates; the larger hydrocarbons, such as propane and butane, are only able to be adsorbed in the small (5^{12}) and/or the large ($5^{12}6^4$) cavities in sII hydrates. An accurate thermodynamics model is vital in comprehensively understanding the stability and the phase transition behaviors of these hydrates under all conditions. In most thermodynamic models, the van der Waals-Platteeuw (vdW-P) model for the solid phase combines an equation of state (EoS) for the fluid phase to simulate the phase equilibrium. Few models[2-9] are capable of describing the various type of phase behavior simultaneously. In this work, we present a simpler approach for modeling of mixture gas hydrates in various type of phase coexisting conditions (the vapor-ice-hydrate (VIH), vapor-liquid-hydrate (VLH), and liquid-liquid-hydrate (LLH) equilibria) simultaneously, based on the previously developed pressure- and temperature-dependent Langmuir adsorption constant in the van der Waals and Platteeuw (vdW-P) model.[2-3] Briefly, the fugacity of a species in the fluid phase is determined by the Peng-Robinson-Stryjek-Vera (PRSV) EOS[10] combined with the predictive COSMO-SAC activity coefficient model[11] through the 1st order modified Huron-Vidal (MHV1) mixing rule,[12] i.e., PRSV+MHV1+COSMOSAC method. This method allows for predicting fluid phase behaviors (e.g., VLE) without any experimental data for the mixture fluids. In the solid hydrate phase, the fugacity is determined from the pressure- and temperature-dependent vdW-P model, which describes the deformation of the occupied lattice with increasing pressures. In this update, we include the guest-guest interactions among the guests encapsulated in different cavities within the solid hydrate phase, coupled with the pressure- and temperature-dependent vdW-P model to determine the Langmuir adsorption constant. This guest-guest interaction is a geometric mean of chemical species-dependent guest-guest interactions with their mixed-gas hydrate occupations. We demonstrate that this approach is capable of modeling CH_4, C_2H_6, C_3H_8, i-C_4H_{10}, and CO_2 of pure and their mixture gas hydrates upon 5 mixed gases from low temperature-low pressure (148.8K and 5.35 x10^2 Pa) to high temperature-high pressure (323.9 K and 4.79 x10^8 Pa) three-phase-equilibrium systems.

Method

The three-phase (VIHE, VLHE, and LLHE) coexisting condition of gas hydrate is determined based on the equality of fugacity of water at temperature T and pressure P among all three phases

$$\bar{f}_w^{I}(T,P,x_i) = \bar{f}_w^{II}(T,P,y_i) = \bar{f}_w^{H}(T,P,z_i) \quad (1)$$

where the superscripts I and II denote for the vapor, liquid, or ice phase; and superscript H denotes the solid hydrate phase; x_i, y_i, and z_i are the compositions within each coexisting phase. The detail could be found in our previous study[2]. Here, the essential of modified van der Waals-Platteeuw model describes the fugacity of water in the solid hydrate phase as

$$f_w^H(T,P,y_i) = f_w^\beta(T,P)exp\left(\sum_m \nu_m \ln\left(1-\sum_l \theta_{ml}\right)\right) \tag{2}$$

where ν_m is the number of cages of type m per water molecule in the hydrate lattice, f_w^β is the fugacity of a hypothetical empty hydrate lattice, given by

$$f_w^\beta(T,P) = P_w^{sat}\varphi_w^{sat}(T)\exp\left(\frac{\int_{P_w^{sat}}^{P} V_w^\beta(T,P)dP}{RT}\right) \tag{3}$$

where φ_w^{sat} is assumed to be unity; V_w^β is the saturated vapor pressure of empty hydrate and P_w^{sat} is the molar volume of the empty lattice[2]. The θ_{ml} is occupancy (fraction) of each cage type m by a guest l and is given by

$$\theta_{ml} = \frac{C_{ml}\bar{f}_l^V(T,P,y_l)}{1+\sum_i C_{mi}\bar{f}_i^V(T,P,y_i)} \tag{4}$$

The coefficient C_{ml} is the Langmuir adsorption constant for guest l in m type cage and is described by the square-well (SW) potential, given as

$$C_{ml} = \frac{V_{ml}}{kT}\exp\left(\frac{\varepsilon_{ml}}{kT}\right) \tag{5}$$

where V_{ml} is the free volume of guest molecule l inside m type cage and ε_{ml} is the depth of the SW cell potential. To consider the distortion of occupied hydrate lattice, a pressure-dependent free volume of encapsulated molecules for the Langmuir constant is as

$$V_{ml}(P) = \frac{V_{ml}^0}{f_{ml}(P)} \tag{6}$$

where V_{ml}^0 is the maximum free volume, and the function $f_{ml}(P)$, whose value increases with pressure, effectively reduces the free volume for gas molecules in the cages at high pressures. An empirical expression $f_{ml}(P) = 1 + A_{ml}\Delta + B_{ml}\Delta^2$ is used, where the symbol $\Delta = P/P_c$ is the reduced pressure and P_c is the critical pressure of water[2]. The parameters A_{ml} and B_{ml} used here are refined by including the mixed-gas hydrate data and only depend on the cage types and not the type of gas molecules.

$$f_{ml}(P) = \begin{cases} 1+2.9967\times10^{-7}\Delta+2.4383\times10^{-7}\Delta^2 & \text{for sI } 5^{12} \text{ cage} \\ 1+2.1773\times10^{-2}\Delta+7.2512\times10^{-2}\Delta^2 & \text{for sI } 5^{12}6^2 \text{ cage} \\ 1+3.6039\times10^{-10}\Delta+7.7857\times10^{-8}\Delta^2 & \text{for sII } 5^{12} \text{ cage} \\ 1+7.2193\times10^{-2}\Delta+2.4735\times10^{-1}\Delta^2 & \text{for sII } 5^{12}6^2 \text{ cage} \end{cases} \tag{7}$$

The Langmuir adsorption constant, considering the guest-guest interaction of the encapsulated gas in the water cages, could be represented as follows

$$C_{ml} = \frac{V_{ml}}{kT}\exp\left(\frac{\varepsilon_{ml}^{GL}}{kT}\right)\exp\left(\frac{\varepsilon_m^{GG}}{kT}\right) \tag{8}$$

where ε_{ml}^{GL} is the interaction between the encapsulated guest molecule l and the water cage, m, and ε_m^{GG} is the guest-guest interaction among guests when the encapsulated guest is occupied in the cage m, given as

$$\varepsilon_m^{GG} = \sum_n\sum_m\sum_i\sum_j \theta_{mi}\theta_{nj}\varepsilon_{minj}^{gg} \tag{9}$$

$$\varepsilon_{mnij}^{gg} = \sqrt{\varepsilon_{mi}^{gg}\varepsilon_{nj}^{gg}} \tag{10}$$

where θ_{mi} is the occupancy of guest i in cage m; θ_{nj} is the occupancy of guest j in cage n; ε_{mi}^{gg} is the guest-guest interaction of guest i in cage m; ε_{nj}^{gg} is the guest-guest interaction of guest j in cage n.

The three-phase coexisting condition is determined as follows. For a given temperature T (or pressure P) and the feed composition with an initial guess for the equilibrium pressure P (or temperature T), either the VIE or VLE or LLE is first determined via isothermal flash calculations. The fluid phase fugacities are then used to calculate water fugacity in both sI and sII in hydrate phase from the vdW-P model. If the water fugacity in the hydrate phase equals that in the fluid phase, then the equilibrium pressure is found for the calculated structure. Otherwise, the pressure is updated, and the calculation is repeated until the equality of water fugacity is attained. The equilibrium pressure and the hydrate structure for the coexisting condition are determined by the lower calculated equilibrium pressure from the above computing process.

The value of the Langmuir adsorption constant is calculated by solving the occupancy from equations (4), (8), and (9) with the Newton-Raphson method. Note that the small guests (i.e., methane. ethane, and carbon dioxide) could occupy the small and large cages of hydrate in both sI and sII; and the large guests (i.e., propane and *iso*-butane) are only able to occupy the large cage of hydrate in sII. The maximum free volume (V_{ml}^{0}) and energy (ε_{ml}^{GL} and ε_{mi}^{gg}), summarized in **Table 1**, were determined by fitting to the experimental VIHE and VLHE data of pure and mixed gas hydrate. For the parameters to describe structure I, the data include methane, ethane, and carbon dioxide pure gas hydrates in the VIHE and VLHE regions, as well as a portion of data of methane-ethane, methane-carbon dioxide, ethane-propane, ethane-carbon dioxide, propane-carbon dioxide binary gas in the VLHE region. For structure II, the data include propane and iso-butane pure gas hydrates in VIHE and VLHE regions, as well as a portion of data of methane-ethane, methane-propane, ethane-propane, propane-carbon dioxide, *iso*-butane-methane, and *iso*-butane-carbon dioxide binary gas hydrates in the VLHE region. The parameters A_{ml} and B_{ml} in Eq. (7) were also determined via the structure I and II data regression process, respectively. Total 564 and 501 experimental data points of pure and binary gas hydrates, respectively, whose structures are known, have been included in the parameter regression process. These fitted parameters are then directly adapted to predict the phase equilibrium of gas hydrates beyond binary gas systems and the phase transitions of mixed-gas hydrates without changing the parameter values.

The following objective function was used for the regression of the parameters in the Langmuir constant

$$\mathrm{RMSE} = \sqrt{\frac{1}{N_{data}} \sum_{j=1}^{N_{data}} \left(\frac{P_j^{\mathrm{calc}} - P_j^{\mathrm{expt}}}{P_j^{\mathrm{expt}}} \right)^2} \tag{11}$$

where P is the equilibrium pressure at some given temperatures, and N_{data} is the number of experimental data used. The quality of predictions of the presented model is evaluated by the absolute average relative deviations in pressure (P), temperature (T), and in hydrate phase composition (z), as follows

$$\%AARD - \xi = \frac{100\%}{N_{data}} \sum_{i} \left| \frac{\xi_i^{\mathrm{calc}} - \xi_i^{\mathrm{expt}}}{\xi_i^{\exp t}} \right| \tag{12}$$

where ξ is the examined property (i.e., P, T, and z).

Results and discussion

The three-phase coexisting condition of pure gas hydrates

Figure 1 illustrates the correlation of the three-phase coexisting conditions of pure gas hydrate of methane, ethane, and carbon dioxide (in sI) and propane and iso-butane (in sII), respectively. The summation of this guest-guest interaction (ε_{ml}^{gg}) with the guest-lattice interaction (ε_{ml}^{GL}) determines the potential energy of guests for stabilizing in the hydrate phase. The distortion of the occupied hydrate lattice framework is counted by a simple polynomial pressure dependence (see Eq. 7), which is capable of describing the three-phase coexisting conditions over a broad range of temperature/pressure regions. The applicability of the model describes the hydrate formation boundary from 5.35 x 10^2 to 4.79 x10^8 Pa (148.8 to 323.9 K). Here, as a demonstration of this approach, **Table 2** summarizes the AARD-Ps (and AARD-Ts at LLHE region) of 5 pure gas hydrates, CH_4, C_2H_6, C_3H_8, *i*-C_4H_{10}, and CO_2. The average accuracies of all pure gas hydrates are 4.55% AARD-P in VIHE region (88 pts), 2.76% AARD-P in VLHE region (367 pts). The overall average of accuracy is 3.11% AARD-P in VIHE and VLHE regions (455 pts), and 0.33% AARD-T in LLHE region (109 pts). The updated model with guest-guest interactions is of the higher accuracy compared with our previous model (3.26 % AARD-P in VIHE and VLHE regions and 0.36 % AARD-T in LLHE region) in pure hydrate systems. Also, turning off the guest-guest interactions in the calculation, the total average of deviations for single-gas hydrate significantly increases to 17.69 % AARD-P in VIHE and VLHE regions, and 0.49% AARD-T in LLHE region. The above illustrates that the consideration of guest-guest interaction boosts the accuracy in the description of the phase equilibrium of single-gas hydrates; that is, it is essential for the potential energy with the guest-guest interaction to better describe the encapsulation of guests in the hydrate phase. The % AARD is better than previous models of Sloan[1] and Klauda[16] and is the result of slightly better predictions for hydrate at high coexisting pressures, especially at LLHE regions.

Table 1. The values of the maximum free volume (V_{ml}^0) and energy (ε_{ml}^{GL} and ε_{ml}^{gg}).

Guest	Cage type	sI			sII		
		$V_{ml}^0 \times 10^3$ (Å^3)	ε_{ml}^{GL}/k (K)	ε_{ml}^{gg}/k (K)	$V_{ml}^0 \times 10^3$ (Å^3)	ε_{ml}^{GL}/k (K)	ε_{ml}^{gg}/k (K)
CH_4	5^{12}	2.626 x 10^{-3}	5169.675	59.438	12.492	2998.320	5.331 x 10^{-6}
	$5^{12}6^2$	18632.317	1456.457	19.974	13388.581	1369.319	56.199
C_2H_6	5^{12}	2.052 x 10^{-11}	4939.894	4.463 x 10^{-7}	2.133 x 10^{-8}	7663.796	1.116 x 10^{-2}
	$5^{12}6^2$	101.19	3603.528	25.028	1.620	5201.343	1.326 x 10^{-7}
CO_2	5^{12}	4.622	2747.680	17.378	2.389 x 10^{-2}	4612.730	9.721 x 10^{-1}
	$5^{12}6^2$	146.765	3235.765	1.990	3.931	3630.572	2.158 x 10^{-5}
C_3H_8	5^{12}	-	-	-	-	-	-
	$5^{12}6^2$	-	-	-	15.863	5112.638	11.979
i-C_4H_{10}	5^{12}	-	-	-	-	-	-
	$5^{12}6^2$	-	-	-	3.787	5547.704	85.258

Table 2. Comparison of the accuracy of different methods in pure gas hydrates.

Guest	Equilibria	No. of data[a]	Temperature range (K)	Pressure range (Pa)	AARD (%)					
					Sloan[1]	Klauda[16]	Delavar[17]	Khosravani[4]	Hsieh[2]	this work
CH_4	V-I-H(sI)	17	148.8-270.9	5.30×10^3 - 2.39×10^6	6.52	2.56	-	1.09	4.85	3.87
	V-L_w-H(sI)	92	273.2-315.1	2.65×10^6 - 2.38×10^8	3.44	2.56	1.56	2.11	3.49	3.85
C_2H_6	V-I-H(sI)	11	200.8-272.0	8.27×10^3 - 4.57×10^5	10.41	3.90	-	3.48	6.52	9.28
	V-L_w-H(sI)	61	273.4-287.4	5.03×10^5 - 3.30×10^6	10.94	3.51	1.21	1.39	4.09	3.98
	L_w-L_{HC}-H(sI)	55	288.0-323.9	3.33×10^6 - 4.79×10^8	20.20	20.40	-	12.68	0.58	0.54
C_3H_8	V-I-H(sII)	21	245.1-272.9	3.95×10^4 - 1.72×10^5	13.67	6.65	-	2.33	3.20	3.22
	V-L_w-H(sII)	70	273.3-278.9	1.72×10^5 - 5.67×10^5	3.32	3.50	1.26	3.45	2.88	2.89
	L_w-L_{HC}-H(sII)	27	278.4-278.9	6.16×10^5 - 1.68×10^7	-	-	-	-	0.05	0.06
CO_2	V-I-H(sI)	22	151.5-271.8	5.35×10^2 - 1.05×10^6	17.10	3.04	-	2.81	3.23	4.92
	V-L_w-H(sI)	121	271.6-283.2	1.04×10^6 - 4.51×10^6	3.22	2.79	1.51	3.06	2.90	1.59
	L_w-L_{CO2}-H(sI)	21	282.9-292.7	4.50×10^6 - 1.86×10^8	22.00	24.60	-	10.1	0.37	0.20
i-C_4H_{10}	V-I-H(sII)	17	241.1-275.1	1.76×10^4 - 1.09×10^5	-	-	-	-	2.87	3.32
	V-L_w-H(sII)	23	273.3-275.1	1.18×10^5 - 1.69×10^5	-	-	0.99	-	0.81	0.91
	L_w-L_{HC}-H(sII)	6	275.4-275.8	2.26×10^5 - 1.43×10^7	-	-	-	-	0.09	0.11
Overall[b]				AARD	8.94	6.85	1.39	4.41	2.70	2.56
Overall[c]	VIHE & VLHE	455		AARD-P	7.24	3.79		3.07	3.26	3.10
	LLHE	109		AARD-T					0.36	0.33

[a]The data used here are the same as the previous work[2].
[b]The accuracies are represented by all equilibria regions.
[c]The accuracies are represented according to the different equilibria regions.

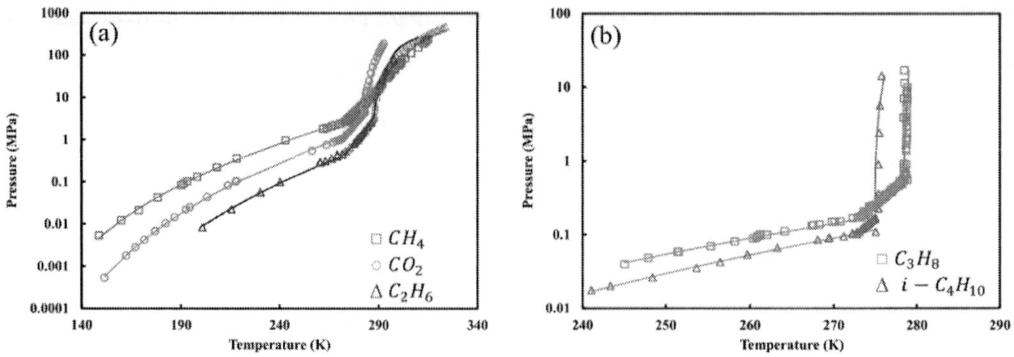

Figure 1. The three-phase coexisting conditions of (a) structure I hydrate of CH_4 (squares), C_2H_6 (triangles), and CO_2 (circles) and (b) structure II hydrate of C_3H_8 (squares) and i-C_4H_{10} (triangles).

The three-phase coexisting condition of mixed-gas hydrates

Combined the concepts of the guest-guest interaction among the encapsulated guests and the pressure effect of lattices deformation of the solid hydrate phase, the updated model can be further applied to the modeling of mixed-gas hydrates. **Table 3** summarizes the AARD-Ps in VLHE and AARD-Ts in LLHE of 16 mixed-gas hydrates containing two or more of the 5 investigated gases in this work (CH_4, C_2H_6, C_3H_8, i-C_4H_{10}, and CO_2), including 11 binary, 2 ternary, 2 quaternary, and 1 quinary mixed-gas hydrates. The average accuracies of all binary gas hydrates are 5.34% AARD-P in VLHE region (538 pts) and 0.28% AARD-T in LLHE region (111 pts). The average accuracies of mixed gas hydrates beyond binary systems are 8.52% AARD-P in VLHE region (43 pts) and 0.36% AARD-T in LLHE region (14 pts). The total average of accuracy is 5.57% AARD-P in VLHE region (581 pts), and 0.29% AARD-T in LLHE region (125 pts). The updated model extends its capability of describing the mixed gas hydrates by including the parameter sets for the encapsulated guests in the structure other than they are hosted in pure gas hydrate. The errors of mixed gas hydrate formation are significantly reduced, compared with the calculation from the previous model (1105 % AARD-P in VIHE and VLHE regions and 3.53 % AARD-T in LLHE region). The guest-guest interaction, again, is essential for the potential energy of encapsulated guests within the solid hydrate phase. When turning the guest-guest interaction off, the total average of AARD-P for single-gas hydrate significantly increases to 28.21 % AARD-P in VIHE and VLHE regions, and 0.64% AARD-T in LLHE region.

Table 3. Comparison of the accuracy of different methods in mixed gas hydrates.

Guest	Equilibria	No. of data[a]	Temperature range (K)	Pressure range (Pa)	AARD (%) Sloan[1]	Klauda[16]	Delavar[17]	Khosravani[4]	Hsieh[2,b]	this work
CH_4-C_2H_6	V-L_w-H(sI, sII)	60	274.2-304.1	8.83×10^5 - 6.86×10^7	4.86	5.50	2.89	3.58	14.13	4.67
CH_4-C_3H_8	V-L_w-H(sII)	80	274.4-304.9	2.45×10^5 - 6.90×10^7	4.55	4.23	3.13	4.61	7151.97	3.85
CH_4-C_3H_8	L_w-L_{HC}-H(sII)	60	279.6-300.2	6.60×10^5 -1.75×10^7	-	-	-	-	4.61	0.15
CH_4-i-C_4H_{10}	V-L_w-H(sII)	86	273.8-305.0	1.27×10^5 - 6.33×10^7	-	-	2.68	-	185.91	7.08
CH_4-i-C_4H_{10}	L_w-L_{HC}-H(sII)	38	275.4-299.3	1.79×10^5 - 1.45×10^7	-	-	-	-	3.54	0.54
CH_4-CO_2	V-L_w-H(sI)	68	273.5-287.6	1.10×10^6 - 1.10×10^7	2.29	3.66	1.75	1.37	7.84	3.46
C_2H_6-C_3H_8	V-L_w-H(sI, sII)	57	273.1-283.3	4.40×10^6 - 2.03×10^6	5.72	8.86	3.11	5.00	16.92	3.91
C_2H_6-C_3H_8	L_w-L_{HC}-H(sI, sII)	13	278.1-284.5	9.10×10^5 - 7.28×10^6	-	-	-	-	0.25	0.10
C_2H_6-CO_2	V-L_w-H(sI)	45	273.5-287.8	5.65×10^5 - 4.08×10^6	3.56	3.60	2.41	2.97	10.86	5.03
C_3H_8-CO_2	V-L_w-H(sI, sII)	91	273.7-286.2	2.21×10^5 - 4.27×10^6	46.7	17.4	7.32	11.38	462.92	7.71
i-C_4H_{10}-CO_2	V-L_w-H(sI, sII)	52	273.7-280.9	1.45×10^5 - 3.18×10^6	-	-	7.9	-	140.14	5.70
CH_4-C_2H_6-C_3H_8	V-L_w-H(sI, sII)	18	275.8-298.1	9.17×10^5 - 2.45×10^7	4.96	1.00	-	1.21	47.62	8.62
CH_4-C_3H_8-CO_2	V-L_w-H(sII)	4	277.2-288.6	1.66×10^6 - 7.24×10^6	-	-	-	-	76.77	4.28
CH_4-C_2H_6-C_3H_8-i-C_4H_{10}	V-L_w-H(sII)	8	273.5-281.9	9.20×10^5 - 2.67×10^6	-	-	2.24	-	151.69	16.82
C_2H_6-C_3H_8-i-C_4H_{10}-CO_2	L_w-L_{HC}-H(sII)	14	280.0-285.7	1.21×10^6 - 1.47×10^7	-	-	-	-	1.89	0.36
CH_4-C_2H_6-C_3H_8-i-C_4H_{10}-CO_2	V-L_w-H(sII)	13	292.7-308.2	1.12×10^7 - 9.96×10^7	-	-	7.63	-	29.54	4.58
Overall	VLHE	582		AARD-P	14.73	8.21	4.28	5.42	1105.08	5.57
	LLHE	125		AARD-T					3.53	0.29

[a]The data used here is available from the NIST Gas Hydrate database[15].

[b]The accuracies are calculated by using the previous model, and the result shows the limited in the description of phase transition.

One enchanting feature of this model is its performance in LLHE region by inheriting the advanced performance of the COSMO-SAC activity model for the fugacity in fluid phases. **Figure 2** shows the LLHE of methane-propane and ethane-propane binary gas hydrates, respectively. The modeled phase boundaries demonstrate the tendency and high accuracy of the equilibrium pressure calculation with various compositions. Another feature of this model is the high accuracy of the mixed-gas hydrates with CO_2 (5.69%), comparing with previous models from Sloan[1] (23.95%), Klauda[16] (10.27%), Khosravani[4] (6.53%), and Delavar[17] (5.28%, this model did not report its result in the VIHE and LLHE regions). Note that although only a fraction of experimental points of binary gas hydrates has been included in the fitting process, the high predictive performance of three-phase coexisting conditions in mixed-gas hydrate with more than two guests supports the generalization of guest-guest interaction parameters.

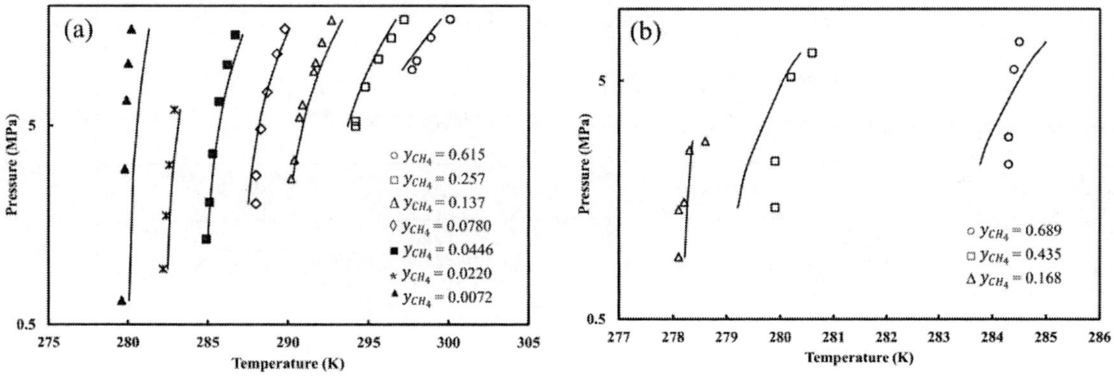

Figure 2. The formative conditions of (a) CH_4-C_3H_8 and (b) C_2H_6-C_3H_8 binary gas hydrates with various compositions. The solid lines are the results of this work.

A phase change behavior (i.e., the structure transition from sI to sII in VLHE region) in binary gas hydrates (CH_4-C_2H_6, C_2H_6-C_3H_8, C_3H_8-CO_2, and *i*-C_4H_{10}-CO_2) has been investigated by using this updated model. For determining the stable hydrate structure, the calculation approach is as follows. The temperature dependence of equilibrium pressures of both sI and sII hydrate could be calculated at a given binary composition (see **Figure 3a**). The lower pressure is identified as the equilibrium pressure at a given temperature, and the hydrate structure is thermodynamically favorable. The transition temperature could be further determined by matching equilibrium pressures to equal in both structures. **Figure 3a** shows the temperature dependence of equilibrium pressures of C_2H_6-C_3H_8 hydrates at $y_{C_3H_8}$ = 0.186 and the prediction of phase change behavior using this model. The modeled phase transition temperature is at 272.9K, which is slightly lower than the reported temperature at T = 273.1-273.8 K[18]. **Figure 3b** illustrates the hydrate structure change of C_2H_6-C_3H_8 hydrates at T = 260-290 K in VLHE. The results support that the binary gas hydrates stabilize in sI at low propane concentrations and exhibit a transition to sII at high propane concentrations. The modeled phase transition temperature increases with propane composition when the equilibrium temperature is below 288K; however, the reverse tendency occurs when it is over 288K.

Similarly, **Figures 3c** and **3e** represent the compositional dependence on equilibrium pressures of C_3H_8-CO_2 and *i*-C_4H_{10}-CO_2 in VLHE with various temperatures, respectively. The prediction of the equilibrium pressures of both systems with CO_2 is higher accuracy (7.71% and 5.70% of AARD-P, respectively) than previous works (see **Table 3**), which allows this model to investigate the phase change behaviors involving CO_2. **Figures 3d** and **3f** illustrate the structural transitions of C_3H_8-CO_2 and *i*-C_4H_{10}-CO_2 binary gas hydrates. For both systems, sII hydrates are thermodynamically favored up to high concentrations of CO_2. The minimum concentration (y_{CO_2}) of phase transition could occur at 0.91 (T=285.6K) and 0.895 (T=284.0K) in C_3H_8-CO_2 and *i*-C_4H_{10}-CO_2, respectively. The above investigated results indicate that the structure of binary gas hydrate maintains the same structure of pure hydrate of the major guest compositions, and the transition boundary is affected by the other companying guest.

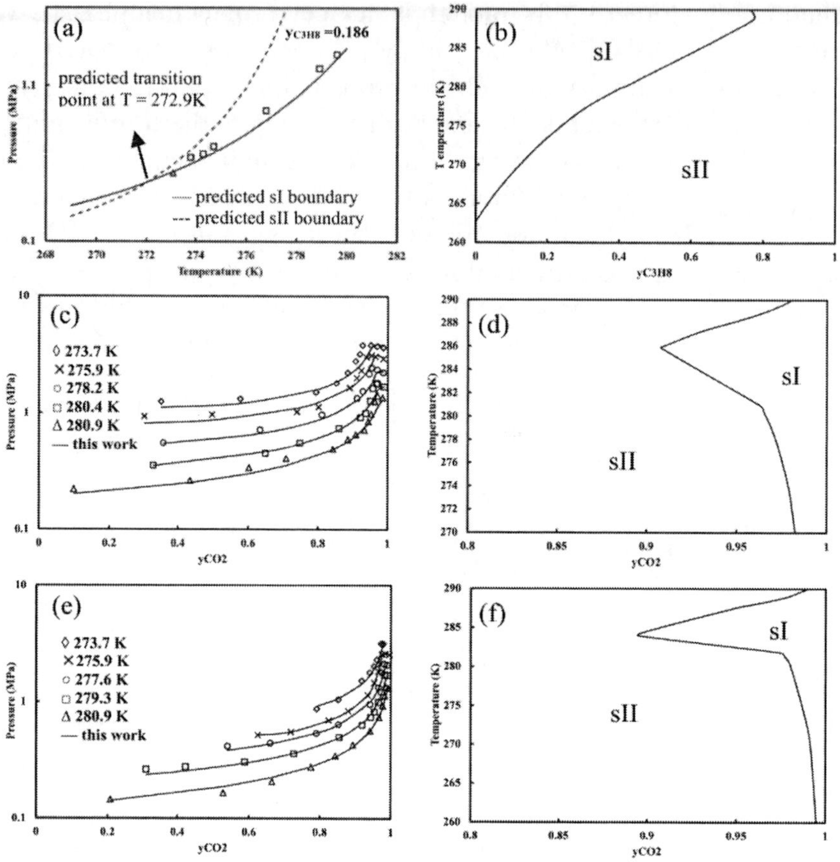

Figure 3. The formative conditions (a) (c) (e) and the predictive phase transitions (b) (d) (f) of C_2H_6-C_3H_8, C_3H_8-CO_2, and *i*-C_4H_{10}-CO_2 binary gas hydrates, respectively, with various compositions in VLHE region.

Exceptionally, CH_4-C_2H_6 hydrates have been reported phase transition behaviors, from sI to sII at y_{CH_4} = 0.72-0.75[19] and from sII back to sI at y_{CH_4} = 0.993-0.994[20] at 274.15K in VLHE, though both pure CH_4 and C_2H_6 hydrates are sI. The updated model shows its capability to demonstrate these phase change behaviors. **Figure 4a** shows the predictive agreement with the compositional dependence on equilibrium pressures of CH_4-C_2H_6 hydrates at T = 274.8-283.2 K and the prediction of phase transition behavior between sI and sII at the temperature region. The result indicates the phase transition from sI to sII is y_{CH_4} ~ 0.662 to 0.742 at T = 274.8-283.2 K. Thought a further investigation, **Figure 4b** illustrates the stable hydrate structure of CH_4-C_2H_6 hydrates at T = 265-305K. According to this predictive phase transition diagram, the phase transition behaviors at T = 274.15K are from sI to sII at y_{CH_4}~0.652, and then from sII to sI at y_{CH_4}~ 0.993. The model predicts the phase boundary transitions matching the experimental report of phase change behavior at the mediate and high methane concentrations.

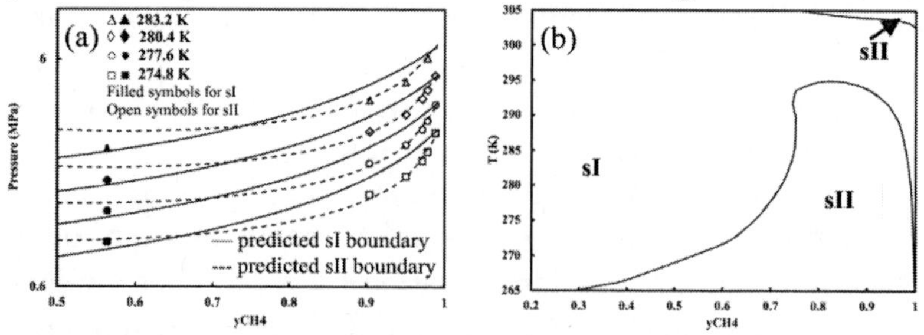

Figure 4. (a) The formative conditions and (b) the predicted phase transition of CH_4-C_2H_6 binary gas hydrates in VLHE region with various vapor mole fractions of methane.

Occupancy

The new model demonstrates its powerful predictive ability in the partitioning behaviors of binary gases in hydrate phase, without including the guest compositions in hydrate phase as a fitting objective. The concentration of methane in hydrate phase of CH_4-C_3H_8, CH_4-*i*-C_4H_{10} binary gas hydrates at VLHE region is in excellent agreement with experiment[21] (see **Figure 5**; water-free mole fractions are presented) in the tendency of temperature and concentration of methane in vapor phase. The average accuracies are 10.60 % AARD-z_{CH_4} and 3.63 % AARD-$z_{C_3H_8}$ in CH_4-C_3H_8 binary gas hydrates (28 pts) as well as 13.09% AARD-z_{CH_4} and 3.45 % AARD-$z_{i-C_4H_{10}}$ in CH_4-*i*-C_4H_{10} binary gas hydrates (19 pts). Moreover, the occupancies of binary gas in the individual hydrate cavity are also predictable with high accuracy in this model. **Figure 6** shows the occupancies of CH_4 and C_2H_6 in the binary hydrate phase in VLHE region at T = 274.2 K. The result indicates that the high quantitively predictive the reported occupancies[21] of methane (4.27 % AARD) in the small cage of sII and qualitative prediction of methane (61.0% AARD) and ethane (11.2 % AARD) in the large cage of sII. A tiny amount of ethane is predicted to occupy both the large cage of sI and sII in this model, matching the experimental observation[22].

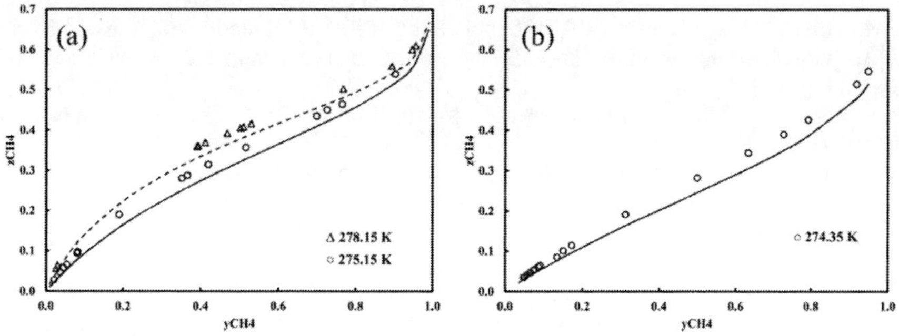

Figure 5. The mole fraction of methane (water-free) in the solid phase of (a) CH_4-C_3H_8 and (b) of CH_4-*i*-C_4H_{10} binary gas hydrates in VLHE. The solid and dashed lines are the predicted results of this work.

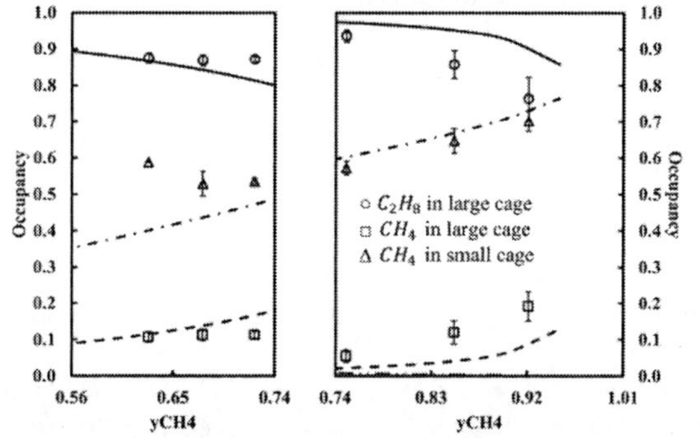

Figure 6. The occupancies of methane and ethane for binary CH_4-C_2H_6 hydrates at T = 274.2K. The solid, dashed, dashed-dotted, and dashed-dotted-dotted lines are the predicted results of this work.

Conclusion

In this work, we investigate the modeling of the phase boundary of mixed gas hydrates over a wide range of conditions: from the VIHE, VLHE, to LLHE. Introducing the guest-guest interaction among the encapsulated molecules, the vdW-P model with explicit inclusion of pressure effect on the Langmuir constant provides a successful description of the water fugacity in the solid hydrate phase. The use of PRSV EoS combined with the COSMO-SAC activity coefficient model provides the fluid phase fugacity without data regression. The universal parameters of guest-guest interaction of each chemical species, contributing to the potential for stabilizing the encapsulated guests within different cavities, are found to accurately describe the formation of the pure and mixed gas hydrates of this study. The model with these guest-guest interactions improves the predictive accuracy, comparing with the calculating result of turning off the guest-guest interactions in single-gas and mixed-gas hydrates. Also, the parameters for the deformation of hydrate lattice are again found to be guest-independent at low to moderate pressures (about 400 MPa); however, to extend the use of the model at higher pressures, the size and shape effects of the guest molecules on the deformation of lattices should be considered to describe the phase behaviors (e.g., retrogression). This updated model extends its ability to predict the coexisting conditions of mixed-gas hydrates, their phase change behaviors, and the occupancy of encapsulated guests in the hydrate phase.

References

1. Sloan, E. D. (1998), "*Clathrate Hydrates of Natural Gases,*" Second ed., New York, NY, Marcel Dekker Inc.
2. Hsieh, M.-K., Ting, W.-Y., Chen, Y.-P., Chen, P.-C., Lin, S.-T., & Chen, L.-J. (2012), "Explicit Pressure Dependence of the Langmuir Adsorption Constant in the van der Waals–Platteeuw Model for the Equilibrium Conditions of Clathrate Hydrates," *Fluid Phase Equilibria,* 325, pp. 80-89.
3. Chin, H.-Y., Hsieh, M.-K., Chen, Y.-P., Chen, P.-C., Lin, S.-T., & Chen, L.-J. (2013), "Prediction of Phase Equilibrium for Gas Hydrate in the Presence of Organic Inhibitors and Electrolytes by Using an Explicit Pressure-dependent Langmuir Adsorption Constant in the van der Waals–Platteeuw Model," *Journal of Chemical Thermodynamics,* 66, pp. 34-43.
4. Khosravani, E., Moradi, G., & Sajjadifar, S. (2012), "Application of PRSV2 Equation of State and Explicit Pressure Dependence of the Langmuir Adsorption Constant to Study Phase Behavior of Gas Hydrates in the Presence and Absence of Methanol," *Fluid Phase Equilibria,* 333, pp. 63-73.
5. Klauda, J. B., & Sandler, S. I. (2000), "A Fugacity Model for Gas Hydrate Phase Equilibria," *Industrial & Engineering Chemistry Research,* 39(9), pp. 3377-3386.
6. Klauda, J. B., & Sandler, S. I. (2003), "Phase Behavior of Clathrate Hydrates: A Model for Single and Multiple Gas Component Hydrates," *Chemical Engineering Science,* 58(1), pp. 27-41.
7. Ballard, A. L., & Sloan, E. D. (2002), "The Next Generation of Hydrate Prediction I. Hydrate Standard States and Incorporation of Spectroscopy," *Fluid Phase Equilibria,* 194, pp. 371-383.
8. Ballard, L., & Sloan, E. D. (2004), "The Next Generation of Hydrate Prediction IV - A Comparison of Available Hydrate Prediction Programs," *Fluid Phase Equilibria,* 216(2), pp. 257-270.
9. Jager, M. D., Ballard, A. L., & Sloan, E. D. (2003), "The Next Generation of Hydrate Prediction - II. Dedicated Aqueous Phase Fugacity Model for Hydrate Prediction," *Fluid Phase Equilibria,* 211(1), pp. 85-107.
10. Stryjek, R., & Vera, J. H. (1986a), "PRSV - An Improved Peng-Robinson Equation of State for Pure Compounds and Mixtures," *Canadian Journal of Chemical Engineering,* 64(2), pp. 323-333.
11. Hsieh, C. M., Sandler, S. I., & Lin, S. T. (2010), "Improvements of COSMO-SAC for Vapor-Liquid and Liquid-Liquid Equilibrium Predictions," *Fluid Phase Equilibria,* 297(1), pp. 90-97.
12. Michelsen, M. L. (1990), "A Modified Huron-Vidal Mixing Rule for Cubic Equations of State," *Fluid Phase Equilibria,* 60(1-2), pp. 213-219.
13. Stryjek, R., & Vera, J. H. (1986b), "PRSV - An Improved Peng-Robinson Equation of State with New Mixing Rules for Strongly Nonideal Mixtures," *Canadian Journal of Chemical Engineering,* 64(2), pp. 334-340.
14. Hsieh, M.-K., Yeh, Y.-T., Chen, Y.-P., Chen, P.-C., Lin, S.-T., & Chen, L.-J. (2012), "Predictive Method for the Change in Equilibrium Conditions of Gas Hydrates with Addition of Inhibitors and Electrolytes," *Industrial & Engineering Chemistry Research,* 51(5), pp. 2456-2469.
15. Kroenlein, K., Muzny, C. D., Kazakov, A., Diky, V. V., Chirico, R. D., Frenkel, M., & Sloan, E. D. (2010), "NIST Gas Hydrate Research Database and Web Dissemination Channel," *Journal of research of the National Institute of Standards and Technology,* 115(2), pp. 85-112.
16. Bandyopadhyay, A. A., & Klauda, J. B. (2011), "Gas Hydrate Structure and Pressure Predictions Based on an Updated Fugacity-Based Model with the PSRK Equation of State," *Industrial & Engineering Chemistry Research,* 50(1), pp. 148-157.
17. Delavar, H., & Haghtalab, A. (2014), "Prediction of Hydrate Formation Conditions Using GE-EOS and UNIQUAC Models for Pure and Mixed-gas Systems," *Fluid Phase Equilibria,* 369, pp. 1-12.
18. Holder, G. D., & Hand, J. H. (1982), "Multiple-phase Equilibria in Hydrates from Methane, Ethane, Propane and water Mixtures," *AIChE Journal,* 28(3), pp. 440-447.
19. Subramanian, S., Kini, R. A., Dec, S. F., & Sloan, E. D. (2000), "Evidence of Structure II Hydrate Formation from Methane+Ethane Mixtures," *Chemical Engineering Science,* 55(11), pp. 1981-1999.
20. Subramanian, S., Ballard, A. L., Kini, R. A., Dec, S. F., & Sloan, E. D. (2000), "Structural Transitions in Methane+Ethane Gas Hydrates — Part I: Upper Transition Point and Applications," *Chemical Engineering Science,* 55(23), pp. 5763-5771.
21. Thakore, J. L., & Holder, G. D. (1987), "Solid Vapor Azeotropes in Hydrate-forming Systems," *Industrial and Engineering Chemistry Research,* 26(3), pp. 462-469.
22. Murshed, M. M., Schmidt, B. C., & Kuhs, W. F. (2010), "Kinetics of Methane-Ethane Gas Replacement in Clathrate-Hydrates Studied by Time-Resolved Neutron Diffraction and Raman Spectroscopy," *The Journal of Physical Chemistry A,* 114(1), pp. 247-255.

608969 Joint (Academic – Industry) Zeolitic Catalyst and Process Development – How Healthy Is the Model?

Abstract:

Driven by efforts to leverage research budgets, introduce more innovation, and improve productivity, many companies have moved to initiate joint catalyst and process development with external organizations. The preferred partners are major research universities, although government laboratories are increasingly engaged. In the US, National Science Foundation (NSF) and funded programs such as NSF's Engineering Research Center ERC) and Industry–University Cooperative Research Centers Program (IUCRC) programs provide incentives for university researchers to collaborate with their industry counterparts. The US Department of Energy (DOE) through its Office of Basic Energy Sciences (BES) Catalysis Science Program has also been an effective facilitator of such relationships. In Europe as well as in China, joint catalyst development programs have been a common practice for many years. There are several recent examples of successful joint university – industry zeolite catalyst and zeolitic catalytic process development. Using information in the public domain, this presentation examines some of the key attributes that have made these joint development programs successful. The presentation also points to practices that may improve the productivity of such relationships. A key question is whether such programs have succeeded too well? Has the intensifying focus on impactful, short term research distorted the mission of academic research to produce the fundamental knowledge and the freedom of academics to pursue research directions that look "interesting" rather than commercially impactful?

609843 Modification of Al Pairing and Distribution in CHA Zeolite

Abstract:

Modifying the aluminum distribution in zeolites could allow one to tune the catalytic, adsorption and/or ion exchange characteristics of the materials. It has been postulated that increased Al pairing in CHA zeolites such that the corresponding Cu-CHA catalysts contain more of the Cu in the Z_2Cu (i.e. Al_2-Cu^{2+}) configuration rather than the Z-Cu(OH) (or Al-$[Cu(OH)]^+$) configuration would lead to improved hydrothermal stability of the catalyst for selective catalytic reduction (SCR)[1]. This is because the Cu^{2+} in the Z_2Cu state is more strongly bound to the zeolite framework. The use of alkali (Na^+, K^+)[2, 3] and Cu^{2+}-containing organic complexes[4, 5] have been suggested as approaches to help alter the Al distribution in CHA zeolites. Each of these techniques appear to have their own limitations – there seems to be an upper limit for Na^+/$TMAda^+$ ratio for the use of Na^+ in the synthesis of CHA while K^+ seems to be leading to the exclusion of $TMAda^+$ from some cages of the CHA zeolite. The use of Cu^{2+}-containing complexes for the synthesis of CHA zeolites seems to require the use of additional alkali to be applicable across a wide range of synthesis conditions. This additional alkali is known to hurt the hydrothermal stability of the resulting zeolite under SCR conditions. An alternate approach to modify Al distribution in CHA zeolites is presented, which may be applicable to a number of other zeolitic systems. The modification of Al distribution is verified by Cu uptake isotherm measurements as well as IR spectroscopy.

References

1. Paolucci, C., et. al. J. Am. Chem Soc. 138, 6028 (2016).
2. Di Iorio, J.R. and Gounder, J. Chem. Mater. 28, 2236 (2016).
3. Di Iorio, J.R., et. al. J. Am. Chem. Soc. 142, 4807 (2020).
4. Ren, L., et. al. Chem. Commun. 47, 9789 (2011).
5. Martinez-Franco, R.et. al. ChemCatChem. 5, 3316 (2013).

602428 Reactive Imbibition of Acidic Hydraulic Fracture Fluids into Unconventional Shales

Abstract:

Unconventional stimulation of gas- and oil-bearing shales by injection of acidic hydraulic fracturing fluids (HFFs) increases shale permeability/porosity and dissolves carbonates, pyrite, and other minerals, which promotes hydrocarbon extraction. In addition to the dissolution of shale minerals, fluid transport also results in the precipitation of secondary minerals (e.g., barite, celestite, gypsum, iron oxides) in the shale matrix, which ultimately reduces permeability and transport. One major consequence of the interaction of shales with HFFs is the overuse of freshwater, resulting in the production of considerable quantities of hard-to-recycle highly saline waters, which threatens watersheds. Improved understanding of the impact of unconventional stimulation on fluid transport is highly desirable as it provides one of the best ways to diminish the environmental impacts of this water-intensive practice.

Here we report unique experimental observations of fluid transport in shales by tracking the rate and extent of imbibition of a bromide tracer in acidic HFFs using synchrotron X-ray fluorescence mapping combined with compositional analysis of the bulk fluid. This approach yields a direct record of time-resolved selective ion diffusion resulting from the penetration of HFFs and attendant mineral transformations. Our experiments highlight for the first time that the penetration of acidic HFFs and the dissolution calcite occur at similar rates in carbonate-bearing shales, suggesting that porosity increases rapidly during imbibition of HFFs and thus promotes fluid penetration. We also show that the variability in mineralogy and chemical reactivity of shales can directly affect reactive imbibition. For example, the presence of major carbonates in the shale matrix reduces reactive imbibition of acidic HFFs due to a rapid neutralization of pH and inhibits fluid penetration. We also found that the presence of Ba in HFFs significantly retards reactive imbibition of acidic HFFs due to rapid precipitation of Ba-bearing minerals, which reduces pososity at the fluid-rock interface. Finally, we show the importance of high ionic strength solutions, similar to produced waters, on the release of colloids and the impact of colloids on ion diffusivity. The main finding from these experimental studies is that the chemical reactivity of unconventional shales regulates the extent and rate of imbibition of HFFs into the shale matrix.

601488 The Impact of Acid-Base Stimulation Sequence on Mineral Stability for Tight/Impermeable Unconventional Rocks: Delaware Basin Case Study

Abstract:

Technological innovations in fracture stimulation of unconventional reservoirs have spurred an energy boom, dramatically lowering CO_2 emissions in the US due to the large-scale substitution of natural gas for coal for electricity generation. However, mineral precipitation due to reactions with injected fluids during unconventional stimulation is a well-recognized problem. Laboratory experiments using a typical hydraulic fracture fluid (HFF) formulation for the Delaware Basin shows that the injection of this fluid initiates two major chemical processes: (i) *mineral dissolution*, creating beneficial secondary porosity; and (ii) *precipitation* of mineral scale (*e.g.*, barite, Fe(III)-(hydr)oxides, etc.) on fracture surfaces and in the shale matrix that can reduce porosity and permeability. These two processes have strong competing effects on the rock that can significantly impact production. Additionally, shale reservoirs are fractured using injections of strong acid (15% HCl) followed with sand slurries containing chemical amendments (pH 7-10) resulting in large pH variations. Consequently, chemical conditions in stimulated reservoirs can depart significantly from simplified model fluids containing only a few targeted chemicals (such as scale inhibitors, biocide, etc.) that have been studied. The goal of this study was to evaluate secondary mineral precipitation and permeability attenuation under "real-world" chemical injection scenarios (formulations, injection sequences, T/P, etc.) specific to the Delaware basin.

Whole cores (1" diameter x 1" length) and ground shale (150-250 μm) from carbonate-rich Bone Spring Formation, Delaware Basin TX, were reacted at 80°C and 85 bar using a HFF recipe and sequential injection typical of Delaware Basin. One subset of ground samples was reacted in glass serum bottles (80°C and ~1 bar) with solution sampling every 72 hours for 3 weeks to evaluate solute evolution with time. We then analyzed these reacted shales and solutions using a variety of laboratory-based and synchrotron-based techniques to characterize both the chemical and spatial distribution of secondary mineral precipitation and identify changes in permeability and mineralogy.

Despite being relatively impermeable (< 5 nD), the HFF was able to penetrate the shale and react with the Fe-bearing phases. The initial pH of the solution is ~0 due to the HCl spearhead. Subsequent additions of borate crosslinked slickwater (pH of 9.2) caused a dramatic increase in solution pH due to a combination of dilution/neutralization by the slickwater and reaction of HCl with the calcite and dolomite the rock. Synchrotron-based X-ray fluorescence mapping coupled with X-ray absorption spectroscopy (both bulk and micro) showed that almost all the iron had been oxidized to Fe(III). These results demonstrate the importance of the acid spearhead and subsequent formulation in order to move the HFF into the rock. These results also improve our understanding of the geochemical reactions occurring in shale reservoirs during fracture stimulation.

609044 Evaluation of Iron Precipitation in the Utica/Point Pleasant Unconventional Play Using Laser Induced Breakdown Spectroscopy (LIBS)

Abstract:

The development of U.S. unconventional resources has had a dramatic impact on global energy supply, with the U.S. becoming the world's leading producer of petroleum and natural gas. This change has been beneficial to the public lowering energy prices, while increasing energy security. Unconventional (or shale) resources differ from conventional reservoirs in a multitude of ways. Unconventional plays undergo significant production drops up to 75% [1] after one year, compared to 5% in conventional wells [2]. Unconventional wells also differ structurally and geologically as they consist of micro-pore and mainly nano-pore structures of interbedded carbonate and hydrocarbon containing shale layers which serve as the source for reactions with fracturing fluid. A majority of the reactions that occur during fracture process happen in the early stages, as the fracturing fluid has yet to come to equilibrium with the reservoir. While some reactions will create the pathways for hydrocarbons to flow through the well; others may cause the shale surface to be coated with insoluble barium, calcium, strontium or iron based solids/precipitate. These insoluble solids tend to clog the fracture and pore networks creating large decreases in production. Prevention of scaling is vital for maintaining well integrity, which is usually achieved by adding scale inhibitors to the fracturing fluid. However in some wells, a scale inhibitor may be incompatible with well geology or ineffective once equilibrium has been reached resulting in scaling and decrease production. Understanding shale-fluid chemistry aids identification of effective scale inhibitor(s) or fracturing fluid formulation modification to improve well performance.

Ohio University (OHIO) is investigating scale formation as significant contributor to unconventional well performance loss, with a particular focus on the Utica/Point Pleasant (UPP) shale play. Shale-fluid interactions are being investigated in bench-scale equipment at reservoir conditions. Previously, it was found iron-based precipitates are forming in UPP wells under experimental trials and iron control additives are effective in maintaining iron in the aqueous phase. This study will evaluate modifications to hydraulic fracturing formula chemistry in order to evaluate the effect on iron precipitation. Pre- and post-trial analysis of the shale surfaces will be performed using Laser-Induced Breakdown Spectroscopy (LIBS), and fluids will be analyzed using Inductively Coupled Plasma Optical Emission Spectrometry (ICP-OES) and Ultra-Violet Visible spectroscopy (UV-Vis). The effect of various commercially available inhibitors and their concentrations will be evaluated. In addition, modeling of the shale-fluid interactions using Geochemist Workbench will be performed and presented.

[1] H. Wachtmeister, L. Lund, K. Aleklett, and M. Höök, "Production Decline Curves of Tight Oil Wells in Eagle Ford Shale," *Nat. Resour. Res.*, vol. 26, no. 3, pp. 365–377, Jul. 2017.

[2] M. Höök, R. Hirsch, and K. Aleklett, "Giant oil field decline rates and their influence on world oil production," *China Energy Effic.*, vol. 37, no. 6, pp. 2262–2272, Jun. 2009.

600100 Improving CO_2-EOR in Shale Using Dilute Amounts of CO_2-Soluble, Wettability-Altering Surfactants

Abstract:

The objective of this work is to determine if the efficacy of CO_2-based EOR techniques for unconventional reservoirs can be enhanced with dilute concentrations of wettability-altering additives. Addition of ionic or nonionic wettability-altering surfactants to brine to improve oil recovery from unconventional formations has been previously studied for water-based hydraulic fracturing operations and subsequent water-based EOR projects. In this work, nonionic surfactants are dissolved in CO_2 (which is a reasonably strong solvent for several types of oil-soluble or water-soluble nonionic surfactants) to attain similar favorable shifts in wettability toward increased water wetness. In essence, this project aims to add favorable shifts in wettability to the suite of mechanisms (e.g. miscibility with oil, oil swelling, oil viscosity reduction, extraction of oil into CO_2, CO_2 diffusion into oil, solution pressure drive upon depressurization) that have been previously attributed to the success of CO_2 recovering oil from shale samples.

The solubility of several nonionic surfactant (which also exhibit water-solubility) was determined at concentrations between 0.1 – 1.0 wt% in CO_2. The cloud point curves for several surfactants were determined at temperatures representative of unconventional formations.

Contact angle measurements were performed on Bakken and Eagle Ford rock samples using dead oil from both formations and synthetic brine. Shale samples were first aged in oil at high temperature and pressure to restore oil-wet characteristics to the rock. The contact angles of water and oil droplets on these samples were determined. Then, shale samples were immersed in water (control test) or an aqueous nonionic surfactant solution containing 0.1 – 1.0 wt% surfactant to verify that the previously reported shifts toward water-wetness occur. Oil-wet shale samples were also immersed in high pressure, single phase CO_2-nonionic surfactant solutions containing 0.1 – 1.0 wt% surfactant, after which contact angles of oil and water droplets on the rocks were determined.

After assessing the change in wettability attributable to the addition of surfactant, lab-scale CO_2 huff 'n puff EOR experiments were conducted on cylindrical, oil-saturated, oil-wet cores immersed in either pure CO_2 or CO_2-surfactant solution. The amount of additional oil recovery attributable to the introduction of dilute amounts of CO_2-soluble surfactants was assessed.

To the best of the authors' knowledge, this is the first attempt to dissolve dilute concentrations of wettability-altering chemicals in CO_2 for EOR in unconventional liquid reservoirs. Although dilute amounts of water-soluble/CO_2-soluble surfactants have been previously dissolved in CO_2 for improved mobility control in conventional formations via the in situ generation of CO_2-in-water foams as the surfactant partitions into the brine, this study focuses on the addition of nonionic surfactants to CO_2 with the sole objective of having the surfactant adsorb onto the rock and favorably alter wettability.

594357 Multi Scale Flow Observation and Stability Measurement for Environmentally Friendly Waterless Fracturing Using Supercritical CO_2 Foam

Abstract:

The energy sector is experiencing a historic boost in oil and natural gas production thanks to development of hydraulic fracturing for unconventional recovery from tight shale formations. However, the process suffers from large dependency on water resources with reported negative environmental impacts. Reduction of freshwater use in hydraulic fracturing via high internal phase emulsions (foams) is a promising approach to protect drinking water resources and enhance ground and surface water quality to avoid large water disposal which may lead to seismic activities. A sustainable approach for reduction of freshwater consumption and produced water disposal is to inject water-less fracturing fluid with minimum aqueous phase volumes obtained from the production cycle itself. The non-aqueous phase, however, could be obtained from an external gas source such as N_2 or CO_2. Application of compressed CO_2 as a non-aqueous phase improves the fluid viscosity, fracture propagation and oil production and it's been introduced as a promising candidate for CO_2 storage and sequestration in shale formations. The mixture of produced water and compressed CO_2 in supercritical state must be stabilized to form homogeneous supercritical CO_2 ($scCO_2$) foam. In this work the multi scale observation and stability measurement of $scCO_2$ foam in different stages of generation, proppant transport, and fracture propagation during a hydraulic fracturing are presented.

Our approach addresses microstructural observation of high internal phase emulsion containing a gas-water lamella stabilized by polyelectrolyte complex nanoparticles (PECNP) and wormlike micelles (WLMs) with electrostatic interactions as defining factors to improve the disjoining pressure over capillary drainage for a better foam stability. The length scale involved in shale production covers multiple orders of magnitude ranging from macro- to micro- scale. Therefore, millimetric view cell observation was coupled with micrometric fluidic visualization to shed light on multi scale observation of physical structure, geometry, dynamic and stability of electrostatically enhanced $scCO_2$ foam. Lamella stability as a result of complexation of two oppositely charged polyelectrolytes with zwitterionic surfactant were investigated in view cell and glass microchips. The electrostatically stabilized interface between the highly concentrated electrolyte and the $scCO_2$ is attributed to carrying the proppants and fast degradation of lamella in the presence of crude oil to maintain fracture conductivity and fast flow back to the surface.

Formation of highly viscous, dry foam, capable of improving fracture propagation, proppant transport, and fracture cleanup were observed in miniature field of view of vertical view cell column where the foam is generated, and stability is influenced by gravity drainage. Foam microstructural graphs were examined to obtain bubble size and size distribution. Subsequently, morphology and structural parameters of foam bubbles including size, size distribution and lamella thickness as well as film drainage and rate of deformation were studied via visualization techniques, post processing software and image analysis. Finally, the hydrodynamic and flow regime parameters were derived to correlate chemical properties of lamella and multiphase flow dynamic and foam physical/geometrical properties. Superior stability of $scCO_2$ foam enhanced with complexes of PECNP and surfactants over a lamella containing sole surfactant was observed in millimetric scale whereas polyhedral structure of dry foam degrades faster in lamellae enhanced with supercharged particles when the foam meets oil. This phenomena occurs due to interaction of oil with the chemical components residing in the lamella such as PECNPs. Synthesized ionic complexes are susceptible to degradation and secession of coiled PEI-DS chains and electrostatic desorption of amine functional groups.

Multiphase flow in fractured medium was emulated using a lab-on-a-chip device. Different etching strategies (wet/dry) for fabrication of pressure resistive microfluidic chips were introduced to study microscale $scCO_2$ bubble geometry, transport and stability in simulated pathways in harsh geological conditions that may range from simplified patterns to complex microcracks based on tomography data. Wet etching technique on glass was performed via UV photolithography, surface layer removal and thermal bonding, whereas, dry etching was conducted with selective laser etching (SLE) inside the glass bulk followed by acid/base wetting for glass debris removal. Micrographs obtained with two etching techniques were analyzed to obtain structural and geometrical parameters for confined $scCO_2$ bubbles in fractured networks. The foam stability and fluid loss in a simulated fracture were investigated and the rate of fluid leak-off was compared for $scCO_2$ foam stabilized with surfactant and PECNP/Surfactant complexes. The microscale observation sheds light on the strong electrostatic compatibility of the synthesized PECNP with zwitterionic surfactants prepared in highly concentrated brine which helps to develop a long term stable $scCO_2$ bubble in fractured medium and effectively addresses the common stability issues with surfactant generated foams suffering low viscosity and proppant carrying capability problems in hydraulic fracturing application.

598993 Interactions of Proppant with Dynamically Complexing, pH-Responsive Viscosity Modifying Agents for Hydraulic Fracture Fluids

Abstract:

Hydraulic Fracturing is now a key technique of extracting unconventional oil and unconventional gas to improve the recovery of oil and gas. One of the essential components of fracking fluid is the viscosity modifying agent, which prevents settling and non-homogenous dispersion of proppant and provides a strong driving force on the proppant to follow the fluid into cracks, fractures, and fissures. It is impossible to acquire adequate and effective transportation of proppants from the ground to the fissure without the system including viscosity modifying agent. Therefore, we report a novel type of viscosity modifying agent based on dynamic binary complexation of a long-chain betaine and amine to actively and reversibly control the settling velocity of proppant. With these viscosity modifiers, the settling velocity of sand could be changed a few orders of magnitudes reversibly by adjusting pH value. The enhanced proppant suspension stability is ascribed to the combination of viscosity effects and attractive intermolecular interactions between the complexes and proppant. Besides, the rheology and interfacial phenomena of the system containing the viscosity modifying agent will be discussed. Furthermore, the proppant settling behavior of these viscosity modifiers has been investigated as a function of temperature and salinity in the conditions relevant to major shale basins in the US to evaluate the potential of applicability in the next-generation hydraulic fracturing fluids.

602559 Characterization of Sorption, Deformation and Mass Transfer in Eagle Ford Shale Samples

Abstract:

When shale rocks are exposed to gases like CO_2 and CH_4, their transport properties can change due to adsorption-induced deformation of their structure, particularly their organic content. The present study investigates the change in mass-transfer and sorption characteristics of Eagle Ford Shale samples when by exposed to a sequence of Helium, Methane and CO_2 at relevant (including stress) reservoir conditions.

In this study, whole cores and trims, extracted from the Eagle Ford Shale formation, are first characterized for their structural properties and sorption characteristics. The core samples are then tested in a sequence of flow-through, pulse-decay and depletion experiments, using Helium, Methane and CO_2 as working gases, to study the interplay between mass transfer, adsorption and deformation.

During the flow-through experiments, we vary the confining pressure in a loading/unloading cycle to investigate the impact (and reversibility) of net stress on permeability at a fixed (average) pore pressure. We vary, in addition, the pore pressure at a fixed confining pressure to delineate the impact of sorption on the observed behavior.

Separate measurements of sorption kinetics and equilibrium (isotherms) are performed on core-trims via thermogravimetric analysis (TGA) in a magnetic suspension balance.

Pulse-decay experiments (loading and unloading) are performed to bridge the independent measurements of permeation and sorption, and we investigate whether pulse-decay behavior can be interpreted from independent flow-through and sorption experiments.

To further facilitate the interpretation of our experimental observations, the deformation, in terms of volumetric strain, is also monitored in situ during the experiments via direct strain gauge measurements.

The overall objective of this experimental effort is to develop an improved understanding of the link(s) between adsorption, deformation and transport in shales, and to explain how pore-level events translate into observed bulk behavior. The improved understanding of these mechanisms will serve to promote accurate estimation of larger-scale shale-gas production behavior.

604792 Blending Optimization of Low-Vapor-Pressure Gasoline-Range Bio-Blendstocks: Linking Economic Values to Fuel Properties

Abstract:

Biomass derived biocrude or bio-oil can be directly upgraded into transportation fuels but subject to high cost penalty and product yield loss associated with deoxygenation. Adding non-traditional blendstocks derived from biomass into gasoline has gained renewed interest since it will avoid expensive deoxygenation processes, reduce greenhouse gas emissions, potentially enhance fuel properties and boost engine efficiencies. A competitive bio-blendstock candidate should have either lower prices or superior properties or both relative to conventional gasoline blending components. Note that gasoline is a mixture of hundreds of compounds specified by properties rather than compositions. The addition of bio-blendstocks, most likely oxygenates, will modify the property mixing rules due to the non-ideal interactions between polar and nonpolar components. In this talk, we will present an equation-of-state-based model for predicting Reid vapor pressure, reduced order non-linear property blending models for computing key properties of final gasoline products, a blending optimization approach to evaluate the economic value of bio-blendstocks to petroleum refiners and to identify the key fuel properties that drive economics, as well as an uncertainty analysis regarding property model and economic assumptions. In additional to ethanol, six low-vapor-pressure bio-blendstock candidates were evaluated in this work: i-propanol, n-propanol, i-butanol, diisobutylene, cyclopentanone, and a mixture of furans.

610149 Solar Fuel Production from H_2o/CO_2 Via Zr-Doped Ceria Based Thermochemical Cycles

Abstract:

Worldwide human development increase with rise in population, which essentially depends on the large amount of the energy and power supply. Currently power generation in many industries (chemical, steel, automobile, and cement) largely depends on supply of fossil fuels like coal, light diesel oil, petrol. These ultimately emits large of CO_2 gas into the atmosphere and caused into upsurge into CO_2 level and accordingly 64% increase in global house effect and related climate changes. Such global warming potential affects the living life on the earth and extremely important to research towards CO_2 capture and conversion carbon neutral technology. In this paper, our focus was on developing carbon neutral and renewable technology to split H_2O/CO_2 into H_2/syngas via two step thermochemical redox cycle. Firstly CO_2 is captured by various methods and used for the two step (reduction and oxidation) thermochemical splitting of CO_2 into CO by using metal oxide as a redox materials. The CO generated can be mixed with the H_2 produced from thermochemical splitting of H_2O. The synthesis gas generated via such thermochemical redox reactions can be directly converted into liquid fuels via Fischer Tropsch process. In this study, investigation was focused to extend the thermochemical splitting capability of Zr-doped ceria. The ceria-zirconia oxides were prepared by using co-precipitation with aqueous ammonia as percipients. Dervied materials were characterized by using various methods. Finally, the redox performance towards thermal reduction (TR) and H_2O/CO_2 splitting reactions was estimated by conducting consecutive thermochemical redox cycles in the thermogravimetric setup (TGA). Also, a detailed solar reactor efficiency analysis will be conducted to estimate the solar-to-fuel energy conversion efficiency by using the real-time experimental data.

603408 Production of Biorenewable Thermal Deoxygenation Oil and Upgrading to Diesel and Jet Fuel Range Hydrocarbons

Abstract:

Thermal DeOxygenation (TDO) is a novel non-catalytic thermochemical process which converts neutralized biomass acid hydrolysate into synthetic crude oil at 450 °C and atmospheric pressure. TDO oil is highly stable and has a very low oxygen content (< 5 wt %) and a low total acid number (TAN) compared to pyrolysis oil. Certain properties of the TDO oil such as broad boiling point range (75-550 °C) and high aromaticity allow it to be compatible with existing petroleum refinery operations and petroleum products. In this study, crude TDO oil was produced from a mixture of levulinate and formate salts using a pilot scale continuous flow reactor, and upgraded and fractionated to produce gasoline, jet fuel, and diesel range hydrocarbons.

Crude oil was hydrogenated to remove residual oxygen and increase hydrogen content of the oil using a commercial Ni/SiO_2-Al_2O_3 catalyst in bench scale continuous flow tubular reactors. The crude and hydrogenated TDO oils were distilled to produce jet fuel (180-250 °C) and diesel (150-325 °C) range hydrocarbons using a fractional distillation apparatus according to American Society for Testing and Materials (ASTM) standards. Four different TDO oil blendstocks (crude TDO oil, Hydrotreated TDO oil, distilled TDO oil, and distilled Hydrotreated TDO oil) were analyzed for chemical and physicochemical properties following ASTM standard methods. Properties of the TDO fuels were compared with ASTM and military fuel standards (MIL-DTL), and blended appropriately to meet the specifications. TDO diesel blendstocks were blended with Ultra Low-Sulfur Diesel (ULSD) in four different volume percent (5 %, 10 %, 15%, and 20%) to produce sixteen different fuel blends. Combustion and emission behaviors of the fuel blends were studied using an air-cooled single-cylinder diesel engine.

Fuel properties such as hydrogen content, specific gravity, viscosity, and net heat of combustion of TDO oil could be improved with hydroprocessing. Jet fuel and diesel fractions of hydroprocessed TDO oils meet all fuel specification with a few exceptions (density, hydrogen content, and heating value for jet fuel and cetane number for diesel). Fuel blends prepared with hydrotreated TDO oil and ULSD exceed or fall within the specified range of fuel specifications. Results show that the TDO oil is a potential renewable fuel as well as a blending agent for making fuel blends with petroleum products.

611819 CO Hydrogenation Mechanism at Varying Partial Pressures over Co-Based Catalysts

Abstract:

Introduction

The catalytic hydrogenation of CO according to Fischer-Tropsch (FT) over Co-based catalysts allows producing long-chain hydrocarbons and oxygenates for the use of transport fuel or as feedstock for lubricants, detergents and plasticizers, respectively. Here we will scrutinize which of the mechanistic suggestions for the reaction are compatible with the experimental results obtained by quantitative chemical transient kinetics (CTK) [1,2].

Materials and Methods

Sudden changes in the chemical composition of the reactant feed are produced to follow the construction of the catalytically active phase (build-up) or its scavenging (back-transient) as a function of time. Data have been quantitatively evaluated at any time to count the amounts of C and O atoms accumulating on the catalyst.

Results and Discussion

Figure 1a demonstrates high surface coverages to be reached during atmospheric CO hydrogenation over Co/MnO_x catalysts. In particular, significant carbon accumulation is encountered for a $H_2/CO=1$ pressure ratio (Fig. 1c). HRTEM shows that Co particles transform into Co_2C for this catalyst [1]. A CO-insertion mechanism occurs at high coverages under these partial pressure conditions since the chain lengthening probability during build-up is directly proportional to the transient CO pressure. Increasing the partial pressure ratio to $H_2/CO=2$ (Figure 1b, d) shows lower carbon accumulation (Figure 1d). Transient chain lengthening probabilities are now directly proportional to both transient CO pressures and accumulated carbon concentrations. We conclude that chain lengthening via CO insertion and C-C coupling are in competition here. Chain growth via CO insertion likely involves formate/carboxylate-derived surface species as well as alkoxy, in agreement with DRIFTS observations [1]. A completely different behavior is observed for Co/SiO_2 catalysts (Figure 1c, f) demonstrating the importance of support effects.

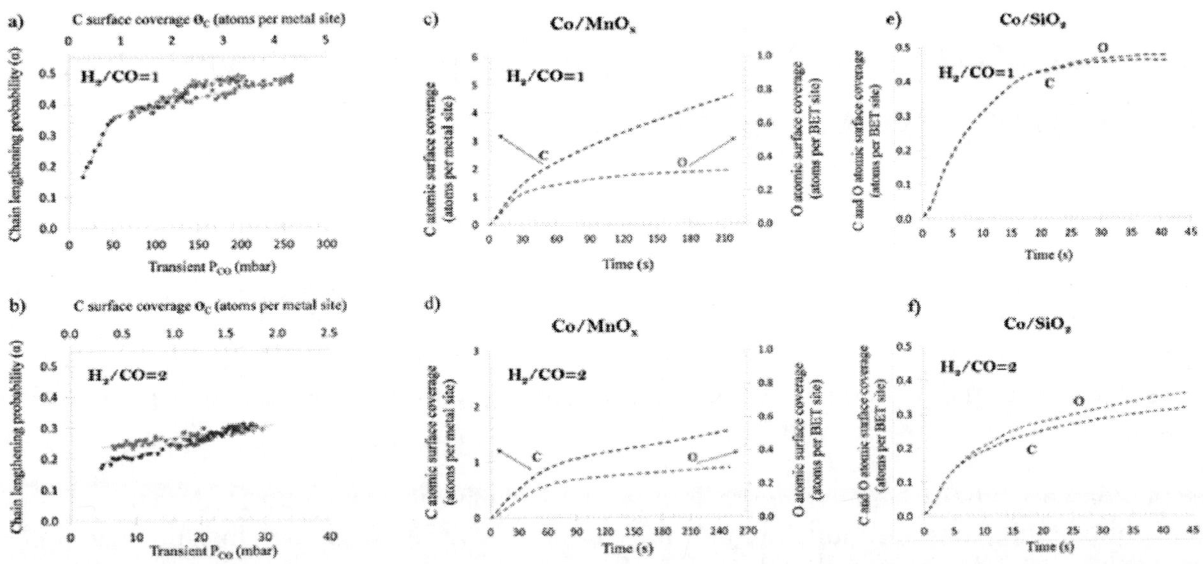

Figure 1. Chain lengthening probabilities (calculated from build-up transients) as a function of the transient CO partial pressure and carbon surface coverage at $p_{H2}/p_{CO}=1$ and $p_{H2}/p_{CO}=2$ for Co/MnO_x (Figures 1a, b). Carbon and oxygen accumulating amounts at the same p_{H2}/p_{CO} ratios for Co/MnO_x (Figures 1c, d) and Co/SiO_2 catalysts (Figures 1e, f).

612265 Metal-Oxide Mediated Redox Scheme for Hydrogen Production from Reducing Fuels

Abstract:

A metal-oxide mediated redox scheme (chemical looping) is presented as an advanced alternative to produce Hydrogen. Previous studies have shown promising results in quantifying the thermodynamic limits of the system at atmospheric pressure, however many applications of H2 require it to be at a higher pressure. The design of a redox scheme at higher pressures provides unique challenges, each of which require careful and holistic consideration.

This presentation will initially summarize previous work on chemical looping for Hydrogen production while highlighting the motivation for the current study. The study will then detail a systematic approach for de-risking technology gaps for scale-up using appropriate experimental studies and thermodynamic simulations. The data obtained from experimental studies and thermodynamic simulations will be used to develop a reactor module design that will be further evaluated using sensitivity studies. The sensitivity studies will use accurate surrogate models constructed from experimental data and black-box optimization principles to identify trade-offs between performance and cost. The overall results will highlight the intricate balance of key operating variables and design parameters that must be achieved to maintain performance advantages of metal-oxide redox schemes at higher pressures.

605383 Metal Loading Effects on the Stability of Ni-Ru Catalysts for Hydrogen Production from Liquid Hydrocarbons at High Pressure and Low Temperature

Abstract:

Hydrogen has been gaining attention as an important energy carrier in the future. Its cost needs to be reduced to allow its market penetration at an acceptable level. Possible solutions are to use a liquid fuel that lowers transportation costs and to reduce the complexity of the current hydrogen production system. A new process that integrates a pre-reformer (PR) and a membrane reformer (MR) can be considered. The PR operates at high pressure and low temperature to produce a methane-rich stream from heavy naphtha. It then enters the MR, where methane steam reforming occurs. Hydrogen separation also occurs due to the high-pressure operation and leads to efficient conversion of methane. The pre-reforming reaction requires a sufficient amount of active metals as well as strong metal-to-support interactions to crack complex hydrocarbons in the fuel at low temperatures (500 ~ 600°C). In order to gain commercial significance, the pre-reforming catalyst needs to have high enough performance to convert the complex hydrocarbons to the methane-rich stream in a stable manner. In the past, we developed our Ni-Ru/CGO catalyst for low-temperature pre-reforming. Thus, we optimized the metal loading of the catalyst for our specific applications. Experiments were performed by changing the nickel and ruthenium contents. The pressure, temperature, steam-to-carbon ratio, and GHSV were 25 bar(g), 550°C, ≤ 3.0, and 100,000 h^{-1}, respectively. Post-experimental analysis on the catalyst samples were performed to investigate the effects of the metal loading. The catalyst had longer duration as its Ni loading increased from 19.5 to 49.5 wt%. Our results can be used as a reference for building our proposed system on an industrial scale.

611151 Hydrogen Production By Liquid Fuel Reforming for on-Board Fuel Cell Applications

Abstract:

Hydrogen production by liquid fuel reforming is promising technology due to the high hydrogen density of the liquid fuels. One of the liquid fuel, diesel, is a common energy source of heavy-duty applications and remote area applications. With the liquid fuel reforming technology, hydrogen can be easily produced on-site from diesel. A fuel cell can utilize the produced hydrogen and generate electricity for various applications.

There are some requirements to apply the liquid fuel reforming technology for on-board fuel cell applications. High mechanical and thermal durability of reforming catalyst, fast start-up time of the reforming system, and complying environmental policies are those requirements that should be considered. In this study, a diesel autothermal reformer integrated with fuel cell was developed and demonstrated with consideration of those aspects.

Introducing a structured reforming catalyst enhanced the mechanical and thermal durability of the liquid fuel reforming system. From the lab-scale tests, the catalytic durability was validated, and the catalyst was scaled-up to the 1 kW_e operation. The structured catalyst showed good durability and stability with even several thermal cycles. Designing novel reforming system configuration with heat utilization optimized the start-up time. By providing a small amount of external heat at the initial heat-up period, steam required for the reforming process could be stably generated, and the reaction temperature could be reached. Moreover, in the developed fuel cell system, catalytic off-gas oxidizer was introduced for complete oxidation of carbon monoxide and proving heat to the fuel cell system.

From the development process of the autothermal reformer and 1 kW_e diesel-driven fuel cell auxiliary power unit, the liquid fuel reforming technology for on-board applications was demonstrated. The developed technology could be applied to broad applications such as backup power and heavy-duty vehicles.

608698 Hydrogen Production from Thermo-Neutral Reforming of Diesel Using Dual Catalyst System

Abstract:

Hydrogen production is gaining increased importance globally for producing clean energy for using in fuel cells. Hydrogen demand in the refining sector is also expected to continue its growth driven mainly by stricter fuel specifications. In addition, hydrogen-based fuel cells for automotive and stationary applications are gaining popularity for various reasons including their higher efficiencies and lower emissions. Use of liquid hydrocarbon fuels to generate hydrogen is being considered as an immediate solution for large scale hydrogen production. Diesel steam reforming is known for posing a major challenge due to its high carbon content, aromatics, and sulfur, which play a major role in catalyst deactivation. Thermo-neutral reforming of diesel with single TNR catalyst had a limited conversion and lower subsequent hydrogen yield. To overcome this problem a dual catalyst system using a multi-components TNR catalyst and a commercial steam reforming catalyst was achieved with complete diesel conversion. A two stage catalyst loading was attempted using TNR catalyst at the top of the catalyst bed and steam reforming catalyst at the bottom of the bed. Relatively, smaller amount (6 cc) of TNR catalyst was loaded on top of larger amount (50 cc) of steam reforming catalyst. The catalyst was reduced with dilute hydrogen for 4 hours prior to the reforming operation. Initial assessment of the catalysts performance showed complete conversion of diesel for a period of 70 hours time-on-stream with above 60 mol% hydrogen yield. Above 20% water conversion was achieved resulting higher hydrogen yield (above 90 l/h). During the experiment higher steam to carbon ratio (3.5) was used to prevent coke formation. The results show promising features of diesel reforming for on-board as well as stationary power generation applications and detail results on catalyst formulation, characterization and testing with diesel fuel will be presented.

611857 Renewable H_2 Production from Steam Reforming of Glycerol (SRG) over Encapsulated Ni/Silicalite-1 Zeolite Supported Catalyst

Abstract:

Introduction

Biodiesel, an environmentally friendly fuel can be produced through catalytic reaction of oil and renewable alcohols; however, biodiesel production generates glycerol as a by-product and is being treated as a waste. Therefore, developing a new route for glycerol conversion into high-added value products is an attractive method to utilize it for economic viability of biodiesel industries. As shown by Equations (1)-(3), various reactions, including SRG, glycerol decomposition and water gas shift,[1] making the system rather complex to be selective to produce H_2 can be promoted. Therefore, the aim of this work focuses on improving the catalytic activity, stability and coke-resistant Ni-based silicalite-1 catalyst by encapsulation process.

Materials and Methods

In this study, Ni-Silicalite-1 zeolite crystals were synthesized using similar approach as reported by Wang [1]. The SRG experiments were carried out at 750 °C for 6 h, and samples of the product gases (H_2, CO_2, CO and CH_4) were subsequently analyzed using gas chromatograph (GC, PerkinElmer, Clarus® 580) equipped with TCD and FID, respectively.

Results and Discussion

A glycerol conversion of ~ 99% was achieved for Ni/Si-1-Encapsulated catalyst, while a lower glycerol conversion of ~85 % was obtained for Ni/Si-1-impregnated between 1 – 2 h, then it dropped to ~80% for 3-6 h as shown in Figure 1. Similarly, the selectivity of H_2 (Figure 2b) was more active (~60%) and stable during the time-on-stream (TOS) as compared to Ni/Si-1-impregnated catalyst shown in Figure 2a.

Significance

The results showed that the Ni/Si-1-Encapsulated catalyst was more active and selective towards H_2 production from SRG and exhibited coke resistance ability at 750 °C.

Reference

[1] Wang N, Sun Q, Bai R, Li X, Guo G, Yu J. In Situ Confinement of Ultrasmall Pd Clusters within Nanosized Silicalite-1 Zeolite for Highly Efficient Catalysis of Hydrogen Generation. Journal of the American Chemical Society 2016;138(24):7484-7.

$$C_3H_8O_3 + 3\ H_2O \rightarrow 3\ CO_2 + 7\ H_2 \qquad \Delta H_{298} = 127.7\ kJ/mol\ (1)$$

$$C_3H_8O_3 \rightarrow 3\ CO + 4\ H_2 \qquad \Delta H_{298} = 251.2\ kJ/mol\ (2)$$

$$CO + H_2O \leftrightarrow CO_2 + H_2 \qquad \Delta H_{298} = -41.2\ kJ/mol\ (3)$$

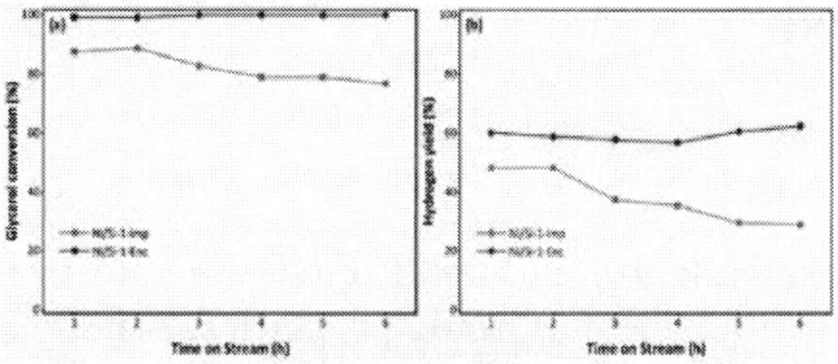

Figure 1. (a) Glycerol conversion, (b) Hydrogen yield. Reaction conditions: P = 1 atm, T = 750 °C, SGFR = 10:1, WHSV = 44,122 mL g_{cat}^{-1} h^{-1} (STP), TOS = 6 h.

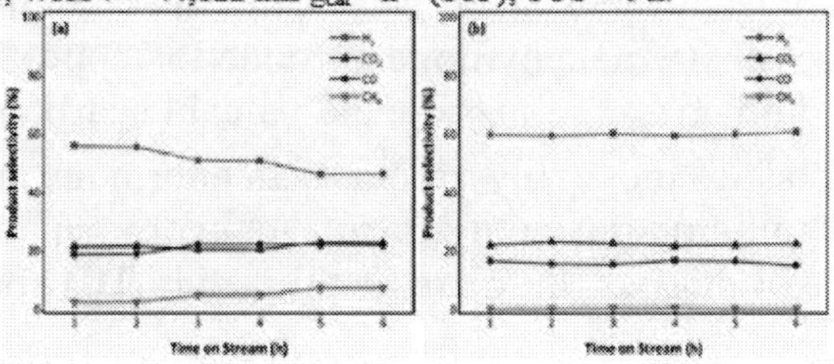

Figure 2. (a) Ni/Si-1-imp, (b) Ni/Si-1-Enc. Reaction conditions: P = 1 atm, T = 750 °C, SGFR = 10:1, WHSV = 44,122 mL g_{cat}^{-1} h^{-1} (STP), TOS = 6 h.

606958 Thermocatalytic Fluidized Bed Co-Pyrolysis of Methane and Plastic Waste into Hydrogen and Carbon Nanotubes over a Multi-Metallic Catalyst

Abstract:

Production of hydrogen at a low cost without polluting the environment has gained prime attention nowadays due to its high energy potential, clean-burning properties. Among all the commercial processes for hydrogen production, the well-established steam methane reforming (SMR) process is a dominant process due to its high efficiency (~65%). However, thermo-catalytic decomposition of methane (TCD) has emerged as a novel alternate hydrogen production approach considering similar energy requirement as SMR along with no CO_x emission and valuable carbon byproducts. The rapidly increasing worldwide demand for plastics inevitably produces an enormous amount of waste plastics which needs to be utilized without polluting the environment. As hydrogen and carbon are the main building blocks of the majority type of plastics, they can be utilized in TCD process for increasing the yields of hydrogen and carbon.

In this context, decomposition of methane and plastics in a fixed bed, as well as fluidized bed reactor (FBR), were investigated. The catalyst used for the process is nickel-based copper-zinc promoted and alumina supported granular powder. By optimizing the concentration of promoters (i.e. copper and zinc), a specific metal crystal size range is obtained which ultimately lead to the high selectivity of the specific form of carbon i.e. carbon nanotubes (CNT). In the current work, the temperature is optimized for maximum hydrogen yield from methane and waste plastics. As the carbon is continuously depositing on the catalyst, the fluidized bed reactor is chosen for the reaction to accommodate spent catalyst regeneration and CNT separation. One of the challenges of the process is maintaining proper hydrodynamic condition in the fluidized bed reactor. The produced carbon is in the form of CNTs which is a valuable by-product which can help in improving the economy of the process. As waste plastic is a major source of carbon, its addition increases the yield of CNTs. At optimum reaction conditions, 75% methane conversion was observed along with the complete conversion of plastics to CNTs and Hydrogen. The obtained CNTs are multi-layered bamboo-shaped carbon nanotubes (Diameter: ~60 nm, Length: 2 microns).

599644 Biomass – Flare Gas Synergistic Co-Processing over Fe / Ni / Pd Doped Mo_2c Supported on Graphene Nanosheets: Experimental and DFT Integrated Approach to Hydrogen Production

Abstract:

Biomass – flare gas synergistic co-processing is a novel energy generation and conversion technology that aims at harnessing abundant renewable energy source: biomass and reduce greenhouse gas emissions from flaring in stranded shale plays. In this study p-cresol is used as a model biomass-lignin molecule for methane assisted reforming studies, then scaled-up to model alkali lignin and eventually hardwood biomass in a detail bottom-up approach to gain the molecular level understanding of the reaction mechanism and ascertain the role of active sites. Novel in-situ catalyst synthesis using biomass precursor was achieved through pyrolysis to yield graphene nanosheets (GNS) supported transition metal and Mo_2C nanoparticles. Initially, experimental work and DFT modelling calculations were performed for methane assisted p-cresol reforming using Fe, Ni, Mo_2C, Fe-Mo_2C, Ni-Mo_2C, Pd-Mo_2C supported on GNS. Detailed mechanistic investigation of methane – p-cresol synergistic reaction experimentally and through molecular simulations helped ascertain the unique reaction pathway occurring on dual active sites on a transition metal doped β-Mo_2C-GNS catalyst. Transition metal doped Mo_2C viz.: Fe-Mo_2C-GNS and Ni-Mo_2C-GNS are equally effective as Mo_2C-GNS for methane dissociation and p-cresol HDO but present significantly lower barrier for H_2 (1.15, 1.13 eV vs. 1.70 eV) and CO (2.87, 2.80 eV vs. 3.67 eV) gas-phase desorption. Dual active sites are required for hydrogen rich syngas production through methane assisted p-cresol reforming as validated by experiments, DFT calculations, and micro-kinetic modelling (MKM). Lignin and hardwood biomass both having higher O:C weight ratio compared to p-cresol (0.46, 1.09 vs. 0.19) were co-processed with methane over FeMo-GNS, NiMo-GNS, and MoPd-GNS catalysts. Transition metal doped Mo_2C was found to be highly active for simultaneous CH_4 activation and extensive hydrodeoxygenation of p-cresol, lignin, and hardwood biomass. Up to 99% hydrogen present in lignin could be valorized as syngas with a concentration of > 65%.

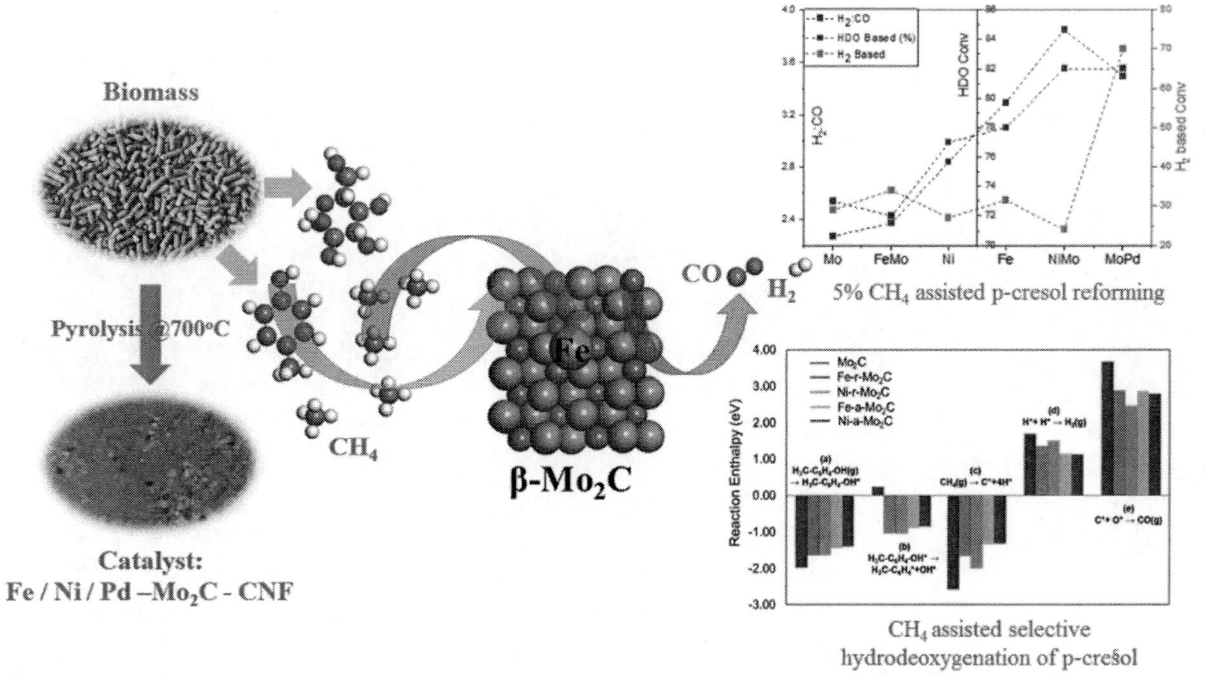

599190 Hydrogen-Enriched Syngas Production from Steam Gasification of Biomass with Eggshell As a CO_2 Sorbent

Abstract:

An investigation into the steam gasification of biomass with an in-situ CO_2 capture was conducted in this study. An enrichment in the hydrogen concentration and a reduction in the carbon dioxide (CO_2) concentration in the gas product was achieved. Sawdust from spruce was utilized as the feedstock while eggshell was used as the CO_2 sorbent material. The gasification experiments were conducted in a bubbling fluidized bed reactor with calcined eggshell as both bed material and CO_2 sorbent.

Activation of the eggshell was performed by calcination at 900°C as thermogravimetric analysis conducted on the eggshell showed that 900°C is sufficient to fully convert the calcium carbonate in the eggshell to calcium oxide. X-ray diffraction peaks also revealed the complete conversion. Effects of important process parameters, steam to biomass ratio (SBR) and calcined eggshell to biomass ratio (CEBR), were studied. Increasing the CEBR promoted the CO_2 uptake via the carbonation reaction and accordingly enhanced hydrogen enrichment. An increase in SBR improved the hydrogen concentration in the gas stream. A maximum hydrogen concentration of 78% and a minimum CO_2 volumetric concentration of 3.3% were obtained in this study at a temperature of 650°C, CEBR of 1.0 and SBR of 1.2. Additionally, results revealed that the water gas shift reaction is the most important gasification reaction in the process.

A SOFC-based Multi-Energy System for the Efficient Production of Hydrogen and Power from Natural Gas for Hydrogen Refuelling Station Applications

ALBERTO CAMMARATA[2], LUCA MASTROPASQUA[1], ALIREZA SAEEDMANESH[1]

JACK BROUWER[1], STEFANO CAMPANARI[2]

[1]National Fuel Cell Research Center, University of California, Irvine, Irvine, CA

[2]Politecnico di Milano, Milano, IT

ABSTRACT

In this work we design a SOFC-based tri-generation system to be installed at a hydrogen refuelling station of 200 kg/d of nominal capacity. The system produces hydrogen, electricity, and cooling power to supply the daily hydrogen demand, all the cooling power demand required for pre-cooling the dispensed hydrogen and all the auxiliary electric loads of the station. We perform stationary zero-dimensional simulations of the system to prove the concept and assess the system potential in terms of overall efficiency and primary energy savings. We test and characterize, via polarization curves, a commercial SOFC short-stack at low fuel utilization factor and with pre-reformed natural gas as input fuel. The experimental characterization is used to calibrate the SOFC model in terms of voltage performance and thermal management of the stack. We then evaluate different system design points by changing the fuel utilization factor in the range 0.35-0.5, while selecting all other system parameters to secure a good plant thermal integration. Our results prove that an overall first law efficiency of about 70% is always obtained, and that the primary energy saving coefficient ranges between 12% and 24% (depending on the design point).

1. INTRODUCTION

The use of a solid oxide fuel cell (SOFC) to simultaneously process hydrocarbon-based fuels - like natural gas - and generate hydrogen, in combination with electricity and thermal energy, represents a promising and versatile technology. The literature provides numerous examples of poly-generation based on SOFCs for a wide range of applications, including small-scale uses [1] and large-scale industrial uses in combination with CO_2 capture [2].

While most hydrogen production from natural gas relies on thermochemical processes like steam methane reforming (SMR) [3], an electrochemical reactor with internal reformation capabilities - like a SOFC - allows co-producing hydrogen, electricity and heating or cooling, within a single unitized unit. In electrochemically-driven internal reforming processes: (i) the amount of high temperature heat produced by the exothermic electrochemical reactions (i.e., hydrogen electro-oxidation) within the fuel cell stack is typically greater than the heat required by the chemical reactions (i.e., steam methane reforming and water gas shift). Therefore, the internal reforming process occurring within the SOFC anodes can be driven by the stack exothermicity, instead of requiring the combustion of additional fuel, as in conventional SMR; and (ii) depending on the comparative magnitude of the yield of the electrochemical reactions (controlled by the cell current and the fuel utilization factor) and the yield of the chemical reactions (controlled by chemical kinetics), the fuel cell can change the conversion path of hydrogen in the anode. At high fuel utilization factors the SOFC operates at a high electricity output. At low fuel utilization factor the SOFC is characterized by a lower yield of electro-oxidation of hydrogen and a higher hydrogen production via internal SMR. Therefore, the SOFC can generate a hydrogen-rich stream at the anode exit. This stream can be subsequently purified and delivered to a compression and storage unit for use in a final application. The SOFC operating condition, in terms of current density and fuel utilization, can be tuned to follow a particular profile of hydrogen and electric/heating demand, making the system flexible and versatile.

We focus here on a case where a tri-generation SOFC-system is coupled with a 200 kg/d hydrogen refuelling station. We assume that the station is equipped with a compressed hydrogen storage tank at 750 bar. On top of supplying the hydrogen demand of the station, the SOFC unit supplies electricity to power all auxiliaries in the refuelling station and cooling load to meet the hydrogen pre-cooling demand upstream the dispenser. Cooling is produced via coupling the heat production from the SOFC system to a low temperature absorption refrigeration unit. The synergy with the cooling load of the pre-cooling unit of the hydrogen station is particularly important due to the high impact that the refrigeration section has on the overall hydrogen cost. For a 200 kg/d refuelling station, Elgowainy et al., [4] predicts that a 13 kW compression refrigeration system is required, consuming approximately 0.3 kWh/kg at full load. This section adds approximately $0.5/kg to the total cost of hydrogen.

Hydrogen purification and compression is made with a small water gas shift (WGS) + pressure swing adsorption (PSA) unit followed by and electrochemical compressor to the station buffer tank. When hydrogen is not required, it can be either stored or the SOFC can move to an electricity-only operating condition by increasing the fuel utilization factor. Once the separation of hydrogen is complete, the residual flue gases are mainly composed of CO_2. Therefore, the system could also be fit to perform carbon separation and production of a pure stream of CO_2.

2. Method

2.1. Plant Layout and Simulations Assumptions

The first modelling assumption is that all components work in stationary conditions. The aim of the simulations is to find several possible stationary design points.

The plant modelling is performed using the GS software, developed at the Department of Energy of Politecnico di Milano since the early 90's [5]. This software is capable of solving zero-dimensional stationary mass and energy balances of many interconnected components, which can form complex plant configurations. The software has been already used by some of the authors for simulating fuel cell systems [6–8].

Figure 1 depicts a schematic of the system that we propose in this work. Natural gas (point no. 1) is first sent through a De-Sulphurisation Unit (DSU), then preheated in a heat exchanger and split in two streams. One (point no. 4) is burnt in a catalytic oxidizer to heat the cathodic air stream to 750°C; the other (point no. 5) is the fuel for the SOFC stack. The plant is designed to operate the SOFC at an average temperature of 750°C. Higher temperatures would increase the risk of local hot-spots, which would increase stack degradation.

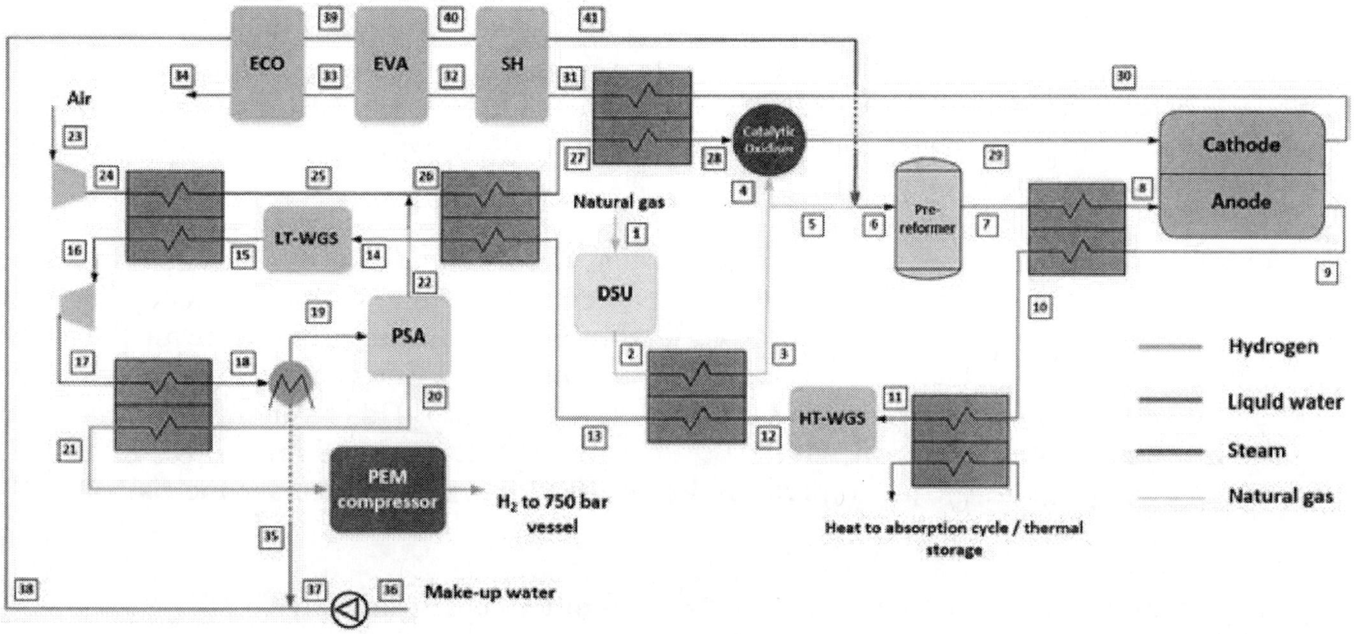

Figure 1: *Plant layout*

The natural gas in the anodic line is mixed with superheated steam from the steam generator (point no. 41). The steam mass flow is constrained to maintain a Steam To Carbon Ratio (STCR) equal to 2.0 at anode inlet to avoid carbon deposition. The degree of steam superheating is selected to ensure 500°C at the inlet of the prereformer (point no. 6). The outlet of the prereformer (point no. 7) is pre-heated using the enthalpy content of the anode exhausts from the SOFC stack (point no. 9).

We selected four different design points at the following low fuel utilization factors (U_f): 0.35, 0.4, 0.45 and 0.5. Higher values of fuel utilization factors are not explored in this work as not of interest for a high hydrogen production system. For each U_f, the cell voltage V_{cell} is chosen to impose a fixed temperature (i.e., 765°C) at the SOFC outlet (equal for anodic and cathodic streams). We use the cell voltage as a parameter to control the degree of exothermicity or endothermicity to fix the stack outlet temperature. For example, a higher cell voltage implies lower electrochemical losses (lower heat generation), which would balance the additional heat produced by H_2 oxidation at higher fuel utilization factors. On the other hand, at lower fuel utilization factors the cell may become endothermic (due to low yield of exothermic hydrogen electro-oxidation reactions); therefore, in our simulation, we selected a low-enough cell voltage to compensate for the low exothermicity of the cell with more heat generation from electrochemical losses to meet the assumption on the stack outlet temperature.

The anode exhaust is cooled down by supplying heat to an absorption refrigeration cycle, which produces enough cooling load to cool the H_2 before injection into the car tanks. We estimated the heat required by the absorption cycle starting from the daily cooling energy for 200 kg of hydrogen, and assuming a COP for the refrigeration system.

$$Q_{cool} = m_{H_2} c_{p_{H_2}} \left(T_{H_2,storage} - T_{tank}\right) \tag{1}$$

$$Q_{abs,cy} = \frac{Q_{cool}}{COP} \tag{2}$$

where H_2 is assumed to be chilled from a storage tank at 25°C and injected in car tanks at -40°C. From equation (1), we estimated that the specific cooling energy required to chill the hydrogen distributed each day is 0.937 MJ/kg_{H_2}. The COP is safely assumed to be 0.62 [21]. Therefore, we found that the specific heat to be delivered is 1.512 MJ_{th}/kg_{H_2} or 3.5 kW_{th} must be supplied for every design case.

Downstream the absorption cycle, a High Temperature Water Gas Shift (HT-WGS) increases the H_2 and CO_2 fractions in the mixture, while decreasing CO and H_2O content. The heat produced during the HT-WGS is used to preheat the input natural gas and the air flow. The preheating process is constrained by a fixed temperature of 180°C at point no. 14, which is the inlet of the Low Temperature Water Gas Shift reactor (LT-WGS). This temperature is required to achieve a good CO conversion within the reactor. The stream is then cooled to 90°C by preheating the input air flow (a temperature lower than 90°C would increase the risk of water condensation).

The stream in point no. 16 is compressed and chilled as required by the Pressure Swing Adsorption (PSA) unit [1]. As a result, water is condensed and separated. Finally, a large fraction of pure H_2 is recovered and compressed to a 750 bar storage tank using a Polymeric Electrolyte Membrane (PEM) compressor. The PSA off-gas, which contains residual H_2 and CO, is mixed with the input preheated air in order to decrease the natural gas converted in the catalytic oxidizer.

We performed the PSA modelling through a Langmuir isotherms-based model previously developed in [2,9]. The model predicts that it is possible to achieve pure H_2 with 95% recovery at 3.5 bar and with reasonable beds size and cycle time.

We estimated the electric power required for H_2 compression based on the work of Kee et al [10]. The following expressions are used for the calculation of the specific energy requirement for an electrochemical H_2 compressor (F and R are the Faraday's and the universal gas constants respectively).

$$\frac{P_{comp}}{\dot{m}_{H_2}} = ASR \left(\frac{2F}{M_{H_2}}\right)^2 \frac{\dot{m}_{H_2}}{A_{mem}} + \frac{RT}{M_{H_2}} ln(\beta) \tag{3}$$

where β is the compression ratio, A_{mem} is the membrane surface and M_{H_2} is the molar mass of H_2. Note that β is the ratio between 750 bar and H_2 pressure after being preheated, which is computed accounting a 2% pressure loss in the heat exchanger starting from 3.5 bar. Assuming an Area Specific Resistance (ASR) of 0.1 Ωcm^2, it is possible to plot the specific work against the parameter $A_{mem}/\dot{m}_{H_2}$. As reported by [10] there is a threshold beyond which it is not worth to further increase A_{mem} to decrease the specific power requirement. In our case, the threshold corresponds to the point $A_{mem}/\dot{m}_{H_2}=10^4$ $m^2kg^{-1}s$, with a specific electric power requirement of 17.67 MJ_{el}/kg.

Some important assumptions have been discussed in this section, the overall set of assumptions is available in Table 1.

Table 1: *Simulations assumptions*

Ambient conditions	NG properties	Chemical reactors
Temperature [°C] 15	Composition [%mol]. CH_4:89.0, CO_2:2.0,	Outlet at full chemical equilibrium
Pressure [bar] 1.01325	C_2H_6:7.0, C_3H_8:1.0, N_2:0.89, C_4H_{10}:0.11	Prereformer inlet temperature [°C] 500
Relative humidity [%] 65	LHV [MJ/kg] 46.482	LT-WGS inlet temperature [°C] 180
		Prereformer and oxidizer $\Delta p/p_{in}$ [%] 3.0
Heat Exchangers	SOFC	WGS reactors $\Delta p/p_{in}$ [%] 2.5
EVA and SH $\Delta p/p_{in}$ [%] hot side: 1.5, cold side: 3.5	Cathode inlet temperature [°C] 750	
ECO $\Delta p/p_{in}$ [%] hot side: 1.5, cold side: 20.0	Outlet Temperature [°C] 765	Miscellaneous
Condenser $\Delta p/p_{in}$ [%] 2.0	Air and fuel channels $\Delta p/p_{in}$ [%] 3	H_2 mass flow rate produced [g/s] 2.315
Others $\Delta p/p_{in}$ [%] hot side: 2.0, cold side: 2.0	Heat loss (fraction of fuel LHV) [%] 3	Heat to absorption cycle [kW] 3.5
Minimum ΔT gas-evaporating water [°C] 10	Anode inlet S/C 2.0	Compressors isentropic efficiency [%] 80
Minimum ΔT gas-water [°C] 15	Anode outlet at chemical equilibrium	Compressors mech.-el. efficiency [%] 94
Minimum ΔT gas-gas [°C] 30		Exhaust stack temperature [°C] 90
Condenser outlet temperature [°C] 35	Pressure Swing Adsorption	DSU pressure drop $\Delta p/p_{in}$ [%] 3
	Required pressure [bar] 3.5	DC/AC conversion efficiency [%] 97
	H_2 recovery [%] 95	PEM compressor inlet temperature [°C] 100

2.2. Key Performance Indicators

We defined a set of Key Performance Indicators (KPIs) to evaluate the overall plant performance of the system. For a "multi-energy" system producing multiple useful products with one input, diverse efficiency measures can be defined. The system considered in this work produces three useful outputs: H_2 for the station, net electricity produced by the plant, and cooling power for H_2. We report the corresponding efficiency definitions in Table 2.

Table 2: *Key performance indicators*

Efficiencies	$\eta_{H_2} = \frac{\dot{Q}_{H_2}^{LHV}}{\dot{Q}_{NG}^{LHV}}$	$\eta_{EL} = \frac{\left(\dot{W}_{EL}^{SOFC} - \dot{W}_{EL}^{PEM}\right)\eta_{DC/AC} - \dot{W}_{EL}^{aux}}{\dot{Q}_{NG}^{LHV}}$
	$\eta_{cool} = \frac{\dot{Q}_{cool}}{\dot{Q}_{NG}^{LHV}}$	$\eta_{I} = \eta_{H_2} + \eta_{EL} + \eta_{cool}$
PES	$PE_{H_2}^{REF} = \frac{\dot{Q}_{H_2}^{LHV}}{\eta_{H_2,@10bar}^{REF}} + \frac{\dot{W}^{10bar \to 750bar}}{\eta_{EL}^{REF}}$	$PE_{EL}^{REF} = \frac{\dot{W}_{EL}^{plant}}{\eta_{EL}^{REF}}$
	$PE_{cool}^{REF} = \frac{\dot{Q}_{cool}}{COP^{REF}\eta_{EL}^{REF}}$	$PES = \frac{PE_{H_2}^{REF} + PE_{EL}^{REF} + PE_{cool}^{REF} - \dot{Q}_{NG}^{LHV}}{PE_{H_2}^{REF} + PE_{EL}^{REF} + PE_{cool}^{REF}}$

In these equations $\dot{Q}_{NG}^{LHV}$ is the thermal input of the system based on the Lower Heating Value (LHV) of the input natural gas. η_I is the overall first law system efficiency.

The Primary Energy Saving (PES) coefficient, which evaluates the relative saving in primary energy realized by the system compared to the separate production of the useful outputs by meaningful reference systems. In our system, the PES is defined as shown in table 2. The terms $PE_{H_2}^{REF}$, PE_{EL}^{REF} and PE_{cool}^{REF} are the primary energies (in kW) required by the reference systems to produce the same amount of H_2, electric power and cooling power produced by the SOFC plant.

In conventional refuelling stations the electricity needed is drawn from the grid. Therefore, we assumed η_{EL}^{REF} as the mean efficiency for generation of grid electricity, estimated equal to 41.59% in the U.S. [11, 12]. We choose a commercial small-scale SMR system (1000 kg/d) as a reference hydrogen production system, with $\eta^{REF}_{H_2,@10bar}$ equal to 67.14% [13]. Note that the system considered produces H_2 at a pressure of 10 bar, hence an additional amount of primary energy is required to compress H_2 to 750 bar. The electric power for this compression is estimated by assuming a conventional inter-cooled compression process. Considering that the reference system for electricity generation would provide power for the additional compression, the efficiency of the reference system for production of H_2 at 750 bar is estimated as 60.32%. Finally, the COP^{REF} for a reference compression refrigeration cycle is 1.22, estimated assuming an ambient temperature of 15°C [4].

2.3. Experimental Setup

We tested a SOFC 6-cell stack to characterize the voltage and thermal performance of the cells operated at low utilization factors. The cells are anode-supported with an active area of 80 cm^2. Figure 2 depicts the test stand installation at the University of California, Irvine and the short-stack as installed within the electric furnace.

(a) *Installation of the short-stack test stand at UC Irvine* **(b)** *Six-cell short-stack installed inside the electric furnace*

Figure 2: *Experimental setup*

The control system on the test stand allows changing the fuel flow rates in accordance to the variation in current density to maintain a constant fuel utilization factor throughout the polarization curve. We performed the polarization curve tests to guarantee that each point on the curve represents a statistically relevant sample of the cell performance at steady state. We refer to [14] for a more detailed description of the experimental method.

2.4. Calibration Procedure

All plant simulations in this work have been performed considering zero-dimensional components. The zero-dimensional SOFC model assumes full chemical equilibrium at the anode outlet, hence CH_4 is fully converted at the considered temperature of 765°C.

Hereafter we show the preliminary calibration results of a one-dimensional, finite-volume, co-flow SOFC model against relevant experimental data. The model is developed following the main equations from [15]. The main chemical and electrochemical model equations are summarized in Table 3. These, together with mass, momentum and energy balances, are solved in each control volume which the cell channels are divided into.

Table 3: *Chemical and electrochemical models used*

Chemical model	$r_{SMR} = k_{f,SMR} p_{CH_4} \left(1 - \frac{\frac{p_{H_2}^3 p_{CO}}{p_{CH_4} p_{H_2O}}}{K_{eq,SMR}(T)}\right)$	$k_{f,SMR} = \gamma_{SMR} exp\left(-\frac{E_{act,SMR}}{RT}\right)$
	$r_{WGS} = k_{f,WGS} p_{CO} p_{H_2O}^{0.5} p_{CO_2} \left(1 - \frac{\frac{p_{H_2} p_{CO_2}}{p_{CO} p_{H_2O}}}{K_{eq,WGS}(T)}\right)$	$k_{f,WGS} = \gamma_{WGS} exp\left(-\frac{E_{act,WGS}}{RT}\right)$
Electrochemical model	$i_{H_2} = i_{0,H_2}\left[\exp\left(\frac{F\eta_{act,H_2}}{RT}\right) - \exp\left(-\frac{F\eta_{act,H_2}}{RT}\right)\right]$	$i_{0,H_2} = \gamma_{H_2} p_{H_2} p_{H_2O} \exp\left(-\frac{E_{act,H_2}}{RT}\right)$
	$i_{O_2} = i_{0,O_2}\left[\exp\left(\frac{F\eta_{act,O_2}}{RT}\right) - \exp\left(-\frac{F\eta_{act,O_2}}{RT}\right)\right]$	$i_{0,O_2} = \gamma_{O_2} p_{O_2}^{0.25} \exp\left(-\frac{E_{act,O_2}}{RT}\right)$

The activation energies $E_{act,SMR}$, $E_{act,WGS}$ are set to 102 kJ/mol and 103.2 kJ/mol [16], respectively. In this preliminary calibration, the pre-exponential factors γ_{SMR}, γ_{WGS}, γ_{H_2}, γ_{O_2} are calibrated, keeping constant all other model parameters. γ_{H_2} and γ_{O_2} are used to define exchange current densities for H_2 electro-oxidation and O_2 electro-reduction respectively, as shown in Table 3.

The values used for E_{act,H_2}, E_{act,O_2}, thermal and electric conductivities, channel geometry and heat transfer coefficients are equal to those indicated in Tables 1 and 4 of reference [15] (apart from interconnect thermal conductivity, which is 5 W/m/K). Finally, Table 4 summarizes the characteristics of the membrane-electrodes assembly (MEA) structure.

Table 4: *MEA components characteristics: thickness [μm], porosity, tortuosity, mean pore radius [μm]*

Component	thickness	ϵ	τ	r_{por}
Anode	240	0.45	3.5	0.5
Cathode	40	0.3	3.5	0.5
Electrolyte	8	-	-	-

The calibration process is performed assuming the temperature of the interconnect (both fuel and air sides) fixed at 750°C to account for the presence of the furnace in the experimental setup. The fuel composition used for the calibration derives from a full equilibrium calculation at 600°C (pre-reformer temperature), starting from a mixture of 33.42% CH_4 and 66.58% H_2O, simulating humidified natural gas. Hence the considered anode inlet composition is H_2 53.13%, H_2O 23.48%, CO 6.87%, CO_2 8.13%, CH_4 8.39%.

Figure 3 depicts the polarization curves of the short-stack as the average voltage of the six cells. This value is used as reference to calibrate the SOFC model. The model agrees with the experimental data by using the following pre-exponential factors in the equations in Table 3: γ_{SMR}=5.55·10^3, γ_{WGS}=1.0·10^2, γ_{H_2}=1.448·10^9, γ_{O_2}=3.095·10^9.

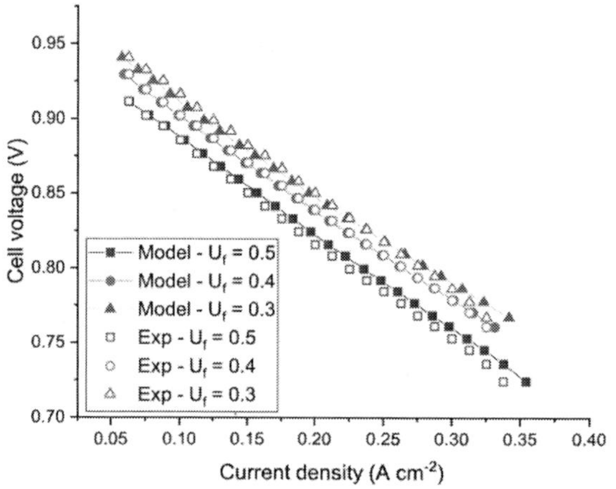

Figure 3: *Experimental and calibration results at $T_{furnace}$=750°C*

3. Results and Discussion

Table 8 (see appendix) summarizes the thermodynamic points for the case U_f=0.4. The plant is capable of respecting the constraints imposed and the resulting temperature distribution proves that a good thermal integration is achieved. Note that the HT-WGS inlet temperature is within the suitable range (i.e., 300°C-450°C) for Platinum catalysts typically used in these reactors [18]. Higher operating temperatures above 500°C are also reported for HT-WGS reactors by [19].

Table 5 compares the SOFC operating conditions for the four design points at different fuel utilization.

Table 5: *Comparison of different U_f: SOFC conditions (V_{cell} in [V]), system input flows and steam produced [g/s]*

U_f	V_{cell}	U_{ox}	NG	Air	Make-up water	Steam
0.35	0.499	0.444	8.88	120	5.8	19.8
0.4	0.601	0.501	9.66	131	4.5	21.4
0.45	0.681	0.557	10.58	145	2.9	23.4
0.5	0.744	0.612	11.70	162	1.0	25.7

As expected, a higher operating voltage is required at higher fuel utilization factors to meet the constraint on the SOFC outlet temperature (i.e., 765°C). At higher voltages less heat is produced due to lower electrochemical losses, which balances the greater heat generation associated to H_2 electro-oxidation within the SOFC. The input natural gas must increase at higher fuel utilization, since the equivalent H_2 flow at the anode outlet must remain constant to produce the same amount of H_2. Consequently, the steam produced by the steam generator must increase to achieve a S/C of 2.0 at the inlet of the anode and the input air must also increase to provide the heat required by the steam generator.

Note that the assumption on fixing the cell outlet temperature across all fuel utilization factors design points may entail excessively reducing the cell voltage. A cell voltage below 0.6V, combined with a high Nernst voltage, would significantly increase the current density possibly causing instabilities in the SOFC stack. This issue constitutes a reason for further and more detailed investigation of the exothermic operation of SOFCs at low utilization factors both theoretically and experimentally.

In Table 5, we also report the variation of the make-up water flow, which decreases at high U_f. In fact, at higher fuel utilization factor, a greater amount of steam is produced inside the SOFC module due to H_2 oxidation, which significantly increases the mass-flow of condensed water (point no. 35). The more significant H_2O production within the SOFC also increases the H_2O content at the inlet of WGS reactors, which explains the slightly better CO conversion achieved at high U_f.

Table 6: *Comparison of different U_f: energy flows [kW] and efficiencies [%]*

U_f	$\dot{Q}_{NG}^{LHV}$	$\dot{Q}_{H_2}^{LHV}$	$\dot{W}_{EL}^{SOFC}$	$\dot{W}_{EL}^{PEM}$	$\dot{W}_{EL}^{aux}$	$\dot{Q}_{cool}$	η_{H_2}	η_{EL}	η_{cool}	η_I
0.35	412.7	277.7	63.0	40.9	18.9	2.17	67.29	0.62	0.53	68.44
0.4	449.9	277.7	93.8	40.9	20.5	2.17	61.85	6.88	0.48	69.21
0.45	492.9	277.7	130.3	40.9	22.4	2.17	56.44	13.08	0.44	69.96
0.5	543.6	277.7	174.0	40.9	24.6	2.17	51.08	19.22	0.40	70.70

Table 6 shows the energy flows and efficiencies resulting from the simulations. The amount of H_2 and cooling power produced by the plant is constant in all simulations, hence the corresponding hydrogen and cooling efficiencies decrease due to the larger input natural gas at higher U_f. On the contrary, the power produced by the SOFC module steeply increase, while electricity for auxiliaries slightly increase. Therefore, the net electric power produced by the plant significantly increase with the fuel utilization.

Table 7: *Comparison of different U_f: Primary Energy flows [kW] and PES [%]*

U_f	$PE_{H_2}^{REF}$	PE_{EL}^{REF}	PE_{cool}^{REF}	PE_{TOT}^{REF}	$\dot{Q}_{NG}^{LHV}$	PES
0.35	460.4	6.2	4.3	470.8	412.7	12.36
0.4	460.4	74.2	4.3	538.9	449.9	16.68
0.45	460.4	154.7	4.3	619.3	492.9	20.56
0.5	460.4	251.2	4.3	715.9	543.6	24.06

Table 7 shows the primary energy flows computed and the resulting savings. All the design points found achieve a PES larger than 10%, which proves the good level of synergistic integration achieved by the proposed tri-generation system. The PES increases with U_f due to the more efficient production of additional electricity by the plant compared to the reference system for electricity generation.

In summary, a relatively large U_f increases the plant first law efficiency and PES due to the very efficient production of additional electric power. Moreover, the significantly larger water production within the SOFC module increases CO conversion in WGS reactors and decrease the required make-up water. Nevertheless, the mass flows generally increase (input air and natural gas, steam generated), which would require larger pipes to limit pressure losses, and larger components in general (heat exchangers in particular). Note that also the PSA size would increase with U_f. From the model applied, it is estimated that two beds of 0.2 m^2 each are required when U_f=0.35, the volume increases to 0.245 m^2 in case U_f=0.5. Note that the SOFC membrane area required at large U_f would steeply increase for three reasons, i) the overall stack current must increase due to significantly larger power produced ii) the operating voltage is larger, which reduces the SOFC current density (A/m^2) iii) the current density is further reduced by the lower Nernst voltage.

4. Conclusion

In this work, a novel tri-generation system for refuelling station applications is proposed. We simulated the system considering stationary operation and zero-dimensional components. The only parameter changed in the simulations is the SOFC U_f, while other parameters are adapted to provide a good thermal integration for the plant. Results show that an overall efficiency η_I of about 70% is always achieved, and the primary energy saving coefficient lies in the range of 12%-24%. Moreover, despite operating the SOFC at relatively high U_f brings several benefits in terms of overall efficiency and PES, components size and cost may increase significantly.

The polarization curves of a SOFC stack operating at low U_f are shown and used to calibrate a one-dimensional SOFC model. The results showed that the model is capable of replicating the experimental data. The calibrated one-dimensional SOFC model will be used to update future plant simulation results. In particular, the one-dimensional model will provide useful information such as the temperature distribution inside the cell and the actual CH_4 conversion. Moreover, an economic analysis will be performed to provide a more thorough comparison between simulations at different U_f, aiming in particular at estimating the Levelized Cost Of Hydrogen (LCOH) produced.

References

[1] P. Margalef, T. Brown, J. Brouwer, S. Samuelsen (2011), "Conceptual Design and Configuration Performance Analyses of Polygenerating High Temperature Fuel Cells", *International J. of Hydrogen Energy*, 36, 10044-10056.

[2] L. Mastropasqua, A. Pegorin, S. Campanari (2020), "Low Fuel Utilisation Solid Oxide Fuel Cell System for CO_2-free Hydrogen Production in Oil Refineries", *J. of Power Sources*, 448, 227461.

[3] G. Collodi, G. Azzaro, N. Ferrari, S. Santos (2017), "Techno-economic Evaluation of Deploying CCS in SMR Based Merchant H_2 Production with NG as Feedstock and Fuel", *Energy Procedia*, 114, 2690-2712.

[4] A. Elgowainy, K. Reddi, D. Lee, N. Rustagi, E. Gupta (2017), "Techno-Economic and Thermodynamic Analysis of Pre-Cooling Systems at Gaseous Hydrogen Refueling Stations", *International J. of Hydrogen Energy*, 42, 29067-29079.

[5] S. Consonni (1992), "Performance Prediction of Gas/steam Cycles for Power Generation", Princeton University, 1992.

[6] S. Campanari, P. Chiesa, G. Manzolini, S. Bedogni (2014), "Economic Analysis of CO_2 Capture from Natural Gas Combined Cycles Using Molten Carbonate Fuel Cells", *Applied Energy*, 130, 562-573.

[7] S. Campanari, L. Mastropasqua, M. Gazzani, P. Chiesa, M.C. Romano (2016), "Predicting the Ultimate Potential of Natural Gas SOFC Power Cycles with CO_2 Capture-Part A: Methodology and Reference Cases", *J. of Power Sources*, 324, 598-614.

[8] L. Mastropasqua, S. Campanari, J. Brouwer (2017), "Electrochemical Carbon Separation in a SOFC-MCFC Polygeneration Plant with Near-Zero Emissions", *J. of Engineering for Gas Turbines and Power*, 140, 013001.

[9] L. Mastropasqua, F. Baia, L. Conti, S. Campanari (2018), "Electrochemical Energy Storage and Synthetic Natural Gas Production based on Reversible Molten Carbonate Cells", *12th International Conference on Energy Sustainability*.

[10] B. L. Kee, D. Curran, H. Zhu, R. J. Braun, S. C. DeCaluwe, R. J. Kee, S. Ricote (2019), "Thermodynamic Insights for Electrochemical Hydrogen Compression with Proton-Conducting Membranes", *Membranes*, 9, 77.

[11] https://www.eia.gov/energyexplained/electricity/electricity-in-the-us.php

[12] https://www.eia.gov/electricity/annual/html/epa_08_01.html

[13] https://www.bayotech.us/products-bayotech-h2-1000

[14] L. Mastropasqua and S. Campanari and J. Brouwer (2017), "Solid oxide fuel cell short stack performance testing - Part A: Experimental analysis and μ-combined heat and power unit comparison", *Journal of Power Sources*, 371, 225 - 237.

[15] S. Campanari, P. Iora (2004), "Comparison of Finite Volume SOFC Models for the Simulation of a Planar Cell Geometry", *Fuel Cells*, 5, 34-51.

[16] L. Mastropasqua, A. Donazzi, S. Campanari (2019), "Development of a Multiscale SOFC Model and Application to Axially-Graded Electrode Design", *Fuel Cells*, 19, 125-140.

[17] S. Campanari, P. Iora (2004), "Definition and Sensitivity Analysis of a Finite Volume SOFC Model for a Tubolar Cell Geometry", *J. of Power Sources*, 132, 113-126.

[18] V. Palma, C. Ruocco, M. Cortese, S. Renda, E. Meloni, G. Festa, M. Martino (2020), "Platinum Based Catalysts in the Water Gas Shift Reaction: Recent Advances", *Metals*, 10, 866.

[19] W. Chen, C. Chen (2020), "Water Gas Shift Reaction for Hydrogen Production and Carbon Dioxide Capture: A Review", *Applied Energy*, 258, 114078.

[20] L. Zhao, J. Brouwer, S. Samuelsen (2014), "Dynamic Analysis of a Self-Sustainable Renewable Hydrogen Fueling Station", *12th International Conference on Fuel Cell Science, Engineering and Technology*.

[21] A. Lavernia, M. Asghari, J. Brouwer (2020), "Absorption Cooling for Data Centers Powered By Solid Oxide Fuel Cell Waste Heat", *ASHRAE Transactions*,126, 204–211,American Society of Heating, Refrigeration and Air Conditioning Engineers, Inc.

Appendices

Table 8: *Thermodynamic points for the case U_f = 0.4 (molar fractions in [%])*

Stream	T [°C]	p [bar]	$\dot{m}$ [g/s]	$\dot{n}$ [mol/s]	Ar	CH_4	CO	CO_2	C+	H_2	H_2O	N_2	O_2	H_2O (L)
1	15.0	1.34	9.659	0.536	0.0	89.0	0.0	2.0	8.11	0.0	0.0	0.89	0.0	0.0
2	15.0	1.30	9.659	0.536	0.0	89.0	0.0	2.0	8.11	0.0	0.0	0.89	0.0	0.0
3	457.8	1.27	9.659	0.536	0.0	89.0	0.0	2.0	8.11	0.0	0.0	0.89	0.0	0.0
4	457.8	1.27	0.916	0.051	0.0	89.0	0.0	2.0	8.11	0.0	0.0	0.89	0.0	0.0
5	457.8	1.27	8.744	0.485	0.0	89.0	0.0	2.0	8.11	0.0	0.0	0.89	0.0	0.0
6	500	1.27	30.175	1.675	0.0	25.79	0.0	0.58	2.35	0.0	71.02	0.26	0.0	0.0
7	374.9	1.23	30.175	1.813	0.0	24.68	0.06	4.29	0.0	12.68	58.05	0.24	0.0	0.0
8	735.0	1.21	30.175	1.813	0.0	24.68	0.06	4.29	0.0	12.68	58.05	0.24	0.0	0.0
9	765.0	1.17	43.108	2.708	0.0	0.02	7.90	11.52	0.0	36.82	43.58	0.16	0.0	0.0
10	475.2	1.15	43.108	2.708	0.0	0.02	7.90	11.52	0.0	36.82	43.58	0.16	0.0	0.0
11	438.9	1.13	43.108	2.708	0.0	0.02	7.90	11.52	0.0	36.82	43.58	0.16	0.0	0.0
12	487.8	1.10	43.108	2.708	0.0	0.0	3.12	16.31	0.0	41.66	38.75	0.16	0.0	0.0
13	366.5	1.08	43.108	2.708	0.0	0.0	3.12	16.31	0.0	41.66	38.75	0.16	0.0	0.0
14	180.0	1.06	43.108	2.708	0.0	0.0	3.12	16.31	0.0	41.66	38.75	0.16	0.0	0.0
15	215.2	1.03	43.108	2.708	0.0	0.0	0.14	19.29	0.0	44.64	35.77	0.16	0.0	0.0
16	90.0	1.01	43.108	2.708	0.0	0.0	0.14	19.29	0.0	44.64	35.77	0.16	0.0	0.0
17	266.1	3.64	43.108	2.708	0.0	0.0	0.14	19.29	0.0	44.64	35.77	0.16	0.0	0.0
18	243.0	3.57	43.108	2.708	0.0	0.0	0.14	19.29	0.0	44.64	35.77	0.16	0.0	0.0
19	35.0	3.50	26.169	1.768	0.0	0.0	0.21	29.56	0.0	68.38	1.61	0.24	0.0	0.0
20	35.0	3.50	2.315	1.148	0.0	0.0	0.0	0.0	0.0	100.0	0.0	0.0	0.0	0.0
21	100.0	3.43	2.315	1.148	0.0	0.0	0.0	0.0	0.0	100.0	0.0	0.0	0.0	0.0
22	35.0	1.20	23.854	0.619	0.0	0.0	0.61	84.35	0.0	9.76	4.58	0.70	0.0	0.0
23	15.0	1.01	131.480	4.557	0.92	0.0	0.0	0.03	0.0	0.0	1.03	77.29	20.73	0.0
24	34.9	1.22	131.480	4.557	0.92	0.0	0.0	0.03	0.0	0.0	1.03	77.29	20.73	0.0
25	120.3	1.20	131.480	4.557	0.92	0.0	0.0	0.03	0.0	0.0	1.03	77.29	20.73	0.0
26	107.6	1.20	155.330	5.177	0.81	0.0	0.07	10.12	0.0	1.17	1.46	68.12	18.25	0.0
27	216.7	1.17	155.330	5.177	0.81	0.0	0.07	10.12	0.0	1.17	1.46	68.12	18.25	0.0
28	426.1	1.15	155.330	5.177	0.81	0.0	0.07	10.12	0.0	1.17	1.46	68.12	18.25	0.0
29	750.0	1.12	156.240	5.198	0.81	0.0	0.0	11.21	0.0	0.0	4.61	67.84	15.53	0.0
30	765.0	1.08	143.310	4.793	0.88	0.0	0.0	12.16	0.0	0.0	4.99	73.57	8.40	0.0
31	559.2	1.06	143.310	4.793	0.88	0.0	0.0	12.16	0.0	0.0	4.99	73.57	8.40	0.0
32	446.1	1.04	143.310	4.793	0.88	0.0	0.0	12.16	0.0	0.0	4.99	73.57	8.40	0.0
33	135.0	1.03	143.310	4.793	0.88	0.0	0.0	12.16	0.0	0.0	4.99	73.57	8.40	0.0
34	90.0	1.01	143.310	4.793	0.88	0.0	0.0	12.16	0.0	0.0	4.99	73.57	8.40	0.0
35	35.0	3.50	16.939	0.940	0.0	0.0	0.0	0.0	0.0	0.0	0.0	0.0	0.0	100.0
36	15.0	1.01	4.492	0.249	0.0	0.0	0.0	0.0	0.0	0.0	0.0	0.0	0.0	100.0
37	15.0	1.71	4.492	0.249	0.0	0.0	0.0	0.0	0.0	0.0	0.0	0.0	0.0	100.0
38	30.8	1.71	21.431	1.190	0.0	0.0	0.0	0.0	0.0	0.0	0.0	0.0	0.0	100.0
39	105.0	1.37	21.431	1.190	0.0	0.0	0.0	0.0	0.0	0.0	0.0	0.0	0.0	100.0
40	107.5	1.32	21.431	1.190	0.0	0.0	0.0	0.0	0.0	0.0	0.0	0.0	0.0	0.0
41	529.2	1.27	21.431	1.190	0.0	0.0	0.0	0.0	0.0	0.0	0.0	0.0	0.0	0.0

596277 Self-Contained Electrochemical Process to Produce Pure Compressed Hydrogen from Hydrocarbons and Steam without an External Energy Supply

Abstract:

This paper develops and applies a physics-based model to predict the behavior of a concept that tightly couples a hydrocarbon-fueled solid-oxide fuel cell (SOFC) and an electrochemical protonic-ceramic hydrogen-separation cell (PSM). The system, called Compound Hydrogen Generation (CH2Gen), is designed to produce separated and compressed hydrogen without any external electrical energy supply.

The SOFC is configured as a typical anode-supported membrane-electrode assembly (MEA), with a relatively thick (order millimeter) porous composite anode, a thin (order 20 microns) oxygen-ion membrane, and a thin (order 20 microns) composite cathode. The role of the PSM is to electrochemically separate and compress hydrogen from reformed methane, such as reported by Malerød-Fjeld, et al. [1] The protonic-ceramic membrane is a doped-perovskite material such as yttrium-doped barium zirconate ($BaZr_{0.8}Y_{0.2}O_{3-d}$, BZY20). Because both sides of the PSM operate in a reducing gas-phase environment, the cell can be symmetric with both electrodes using materials such as Ni-BZY.

The SOFC and the PSM could each be operated alone, but the SOFC would need to be connected to an external load and the PSM connected to an external power supply. The innovation with the CH2Gen is associated with the electrical and gas-flow connections between the SOFC and the PSM.

The present paper considers two alternative gas-handling coupling strategies. In both cases, the anode (negatrode) of the SOFC is connected electrically to the cathode (positrode) of the PSM. The cathode (positrode) of the SOFC is connected directly to the anode (negatrode) of the PSM. An important consequence of the electrical coupling is that the SOFC and PSM must operate at the same voltage and current.

Parallel configuration

Figure 1 illustrates a *parallel* CH2Gen implementation. In this case, the SOFC and the PSM share the same fuel chamber, with catalytic fuel reforming within both the SOFC anode and the PSM positrode structures. As a result of oxygen-ion transport through the SOFC membrane, oxygen enters the fuel chamber. As a result of proton transport through the PSM membrane, hydrogen leaves the fuel chamber. As illustrated in Fig. 1, the air chamber, the fuel chamber, and the hydrogen-collection chambers are all represented as perfectly stirred reactors [2].

Although not strictly necessary, both the SOFC and the PSM should operate at the same temperature. Considering catalytic steam reforming, protonic membrane performance, and oxygen-ion membrane performance, operating temperatures should be around 600--700 °C. Thus, at relatively low operating temperatures, the SOFC should likely be based on a doped-ceria (e.g., $Ce_{0.9}Gd_{0.1}O_{2-d}$, GDC10) membrane rather than a yttria-stabilized (YSZ) membrane.

Serial configuration

In a *serial* CH2Gen implementation, the SOFC and PSM are physically separated, with exhaust gas from the SOFC anode chamber feeding directly to the PSM anode chamber. Electrically, however, the SOFC cathode is wired directly to the PSM anode. As illustrated for the parallel implementation, all the gas chambers are idealized as perfectly stirred reactors (PSR).

The SOFC is fed by a hydrocarbon-steam mixture and air. Steam reforming proceeds within the catalytic surfaces of the porous composite SOFC anode (e.g., Ni-BZY). Hydrogen is electrolyzed by oxygen ions O^{2-} emerging from the ion-conducting membrane (e.g., GDC). The gas-phase composition of the SOFC chamber depends on the fuel feed composition and rates, reforming rates, oxygen-ion flux, and gas-phase residence time.

The exhaust of the SOFC anode chamber is fed directly to the negatrode-side of the PSM. In addition to further reforming on the PSM composite electrode to produce H_2, electrochemical processes deliver protons to the proton-conducting membrane. Because the PSM membrane is polarized (i.e., same voltage as the SOFC), protons migrate across the membrane to deliver H_2 into the hydrogen-collection chamber. Assuming a back-pressure regulator on the exhaust of the H_2-collection chamber, the H_2 may be electrochemically compressed.

Operational constraints

The CH2Gen electrical configuration requires that the SOFC and PSM operate at the same voltage and current (electrical and ionic). A further practical constraint is that the exhaust from system be dominantly CO_2 and H_2O, with only trace levels of H_2, CO, or other hydrocarbon side products.

Assume that the hydrogen-collection chamber is held at an elevated pressure (e.g., 5 bar) and that the PSM fuel chamber has only small H_2 partial pressure. Thus, there is a strong electrochemical driving potential to transport H_2 from the hydrogen-collection chamber back to the PSM fuel chamber. The operating voltage for the PSM (and thus the SOFC) must be sufficiently high as to force low-concentration H_2 into the elevated-pressure H_2-collection chamber. In other words, the SOFC voltage must be greater than the open-circuit voltage associated with the PSM MEA.

Mathematical models

The CH2Gen system tightly couples the SOFC with PSM through the fuel flow and the electrical connections. The present model treats the SOFC and PSM as button cells with the gas compartments modeled as perfectly stirred reactors (PSR). The electrochemical button-cell models build upon previous models for fuel cells, electrolysis cells, membrane separations, etc. [3-12]. The SOFC and PSM models are coupled with an iterative algorithm that forces the SOFC and PSM to operate at the same voltage and current.

Summary and conclusions

A model is developed and applied to evaluate the feasibility of a novel electrochemical cell architecture that is designed to reform hydrocarbons, separate and electrochemically compress H_2, and emit only CO_2 and H_2O---all without any external electrical power. The cell combines a solid-oxide fuel cell (SOFC) with a protonic-ceramic separation membrane (PSM) in a unique way. The negatrode of the SOFC is wired to the positrode of the PSM and the negatrode of the PSM is wired to the positrode of the SOFC. In this way, the SOFC provides the power needed to drive the PSM.

Because of the unique electrical connection, both the SOFC and the PSM must necessarily operate at the same voltage and current. Two alternative gas-connection architectures are considered. In the parallel case, the SOFC and PSM share the same fuel chamber. In the serial case, the exhaust of the SOFC fuel chamber feeds the downstream fuel chamber of the PSM. In both cases, all the gas flows are modeled as perfectly stirred reactors.

The results establish the feasibility of the concept, called CH2Gen. Assuming that the concept is viable, process scale-up will likely require alternative gas-handling approaches. As is typical in fuel-cell technology, channel architectures will likely be appropriate. However, the next steps in technology development are beyond the scope of the present feasibility study.

References

1. H. Malerød-Fjeld, D. Clark, I. Yuste-Tirados, R. Zanón, D. Catalán-Martinez, D. Beeaff, S.H. Morejudo, P.K. Vestre, T. Norby, R. Haugsrud, J.M. Serra, and C. Kjølseth. Thermo-electrochemical production of compressed hydrogen from methane with near-zero energy loss. *Nat. Energy*, 2:923–931, 2017.

2. R.J. Kee, M.E. Coltrin, P. Glarborg, and H. Zhu. *Chemically Reacting Flow: Theory, Modeling and Simulation.* Wiley, Hoboken, NJ, 2nd edition, 2018.

3. E.S. Hecht, G.K. Gupta, H. Zhu, A.M. Dean, R.J. Kee, L.Maier, and O. Deutschmann. Methane reforming kinetics within a Ni- YSZ SOFC anode. *Appl. Catal. A*, 295:40–51, 2005.

4. H. Zhu and R.J. Kee. Modeling distributed charge-transfer processes in SOFC membrane electrode assemblies. *J. Electrochem. Soc.*, 155:B175–B729, 2008.

5. R.J. Kee, H. Zhu, B.W. Hildenbrand, E. Vøllestad, M.D. Sanders, and R.P. O'Hayre. Modeling the steady-state and transient response or polarized and non-polarized proton-conducting doped-perovskite membranes. *J. Electrochem. Soc.*, 160:F290–F300, 2013.

6. E. Vøllestad, H. Zhu, and R.J. Kee. Interpretation of defect and gas-phase fluxes through mixed-conducting ceramics using Nernst–Planck–Poisson and integral formulations. *J. Electrochem. Soc.*, 161:F114–F124, 2014.

7. H. Zhu, S. Ricote, W.G. Coors, and R.J. Kee. Interpreting equilibrium-conductivity and conductivity-relaxation measurements to establish thermodynamic and transport properties for multiple charged defect conducting ceramics. *Faraday Discuss.*, 182:49–74, 2015.

8. H. Zhu and R.J. Kee. Membrane polarization in mixed conducting ceramic fuel cells and electrolyzers. *Int. J. Hydrogen Energ.*, 41:2931–2943, 2016.

9. H.Zhu and R.J. Kee. Modeling protonic-ceramic fuel cells with porous composite electrodes in a button-cell configuration. *J. Electrochem. Soc.*, 164:F1400–F1411, 2017.

10. H. Zhu, S. Ricote, C. Duan, R.P. O'Hayre, D.S. Tsvetkov, and R.J. Kee. Defect incorporation and transport with $BaZr_{0.8}Y_{0.2}O_{3-d}$ (BZY20) proton-conducting membranes. *J. Electrochem. Soc.*, 165:F581–F588, 2018.

11. H. Zhu, S. Ricote, C. Duan, R.P. O'Hayre, and R.J. Kee. Defect chemistry and transport within dense $BaCe_{0.7}Zr_{0.1}Y_{0.1}Yb_{0.1}O_{3-d}$ (BCZYYb) proton-conducting membranes. *J. Electrochem. Soc.*,

165:F845–F853, 2018.

12. C. Duan, R.J. Kee, H. Zhu, N. Sullivan, L. Zhu, L. Bian, D. Jennings, and R.P. O'Hayre. Highly efficient reversible protonic ceramic electrochemical cells for power generation and fuel production. *Nature Energy*, 4:230–240, 2019.

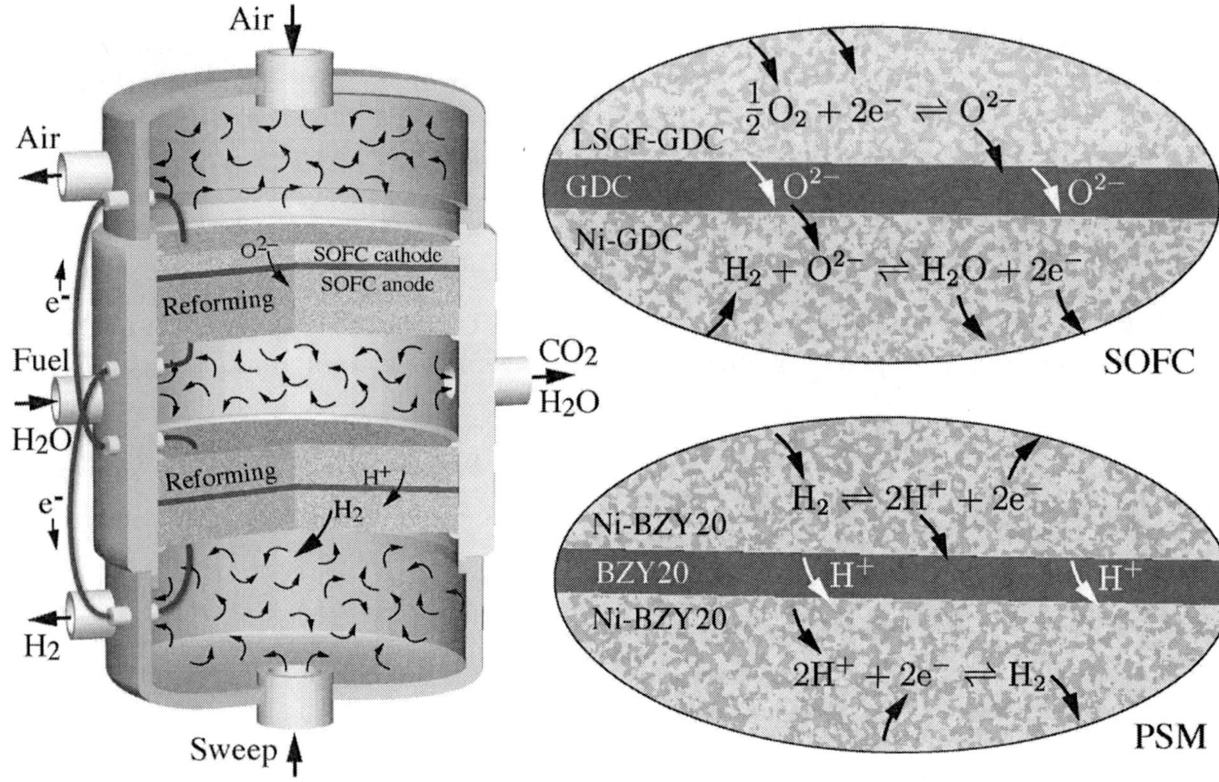

605193 Probing the Spaces of Microporous Materials for Enhanced Molecular Accessibility and Catalytic Control

Abstract:

The confining environment of microporous materials provide additional control for modifying the energetics of adsorbed species and the accessibility of reactants and products. However, small changes in molecular size and structure comparable to the pore or window size of the selected porous material yield large changes in molecular behavior[1]. In this work, the accessibility, adsorption, and transport of hydrocarbons are described using in situ characterization of the position of tunable adsorbing amines with time. Prediction of molecular sieving via three methods of describing rigid molecular size are directly compared with experimentally-measured molecular accessibility obtained via both reactive gas chromatography and in situ infrared spectroscopy[2]. These results are expanded to permit characterization of the different transport time scales of hierarchical microporous materials.

[1] P.J. Dauenhauer, Omar Abdelrahman, "A Universal Descriptor for the Entropy of Adsorbed Molecules in Confined Spaces," ACS Central Science 2018, 4(9), 1235-1243. DOI: 10.1021/acscentsci.8b00419

[2] Omar A. Abdelrahman, K.P. Vinter, L. Ren, D.Xu, R.J. Gorte, M. Tsapatsis, P.J. Dauenhauer, "Simple Quantification of Zeolite Acid Site Density by Reactive Gas Chromatography," Catalysis Science & Technology 2017, 7, 3831-3841. DOI: 10.1039/C7CY01068K

[3] Y. Guefrachi, G. Sharma, D. Xu, G. Kumar, K.P. Vinter, O.A. Abdelrahman, S.A. Hassan, P.J. Dauenhauer, A. Navrotsky, W. Zhang, M. Tsapatsis, "Steam-Induced Coarsening of Single-Unit-Cell MFI Zeolite Nanosheets and its Effect on External Surface Bronsted Acid Catalysis," Angewandte Chemie International Edition 2020, DOI: 10.1002/anie.202000395

601378 Investigation of Potential Active Sites for the Methane Oxidation Reaction on Pd/SSZ-13

Abstract:

It is widely reported that palladium oxide is the active site for methane oxidation on Pd-zeolite catalysts[1], yet there is a lack of understanding of the morphology of the active site and how it interacts with the zeolite structure on the molecular level. This theoretical study aims to identify possible active sites that may exist within Pd/SSZ-13 on the molecular level. A model with two aluminum atoms per unit cell has been chosen to simulate a Si/Al ratio of 17. To ensure that realistic active site candidates are found, a sequence of ordered calculations are performed to yield stable/low energy structures with two aluminum atoms. First, 25 geometrically distinct dual aluminum placements on SSZ-13 are investigated and the lowest energy structure is found (Fig 1A). From this lowest energy aluminum arrangement, all possible dual palladium arrangements are then investigated to find the lowest energy palladium structure. From this final structure, oxygen atoms and palladium atoms are then added to form numerous Pd_xO_x species that may serve as active sites. Lastly, dissociative methane adsorption is investigated for each site to further evaluate it for methane activation (Fig 1B). All geometries are optimized using the PBE functional. Identifying the active sites in this work is critical in the design of zeolite catalysts for the combustion of methane.

Figure 1: (A) Relative energies of 25 distinct dual aluminum placements on SSZ-13 versus distance between aluminum atoms (energies referenced to the lowest energy structure). PBE optimized structures evaluated at single point with HSE06 functional for energy. (B) Dissociative adsorption of methane on a 'bridged' dual palladium oxide site.

References

1. Gannouni, A.; Albela, B.; Zina, M. S.; Bonneviot, L., *Applied Catalysis A: General* **2013,** *464-465*, 116-127.

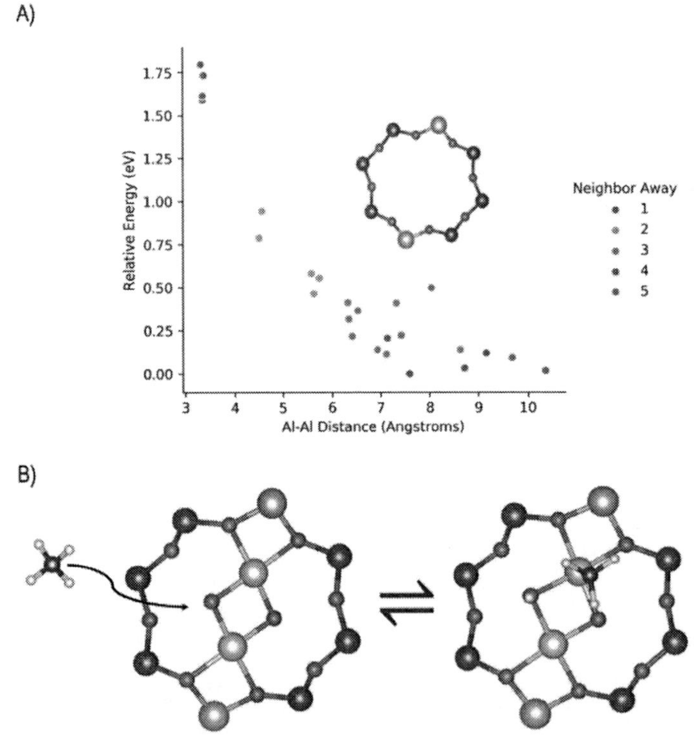

Practical Enhanced Oil and Gas Recovery (EOR/EGR) for Unconventional Wells and Reservoirs - a Next Generation Huff-and-Puff Treatment Combining Gas with Nanoparticles

Authors – R. Watts, Messer Americas; Y. Ahmad, Nissan Chemical America; M. Cantrell, Nitro-Lift Technologies

ABSTRACT

Unconventional reservoirs unlocked through the combination of horizontal drilling and fracturing have revolutionized the exploration and production of hydrocarbons. In the United States, this success has opened the way to the export of these hydrocarbons along with a renaissance in the petrochemical processing industry due to the supply of inexpensive energy and feedstocks. This Unconventional Revolution, now some 10+ years in the making, is still in its infancy. Next generation challenges for economic production, as well as generating free cash flow, include the need for improved hydrocarbon recovery. Current initial drilling and completion practices (aka primary recovery) claim in the range of 6-10% of the oil and up to 25% of the gas in place. Significant opportunities exist to improve upon secondary and tertiary recovery practices for unconventional reservoirs.

One of the more recent methods of enhanced hydrocarbon recovery on existing unconventional wells includes utilizing two injectants, gases and nanoparticles in a huff-and-puff (HNP) style treatment. Gases such as carbon dioxide (CO_2) and nitrogen (N_2) have a long history of use for improving recovery. Nanoparticles, a more recent innovation for recovery enhancement, have demonstrated that ability in fracturing formulations, as well as in remedial applications. A more recent practice, combining the two in a traditional HNP-style injection in existing wells, has demonstrated a synergistic treatment that provides an improved incremental production of hydrocarbons over an extended period. Relative costs of these new nano/gas or nanoparticle assisted HNP, referred to as nano HnP, treatments are significantly less than traditional methods (i.e., refracturing), providing a greater economic return for operators looking to enhance recovery from their existing wells. Most importantly, this new nano HnP EOR/EGR method demonstrates the opportunity for low capital expenditure, low unit cost of production (BOE/Mcfe), along with practical scalability needed for full field re-development to maximize recovery from existing unconventional wells.

INTRODUCTION: Enhancing Recovery – Practical EOR/EGR pre-Unconventional Revolution

Primary, secondary and tertiary recovery represent the three phases of continuously improving recovery of hydrocarbons from existing reservoirs. The time horizons for each phase varies and collectively may span decades over a producing reservoir's life. Figure 1 summarizes each of the phases, common methods, time horizons and overall recovery factor typical for conventional resources as practiced over the last several decades.

As in all phases of recovery, the methodology deployed depends upon several key factors. To make any enhanced recovery method practical, three key components must exist in the method:

1. Enhance recovery
2. Be economical
3. Have scalability for field re-development

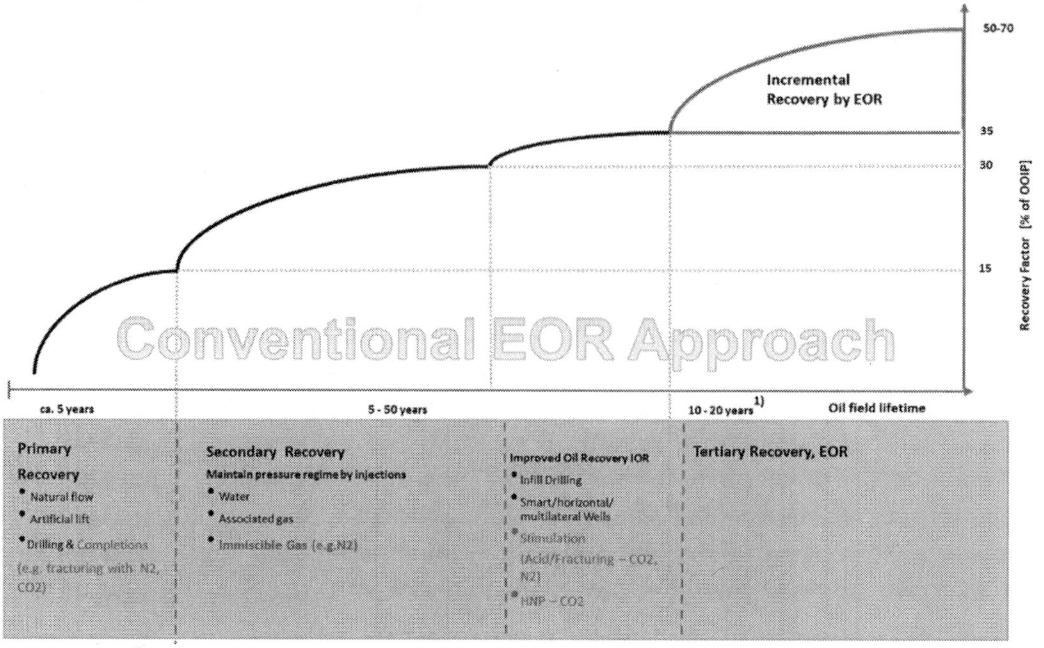

Figure 1. *Recovery Factor as a percent of Original Oil in Place (OOIP) as progress in time from primary to tertiary recovery otherwise known as EOR or enhanced oil recovery.*

In the US, the methods commonly used for enhancing oil recovery have varied from chemical to gas injection. Figure 2 highlights the amount of annual oil production by various EOR methods over time. [1]

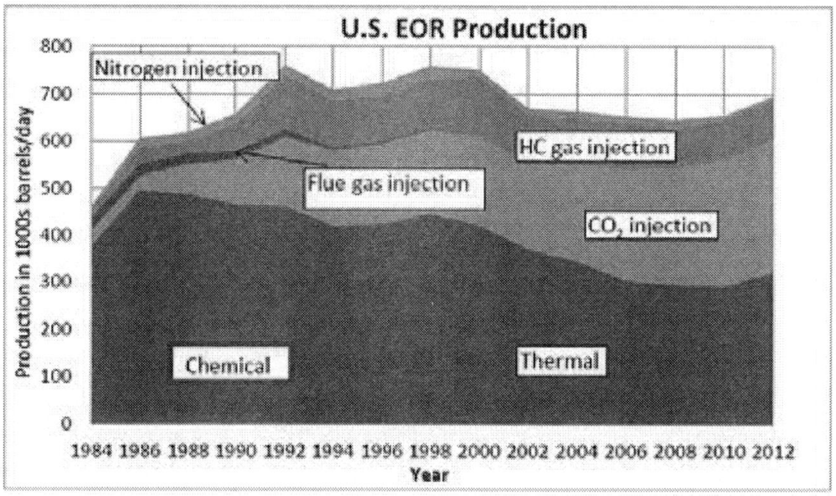

Figure 2. *Annual production of oil by EOR method over time.*

Since the mid-1980's, CO_2 injection for EOR has grown into the "gold standard" for enhancing oil recovery due to its practicality. Miscible CO_2 floods or CO_2 WAG (water alternating gas) floods, shown in Figure 3, provide excellent recovery, and under the right conditions, effective costs. CO_2 improves oil recovery by lowering interfacial tension, swelling the oil, reducing oil viscosity and by mobilizing the lighter components of oil. Figure 4 provides an example of the recovery from a CO_2 project in the Denver Unit of the Wasson Field in West Texas.

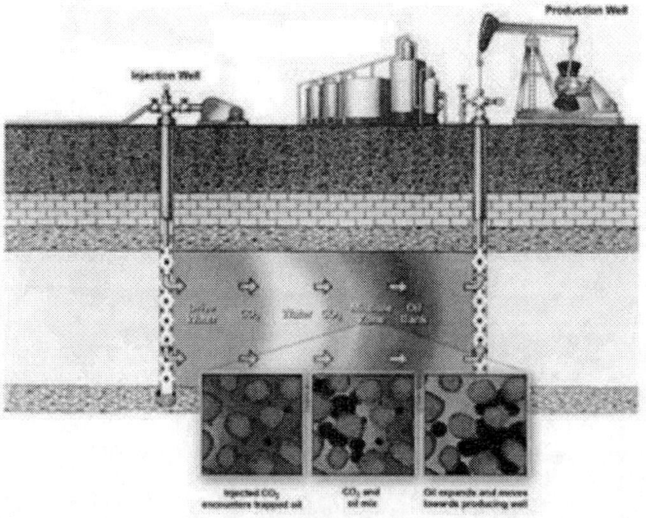

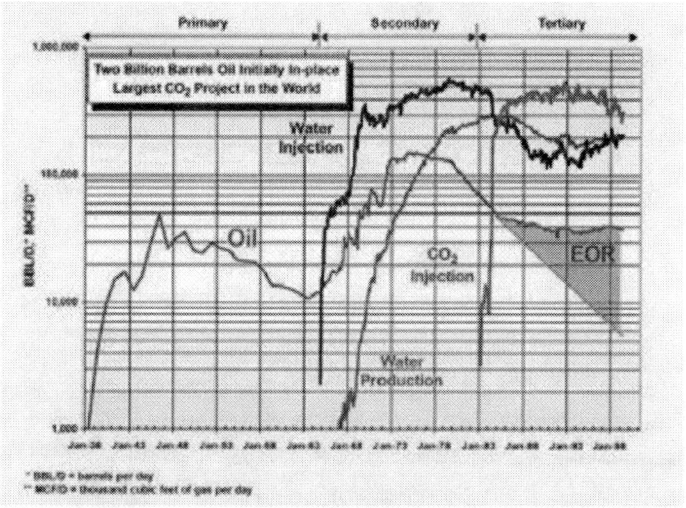

Figures 3 and 4. CO_2 WAG schematic. Recovery from the Denver Unit of the Wasson Field in West Texas. CO_2 injection as the Tertiary or EOR method.

Background

Even as the gold-standard, CO_2 EOR economics are defined by the CO_2 source options and their proximity to the field of injection. Figure 5 highlights the main sources of CO_2 for US EOR floods. Note however, that the illustrative economics, in Table 1, from a National Energy Technology Laboratory report relies upon a much higher oil price than what is true for today's environment. In addition, the cost per Mcf of CO_2 is relatively high compared to the typically targeted CO_2 price of $1/Mcf for floods.

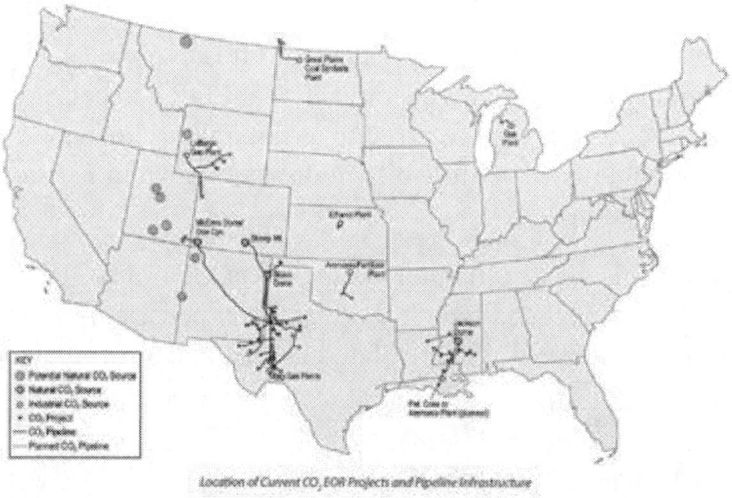

Illustrative Costs and Economics of a CO_2 EOR Project

Oil Price ($/Barrel)	$70
Gravity/Basis Differentials, Royalties and Production Taxes	($15)
Net Wellhead Revenues ($/Barrel)	$55
Capital Cost Amortization	($5 to $10)
CO_2 Costs (@ $2/Mcf for purchase; $0.70/Mcf for recycle)	($15)
Well/Lease Operations and Maintenance	($10 to $15)
Economic Margin, Pre-Tax ($/Barrel)	$15 to $25

Figure 5 and Table 1. Active CO_2 EOR Projects and Pipelines as of 2010. Illustrative CO_2 EOR Project Economics based upon $70/barrel oil price.

Just over 100 CO_2 floods had occurred in the US by 2015 (Table 2)[2], but over 1000 fields represent technically recoverable candidates (Figure 6). And while many efforts are underway to encourage more favorable CO_2 economics, with federal government incentives or with technological advances, the challenges remain.

Number of projects	Lithology	Porosity (percent)	Perm. (md)	Depth (feet)	Gravity (°API)	Viscosity (cp)	Temp. (°F)
				Miscible			
42	ss.	7–26	16–280	1,600–11,950	30–45	0.6–3.0	82–257
2	ss./ls.-dol.	10	4–5	5,400–6,400	35	1	170–181
41	dol.	7–5	2–28	4,000–11,100	28–42	0.6–6.0	86–232
12	dol./ls.	3–12	2–5	4,900–6,700	31–44	0.4–1.8	100–139
6	ls.	4–20	5–70	5,600–6,800	39–43	0.4–1.5	125–135
1	dol./trip. chert	13.5	9	8,000	40	NA	122
7	tripolite	18–24	2–5	5,200–7,500	40–44	0.4–1.0	101–123
1				inadequate data			
				Immiscible			
8	ss.	17–30	30–1,000	1,500–8,500	11–35	0.6–45	99–198
1	dol.	17	175	1,400	30	6	82

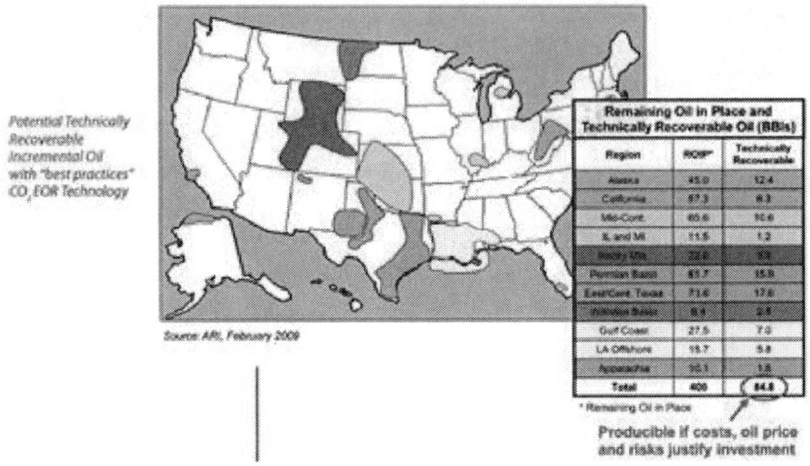

Table 2 and Figure 6. *CO_2 flood projects in the US as of 2015. Technically Recoverable Oil with CO_2 EOR.*

Practical Enhanced Oil and Gas Recovery (EOR/EGR) for Unconventionals

Unconventional EOR as currently practiced and published, is not practical. It is evolving but very limited in its useful application. The most widely implemented unconventional EOR projects are in the Eagle Ford utilizing the method of natural gas cyclic injection, referred to as natural gas huff and puff (NGHNP). Its limited success has been in the Eagle Ford volatile oil window as shown in Figure 7 and requires installation of a large natural gas compression unit.

Grinestaff et al published an economic study this year of the more than 30 Eagle Ford NGHNP pilot projects implemented with a capex cost of ~\$1mm/well and monthly operating expenses of ~\$150k/month.[3] The practice consists of screening candidate wells in isolated groups that are contained

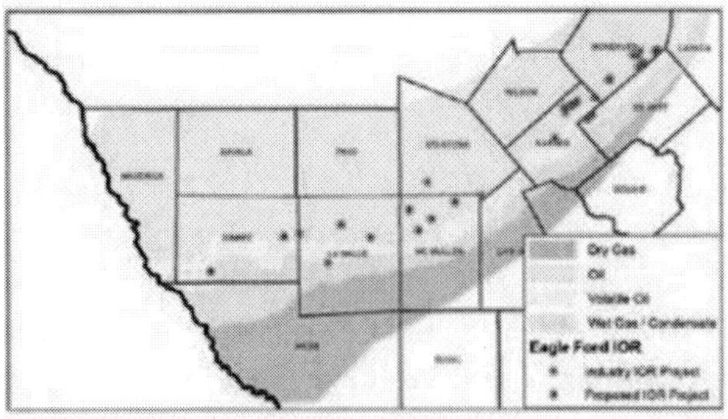

Figure 7. *Eagle Ford map of production windows and IOR (NGHNP) projects. Image of NG compressor.*

in order to successfully build pressure needed to optimize recovery. With that, some of the wells are to be used as injectors and others as producers and these wells may be reversed on a cyclic basis. Production response typically takes 2-3 months. Initial reported unit cost of production equivalents has been as low as $6/barrel of oil. However, the practical application of NGHNP is limited and successful economics were reported as requiring $60 per barrel price of oil over a 10-year investment horizon. (Table 3.)

Table 3—Baker-Deforest & Mitchell Economic Summary

Metric	Baker-Deforest	Mitchell
KPI	Oil-$60, Gas-$2.50	Oil-$60, Gas-$2.50
Injection Rate, MMcfpd	28.6	23.0
Compressor Capacity, MMcfpd	35.0	35.0
Primary EUR/well (MBO)	373	312
IOR Ratio (10 yrs)	1.4	1.3
Capex, M$	15,548	11,147
Gas Purchase, M$	12,198	11,017
Peak Neg. CF, M$	($18,142)	($12,680)
Cashflow, M$	$51,799	$17,051
NPV10, M$	$33,633	$12,078
Disc ROI	2.2	1.1
Undisc ROI	3.3	1.5
IRR,%	64%	62%
Payout, yrs	1.6	1.2

Table 3. *Eagle Ford NGHNP EOR Pilot Project Economics*

Fengshuang et al published last year a comprehensive comparison study of various lab, simulation, and field injection studies done utilizing gas injection in shale reservoirs. None have expanded into the realm of practicality.[4]

Baldwin et al reported in their recent paper the methodology they used for approaching unconventional EOR in the Eagle Ford.[5] The steps and time horizon, shown in Table 4, are very much like a conventional EOR project approach.

Time Line	Work Phase
4-12 weeks	Screening – Candidate Wells
3 months	Feasibility Study – including lab work
6 months	Pilot Design – including simulation
6 months	Pilot Test
Years	Implement full-scale program

Table 4. *Unconventional EOR NGHNP in the Eagle Ford from concept to field re-development.*

nano HnP for Unconventional Reservoirs – 1. technically enhances recovery

A new method for enhancing recovery out of unconventionals was discussed by *Watts et al* utilizing a combination of gas and nanoparticles in a HNP style treatment. (Figure 8).[6] This nano HnP treatment effectiveness, reported in the Eagle Ford, demonstrated the key criteria for being a practical EOR treatment for unconventional reservoirs.

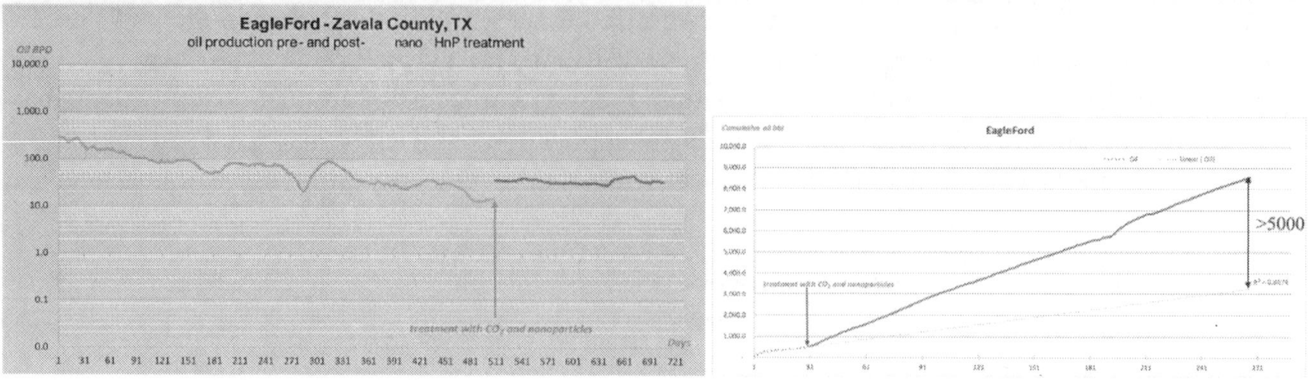

Figure 8. *Eagle Ford well treated with a CO_2 nano HnP showing continued enhanced production over 200 days post-treatment.*

The synergistic gas and nanoparticles well treatments have demonstrated an ability to provide technically recoverable, economical and scalable unconventional enhanced recovery. The nano HnP treatments have exhibited:

- Practical recovery validated in the field, lab and in simulation work
- The ability to work with a variety of gases
- That they require significantly less gas than traditional gas HNP

In the Eagle Ford pilot, the horizontal well's incremental production paid for the full cost of treatment within ~90 days.

Functionally modified silicon dioxide nanoparticles used in the nano HnP treatments have demonstrated the ability to reduce interfacial tension (IFT) as well as modify the wettability of the rock, as shown in Figure 9.[7]

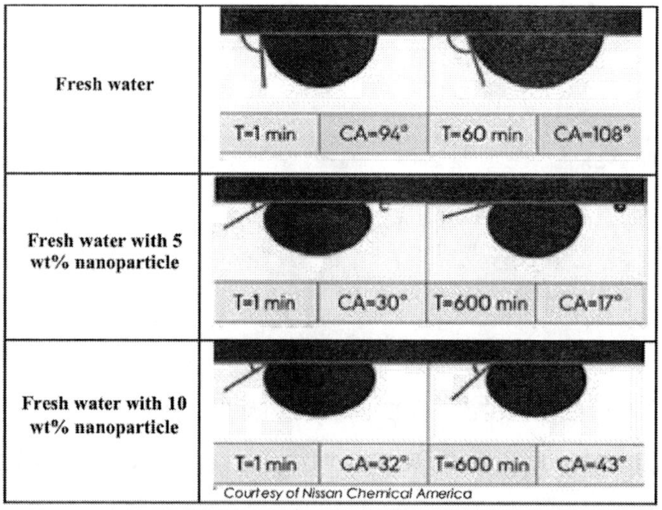

Note: There is some variability in contact angle measurements due to instrument error. As with all lab testing, nanoparticle concentration optimization is crucial in getting accurate results.

Figure 9. *Nanoparticles used in nano HnP treatments demonstrate ability to reduce contact angle and IFT on Montney core plugs.*

In a simulation study of the nano HnP treated well in the Eagle Ford recently published by *Zheng et al*, history matching of the production results in the unconventional HNP simulator concluded that a relative permeability adjustment was required to properly account for the improved well productivity.[8] Neither the small volume of gas nor just water run in a simulated comparison was effective when compared to the improved production created by the nano HnP treatment. (Figure 10)

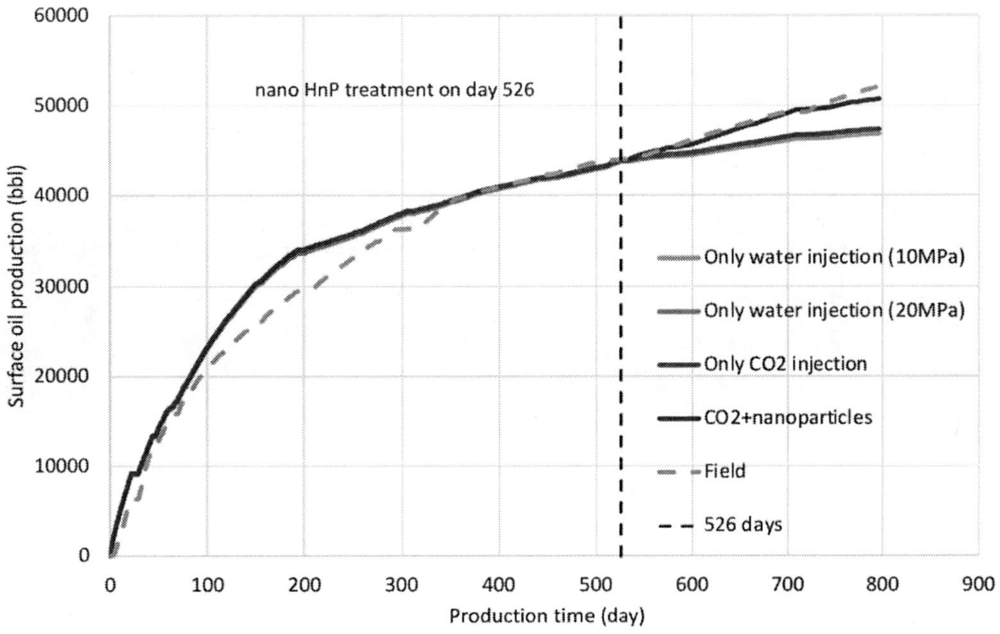

Figure 10. *Simulated Results of Eagle Ford compared to field results of horizontal well treated with CO_2 and nanoparticles (nano HnP).*

nano HnP for Unconventional Reservoirs – 2. economical enhanced recovery

Case Study 1: N_2 nano HnP treated Woodford shale horizontal well

Another example of the economic practicality of EOG/EGR aspect of nano HnP treatments, comes from a case study on the production and economics of horizontal wells treated in the Woodford shale. For the first horizontal well, a 2-year-old unconventional well, a nitrogen (N_2) nano HnP treatment was utilized. Prior to treatment, the well was producing ~1,300 Mcfd and ~6 Bpd. Initial production (IP) of the well when first completed was ~2,600 Mcfd. After the N_2 nano HnP treatment, the well production was greater than its IP for ~60 days and has stayed over 2,200 Mcfd for 150 days. (Figure 11). For this well, the operator's treatment payback was about three months. Fourteen months post- nano HnP treatment, well production reportedly continues to be in the 1,800-2,000 Mcfd range.

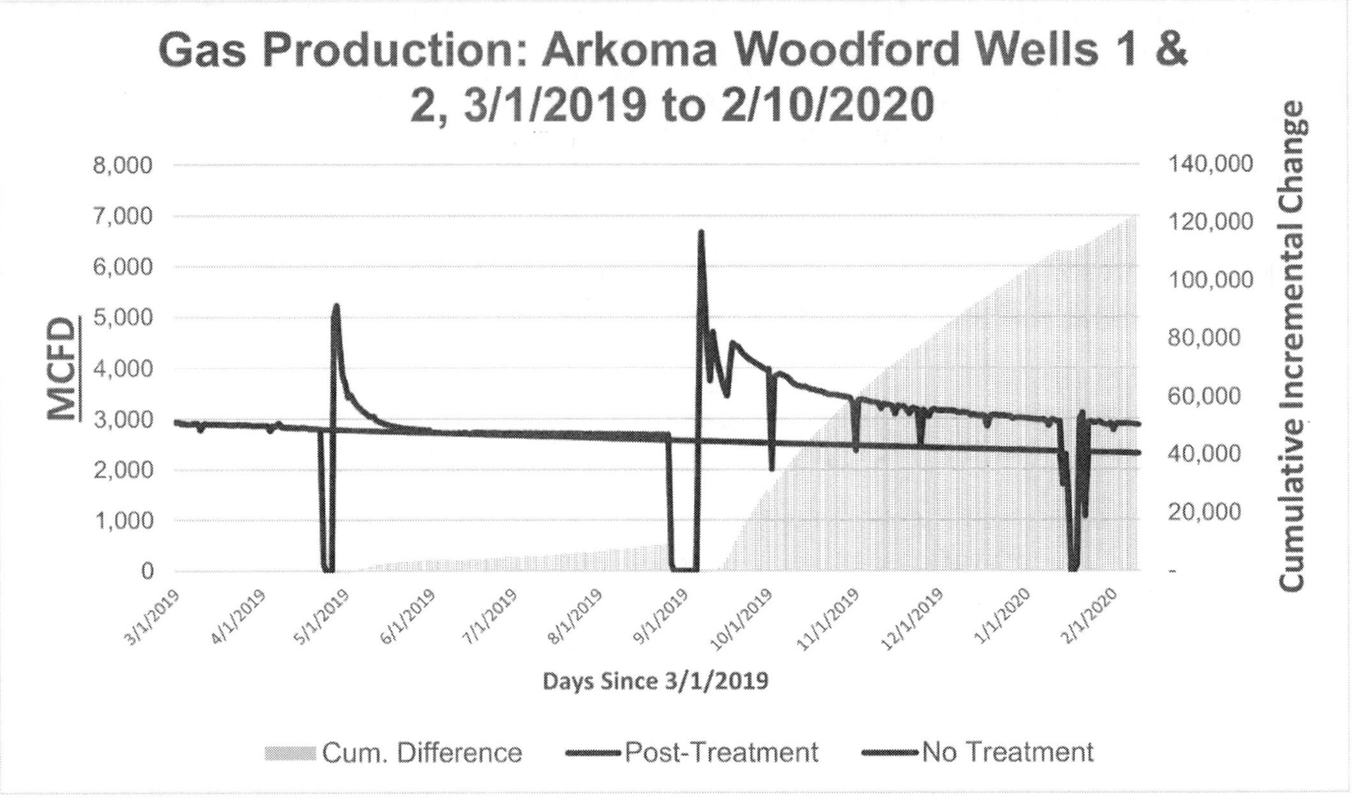

Figure 11. *Woodford horizontal well treated with N_2 nano HnP had post-treatment production exceeding IP.*

Case Study 2: Three horizontal wells in Woodford shale treated with NG nano HnP

In this next example, the economic practicality of EOG/EGR natural gas (NG) nano HnP treatments is demonstrated in three horizontal wells in the Woodford shale.[9] The average incremental production up lift for each of the wells was ~1,200 Mcfd. Payback for the treatments was estimated between 90 to 140 days. In comparison to a new completion, Table 5, the nano HnP represents a unit cost of production of about one third of a new completion unit cost of production as experienced by operators in the region. For the cost of one new completion, 20 nano HnP treatments could be done for a lower total cost with paybacks in about 1/4th of the time.

New Completion

Avg. Mcfd for 1 Year*	**3,500**
Days	365
Annual Production	1,277,500
no. of wells drilled	1
Total Gas Production	1,277,500
Est. Cost	$3,500,000
$-per-Mcf	**$2.74**

Paybacks

$/Mcf	Producer Profit	1 New Completion
$2.00/Mcf	$1.50/Mcf	1.8 Years
$1.75/Mcf	$1.25/Mcf	2.2 Years
$1.50/Mcf	$1.00/Mcf	2.7 Years

nano HnP Treatment

Avg. Mcfd for 1 Year*	**500**
Days	365
Annual Production	182,500
no. of wells treated	**20**
Total Gas Production	3,650,000
Cost	$3,000,000
$-per-Mcf	**$0.82**

Paybacks

$/Mcf	Producer Profit	20 nanoActiv® Treatments
$2.00/Mcf	$1.50/Mcf	0.5 Years
$1.75/Mcf	$1.25/Mcf	0.5 Years
$1.50/Mcf	$1.00/Mcf	0.7 Years

*Current new drill and completion costs in the Arkoma basin are ~$3.5MM per well.

Table 5. *Economics of a nano HnP many times more favorable than a new drilling and completion.*

nano HnP for Unconventional Reservoirs – 3. scalability for field re-development

The process for going from the screening of candidates to full field re-development for nano HnP has many similarities to traditional EOR projects and the more recent NGHNP projects for EOR in the Eagle Ford. The major difference is the practicality of the method of treatment, both in terms of technical capabilities, along with the flexibility and simplicity of the treatments to allow feasible scalability for unconventional reservoirs. No special equipment is needed for these treatments and can be readily executed by a certified service provider.[10] Figure 12 below summarizes the steps to optimize field wide implementation of nano HnP. Each step, focusing on the technical, economical, and scalability of an unconventional enhanced recovery program, consists of the following:

1. ***Screening phase*** – This begins with a focus on the techno-economics of success including the identification of gas source options to go from proof of concept to pilot programs to field-wide program expansion. Since the quantities of gas required for a nano HnP treatment are significantly less than a gas only HNP, and different gases may be used for nano HnP treatments in the same field, gas sources may vary. These sources may include a variety of nearby commercially available plants and field gas and requires a proper understanding of the options for expansion.

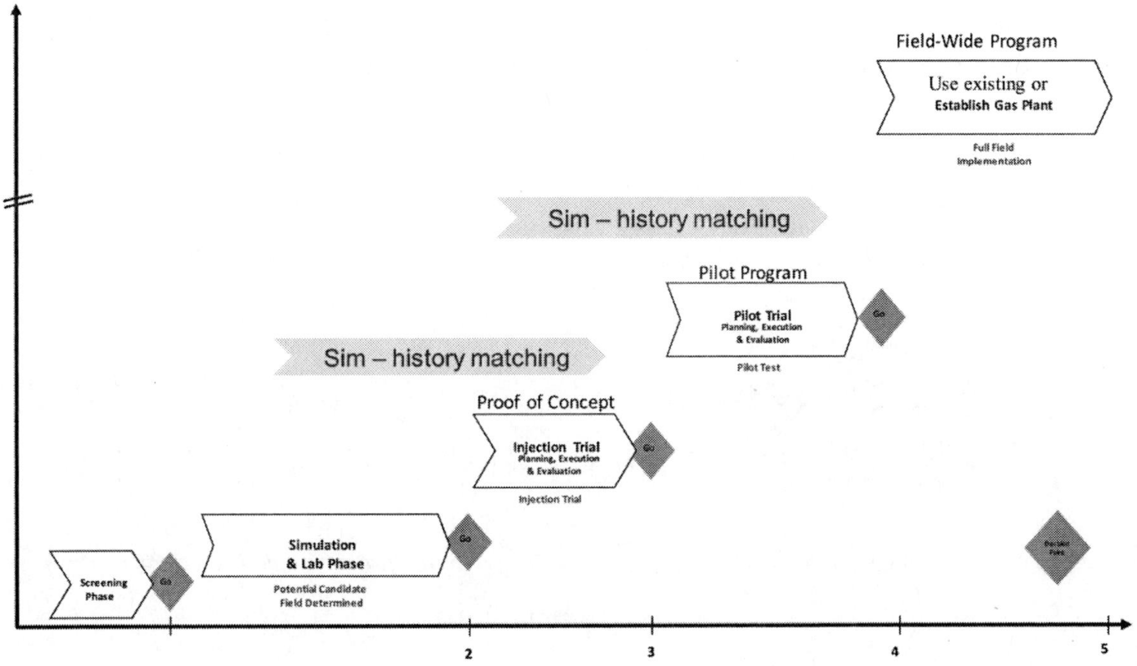

Figure 12. *nano HnP Field Re-development steps for practical enhanced recovery in unconventionals.*

2. ***Simulation and lab phase*** – These phases are utilizing unconventional HNP testing facilities and simulation tools more recently available for commercial applications. Depending on the gas source options, the lab nano HnP test may be run using more than one type of gas. Each type of gas nano HnP test will provide insight into the reservoir response for relative permeability expected

adjustments that can then be input into an unconventional reservoir HNP simulator. This will allow for optimizing initial field treatment designs based upon relative production improvement versus costs of treatments. Securing of gas sources for field implementation and scale-up should begin in earnest at this time.

3. ***Field Injection Trial*** – This step helps ensure well connectivity in identified candidates is understood enough so treatment design can be optimized. Nano HnP well treatments can be optimized around various levels of communication. Monitoring of production results post-treatment provide additional insight into reservoir response, to further refine treatments for technical and economic optimization using the HNP simulator. The securing of gas sources work should continue to ensure there is enough dedicated supply ready for expansion.

4. ***Pilot Program*** – A further expansion and refinement of work from the injection trial.

5. ***Field-wide Program*** – This is execution of the greater scalability plan that comes as a result of the securing of gas sources and optimization of treatment designs in the previous phases.

   Following the steps above, provides the path to a practical unconventional reservoir enhanced recovery field re-development program. The economics of a successful nano HnP program lead to low unit cost of production (Table 6) and scalability to optimize free cash flow with lower capital risk.

Method	CapEx, MM	BOE (Unit Cost)	Payback (varies by NRI)	No. of Wells (Equivalent CapEx)
New Drilling & Completion	$6-10	$49 avg.	540-730 days	1
Re-frac (limited)	$2-3	$13-16	365-540 days	3-5
NG HNP (limited)	$1	$6	365 days	6-10
nano HnP	<$0.25	<$3/BOE or <$1/Mcfe	<100 days	>24-40

NRI = net revenue interest

Table 6. *nano HnP summary economics compared to other currently practiced methods of recovery*

Conclusions

Practical enhanced recovery solutions for unconventional reservoirs have, in recent history, eluded oil and gas operators. But with the current economic emphasis away from growth to creating value by improving free cash flow, an option now exists for operators utilizing the combined technology of gas and nanoparticles in a nano HnP treatment. As defined, practical enhanced recovery is optimized around recovery, economics, and scalability. The flexibility in gas type and volumes along with the ease of implementation for nano HnP treatments means that the path to a field re-development enhanced recovery program readily exists. These nano

HnP treatments have demonstrated all the pre-requisites for a practical enhanced recovery solution for unconventionals as the treatments are:

- Technically recoverable as a result of nano gas synergy reducing IFT, altering wettability/relative permeability and clearing pathways/obstructions
- Economical and flexible, depending on dosage of treatments/operator economics/gas source options (i.e., CO_2, N_2, NG)
- Scalable with defined steps, methods, and practices in parallel with securing gas supply for trials to expanding field-wide

REFERENCES

1. NETL, Mar 2010, *Carbon Dioxide Enhanced Oil Recovery Untapped Domestic Energy Supply and Long Term Carbon Storage Solution.*
2. Verma, M., 2015, *Fundamentals of Carbon Dioxide-Enhanced Oil Recovery (CO2-EOR)—A Supporting Document of the Assessment Methodology for Hydrocarbon Recovery Using CO2-EOR Associated with Carbon Sequestration,* U.S. Department of the Interior, U.S. Geological Survey.
3. Grinestaff, G., et al, 2020, *Evaluation of Eagle Ford Cyclic Gas Injection EOR: Field Results and Economics*, SPE-200427-MS.
4. Fengshuang D. et al, June 2019, *A Review of Gas Injection in Shale Reservoirs: Enhanced Oil/Gas Recovery Approaches and Greenhouse Gas Control,* Mining and Minerals Engineering, Virginia Polytechnic Institute and State University.
5. Baldwin, A. et al, 2020, *A Methodological Workflow for Assessment and Design of a Huff-N-Puff Hydrocarbon Gas Injection Pilot Test as an EOR Technique for Eagle Ford Shale Oil Reservoirs*, SPE-200415-MS.
6. Watts, R. et al, 2019, *Improved Oil Recovery (IOR) Method for Unconventional Reservoirs with a next Generation Huff-and-puff Treatment Combining Gas with Nanoparticles for Enhancing Recovery,* AIChE Annual Meeting.
7. Yuan, L. et al, 2020, *Chemical EOR While Fracturing: An Experimental Study for Evaluating Nanoparticle Additives,* SPE-200382-MS.
8. Zheng, S. et al, 2020, *A Nanoparticle Assisted CO2 Huff-n-puff Field Test in the Eagle Ford Shale,* SPE-200375-MS.
9. NitroLift 2020, *Economic Analysis: New Completion vs. a nanoActiv® Treatment Program*, RecoverMoreOil.com.
10. The nanoActiv® HnP (referred to as nano HnP in this Paper) treatment is proprietary to Messer and Nissan Chemical. This treatment should only be provided by Messer or a Messer Certified Service Provider. nanoActiv is a registered trademark of Nissan Chemical America Corporation.

602808 Understanding the Impact of Formation Properties on Stress Shadow in Multistage Hydraulic Fracturing through Modeling and Simulation

Abstract:

Multistage hydraulic fracturing (MSHF) is a standard technique used to increase reservoir contact with a horizontal wellbore to produce hydrocarbons from tight, unconventional oil plays. A typical Bakken completion consists of 50 stages, with three to four perforation clusters per stage. An induced fracture will propagate from an initiation point, often intersecting natural fractures or other plains of weakness, enhancing connectivity between the wellbore and the reservoir. The geometry and propagation of hydraulic fractures within a fracture stage are influenced by the distribution of reservoir properties (e.g., Young's modulus and Poisson's ratio). Microfractures around the hydraulic fracture (macrofracture) are generated by fluid leak-off during stimulation. The microfractures improve overall fracture conductivity and extend the stimulated reservoir volume (SRV) that can result in short- and long-term increases in well productivity and hydrocarbon recovery factor.

The magnitude and orientation of principal stresses affect the geometry of induced fractures. A hydraulic fracture will preferentially propagate perpendicular to the minimum principal stress. As closely spaced hydraulic fractures develop and interact with each other, the stress in the zone being stimulated accumulates, resulting in a stress shadow. The stress shadow modifies the contrast between principal horizontal stresses ($S_{H\_max} - S_{H\_min}$), influencing fracture development (i.e., orientation and geometry). The impact of stress shadows can be managed by controlling the spacing of stimulations and taking into account the reservoir properties in the MSHF design.

To investigate the effect of reservoir properties and stress contrast on stress shadow, a three-stage fracture model was generated in a three-dimensional lattice-based simulator. The models were built using Bakken rock mechanical properties and stress tensors. The three-stage fracture models were simulated in two phases to understand the effect of each parameter. Results suggest that more effective MSHF can be designed by managing fracture spacing. The simulation implies that fracture spacing is affected by reservoir properties, principal stresses, and fracture geometry. Principal stresses and Young's modulus play a more important role in fracture propagation and stress shadow development than Poisson's ratio.

602266 Modeling Fracture Propagation and Fluid Flowback in Hydraulic Fractures Created Using High Viscosity Shear Thinning Gels

Abstract:

The contribution of shale gas to the total natural gas production in United States continues to increase. Nonetheless, the growing demand for natural gas calls for further increase in production of hydrocarbons from unconventional reservoirs, which is currently being carried out using hydraulic fracturing process[1]. Amongst the numerous parameters that govern the production of shale gas from a fractured reservoir, the crucial ones are average fracture conductivity and propped fracture length. But the use of slick water as fracturing fluid produces fractures with low average fracture conductivity and low propped fracture length. These limitations of slick water can be overcome by using Non-Newtonian gels as fracturing fluids[2-3].

The use of high viscosity Non- Newtonian gels as fracture fluid has many advantages like higher average fracture conductivity, longer propped fracture length and low wastewater production[4]. While the most commonly used polymer gels in hydraulic fractures are shear-thinning fluids, there are some common drawbacks associated with these gels. These gels penetrate into the formation causing formation damage[5-6]. The low shear rate at the fracture surface causes these gels to form a filter cake at the fracture face causing fracture damage[7-8]. Also, the high yield stress of these gels makes their flowback during oil and gas production very difficult[9-10]. Nonetheless, effective production of oil and gas from these fractures can still be obtained if proper flow back of the gel occurs at the inception of hydrocarbon production. Therefore, modeling fluid flow back in these damaged fractures is essential for production planning of unconventional fractured reservoirs.

In this work, a fracture propagation model was first developed by considering the Non-Newtonian flow of shear-thinning fluid inside the fracture. The proppant loading along with the proppant settling velocities were obtained from experimental results[11]. The fracture geometry and fracture damage were obtained for a specified operation time from the model, which were then used as the initial condition to a flowback model to describe the flow of reservoir fluids in the fracture. The flowback model was then validated with experimental results obtained for shear-thinning fluids. The application of such a Non-Darcy flow model where a Non-Newtonian fluid displaces a Newtonian fluid is useful in predicting the oil and gas production from unconventional fractured reservoirs under formation damage. Finally, a sensitivity analysis was performed with respect to the power-law index and coefficient of the viscosity equation to determine the optimal viscosity of the gel to be used for achieving the maximum fluid flowback amount.

Literature Cited:

[1] Wang, Q., et al., Natural gas from shale formation–the evolution, evidences and challenges of shale gas revolution in United States. Renewable and Sustainable Energy Reviews, 2014. 30: p. 1-28.

[2] Brannon, Harold Dean, and Charles Elmer Bell. "Eliminating Slickwater Fracturing Compromises for Improved Shale Stimulation." *SPE Annual Technical Conference and Exhibition*, 2011,

[3] Clark, J.b. "A Hydraulic Process for Increasing the Productivity of Wells." *Journal of Petroleum Technology*, vol. 1, no. 01, Jan. 1949, pp. 1–8.

[4] Civan, Faruk. "Analyses of Processes, Mechanisms, and Preventive Measures of Shale-Gas Reservoir Fluid, Completion, and Formation Damage." *SPE International Symposium and Exhibition on Formation Damage Control*, 2014.

[5] Barati, Reza, and Jenn-Tai Liang. "A Review of Fracturing Fluid Systems Used for Hydraulic Fracturing of Oil and Gas Wells." *Journal of Applied Polymer Science*, vol. 131, no. 16, July 2014

[6] Kim, C.m., and J.a. Losacano. "Fracture Conductivity Damage Due to Crosslinked Gel Residue and Closure Stress on Propped 20/40 Mesh Sand." *SPE Annual Technical Conference and Exhibition*, 1985.

[7] Wang, John Yilin, et al. "Modeling Fracture Fluid Cleanup in Tight Gas Wells." *SPE Hydraulic Fracturing Technology Conference*, 2009.

[8] Wang, Yilin, et al. "Simulation of Gel Damage on Fracture-Fluid Cleanup and Long-Term Recovery in Tight-Gas Reservoirs." *SPE Eastern Regional/AAPG Eastern Section Joint Meeting*, 2008.

[9] Maloney, Daniel R., et al. "Non-Darcy Gas Flow Through Propped Fractures: Effects of Partial Saturation, Gel Damage, and Stress." *SPE Production Engineering*, vol. 4, no. 04, 1989, pp. 417–422.

[10] Balhoff, Matthew T., and Matthew J. Miller. "An Analytical Model for Cleanup of Yield-Stress Fluids in Hydraulic Fractures." *SPE Journal*, vol. 10, no. 01, Jan. 2005, pp. 5–12.

[11] Yegin, Cengiz, et al. "Novel Hydraulic Fracturing Fluids with Improved Proppant Carrying Capacity and PH-Adjustable Proppant Deposition Behavior." *Journal of Petroleum Science and Engineering*, vol. 145, 2016, pp. 600–608.

MULTI-STAGE HYDRAULIC FRACTURING SIMULATION WITH FLUID AND HEAT FLOW, AND GEOMECHANICAL EFFECTS

Philip H. Winterfeld and Yu-Shu Wu, Colorado School of Mines, Golden, CO

Biography

Philip H. Winterfeld is Research Associate Professor in the Petroleum Engineering Department at Colorado School of Mines.
Yu-Shu Wu is Professor, Petroleum Engineering Energi Simulation Chair, in the Petroleum Engineering Department at Colorado School of Mines.

Abstract

Multi-stage hydraulic fracturing and horizontal wells are the technologies responsible for enhanced oil and gas production from unconventional reservoirs. In fracturing operations, fractures are created in series along a horizontal well and the stress field is locally altered by these created fractures, affecting the orientation and propagation of new, adjacent fractures that are created next.

New fractures tend to propagate in a complex manner due to local stress field redistribution during fracturing. Here, we develop a modeling approach to simulate realistic 3-D fracturing geometry while incorporating the effects of existing fractures (or stress shadow) and elevated formation pressure due to injected fracturing fluids on new fracture development. Our model is a general three-dimensional one that accounts for fluid and proppant flow, and heat transfer in the well, the rock mechanics of fracturing including fracture initiation, growth, and propagation, and fluid and proppant flow, and heat transfer in the fracture. The model is capable of simulating multiple fractures that would occur during fracturing of a horizontal well in an unconventional reservoir and includes geomechanical effects such as the stress shadow effect, the effect on fracture growth from superposition of induced stresses by near-by fractures.

In addition to the model formulation, we include a series of simulations using analytical solutions, problems from the literature, and field cases for model verification and demonstration of its capabilities.

Introduction

One of the earliest hydraulic fracturing simulators was formulated by Perkins and Kern[1], who applied the plane strain crack solution by Sneddon[2] to obtain the PK Model. This model was later extended by Nordgren[3] and was called the PKN model, applicable to long fractures with constant height and elliptical cross-section. Khristianovic and Zheltov[4] and Geertsma and de Klerk[5] both developed the KGD model, applicable to short fractures under plane strain. These simple models were subsequently modified, for example extended to power law fluids[6]. Later models[7] include pseudo three-dimensional models, which were an extension of the work of Simonson[8] to multiple layers. These models were cell-based and lumped. In the lumped models, the fracture vertical cross section consists of two half-ellipses whose dimensions change according to the amount of fluid in the fracture. In the cell-based models, the fracture is represented as a series of PKN-like cells that have varying height. Planar three-dimensional models represent the fracture as a two-dimensional grid that may be fixed or moving and the planar fracture is perpendicular to the minimum stress direction. In general three-dimensional models, this orientation requirement is relaxed and the fracture may be three-dimensional.

We present here a general three-dimensional hydraulic fracturing model that accounts for fluid and proppant flow, and heat transfer in the well, the rock mechanics of fracturing including fracture initiation, growth, and propagation, and fluid and proppant flow, and heat transfer in the fracture. The model is capable of simulating multiple fractures that would occur during fracturing of a horizontal well in an unconventional reservoir and includes geomechanical effects such as the stress shadow effect, the effect on fracture growth from superposition of induced stresses by near-by fractures. In addition, we present analytical solutions, cases from the literature, and field cases to demonstrate the capabilities of the simulator.

Simulator formulation

Slurry properties

Slurry, the fluid used in hydraulic fracturing, is composed of conserved proppant and fluid components. Two important physical properties of slurry are density and viscosity. The slurry density, ρ_{sl}, is a weighted sum of the solid and fluid component densities:

$$\rho_{sl} = \sum_p c_p \rho_p + \left(1 - \sum_p c_p\right) \sum_f x_f \rho_f \tag{1}$$

where subscripts p and f refer to proppant and fluid components, respectively, c is proppant component volume fraction, ρ is density, and x is fluid component volume fraction in the fluid portion of slurry. The proppant component density is constant and the fluid component density is slightly compressible:

$$\rho_f = \rho_{f,0}\left(1 + C_P\left(P - P_{ref}\right) - C_T\left(T - T_{ref}\right)\right) \tag{2}$$

where $\rho_{f,0}$ is fluid component density at reference pressure and temperature, P_{ref} and T_{ref}, respectively, and C_P and C_T are fluid component compressibility and thermal expansion coefficient, respectively.

Slurry viscosity, μ_{sl}, has contributions from fluid and proppant components. This dependence is expressed as the fluid viscosity, μ_{fl}, multiplied by a correction factor, $f(c_p)$, that is a function of proppant volume fraction:

$$\mu_{sl} = \mu_{fl} f(c_p) \tag{3}$$

A number of expressions have appeared in the literature for these corrections; one example[9] is the following exponential relation:

$$\mu_{sl} = \mu_{fl}\left(1 - \frac{\sum_p c_p}{c_{\max}}\right)^{-n} \quad (4)$$

where c_{max} is the maximum proppant volume fraction, at which slurry is a rigid porous medium, and n is an exponent typically between 1.0 and 2.5. The viscosity of the fluid portion of slurry depends on fluid component volume fraction and one common fluid viscosity expression is the exponentially weighted product shown below:

$$\mu_{fl} = \prod_f \mu_f^{x_f} \quad (5)$$

Fracture mass conservation equations

Slurry fluid and proppant components are conserved. We consider a fracture to be a thin slit in the xz-direction and the integrated mass conservation equation averaged over that width for fluid is the following, with source/sink terms for fluid leakoff into the surrounding formation and fluid flow between the well and the fracture:

$$\frac{\partial}{\partial x}\left[\rho_f x_f\left(1-\sum_p c_p\right)\vec{v}_f w\right] + \frac{\partial}{\partial z}\left[\rho_f x_f\left(1-\sum_p c_p\right)\vec{v}_f w\right] + \frac{\partial}{\partial t}\left[\rho_f x_f\left(1-\sum_p c_p\right) w\right] + \rho_f x_f q_{leak} + \rho_f x_f\left(1-\sum_p c_p\right) q_{wf} = 0 \quad (6)$$

where subscripts *leak* refers to fluid leakoff[10] and *wf* refers to flow between the well and fracture. The conservation equation for a proppant component is:

$$\rho_p \frac{\partial}{\partial x}\left(c_p \vec{v}_p w\right) + \rho_p \frac{\partial}{\partial z}\left(c_p \vec{v}_p w\right) + \rho_p \frac{\partial}{\partial t}\left(c_p w\right) + \rho_p c_p q_{wf} = 0 \quad (7)$$

Unlike fluid, proppant does not leak off through the fracture face.

Because a fracture occupies a relatively small range of depth, we consider fluid component density in Equation 6 to be constant. Then, we sum Equations 6 over fluid components and Equation 7 over proppant components, and add the results to obtain an overall slurry conservation equation:

$$\frac{\partial}{\partial x}\left(\vec{v}_{sl} w\right) + \frac{\partial}{\partial z}\left(\vec{v}_{sl} w\right) + \frac{\partial w}{\partial t} + q_{leak} + q_{wf} = 0 \quad (8)$$

The coefficient of w in both x- and z-derivatives in Equation 8 is the slurry velocity, v_{sl}, a weighted sum of fluid and proppant velocities:

$$\vec{v}_{sl} = \left(1-\sum_p c_p\right)\vec{v}_{fl} + \sum_p c_p \vec{v}_p \quad (9)$$

Slurry flow through the fracture, a slit of width w, is assumed to be that for laminar flow:

$$\vec{v}_{sl} = \frac{w^2}{12\mu_{sl}} \nabla\left(P + \gamma_{sl} z\right) \quad (10)$$

where γ_{sl} is the slurry gradient. For fluid rheology other than Newtonian, the slurry viscosity in Equation 10 is considered to be an effective one.

Proppant component and fluid velocities generally differ because proppant is denser than fluid and consists of discrete granules that can interact with each other and with the fracture wall. Proppant z-direction velocity, $v_{p,z}$, is assumed to be the sum of the slurry velocity plus a proppant settling velocity that accounts for the difference in fluid and proppant component densities:

$$v_{p,z} = v_{sl} + v_{p,stl} \tag{11}$$

where $v_{p,stl}$ is proppant component settling velocity. Proppant settling velocity is based on a modified Stoke's law, and one example is from Friehauf[11], where inertial effects, the effect of interfering proppant particles, and the effect of the fracture wall are accounted for. For the x-direction, the proppant velocity, $v_{p,x}$, is affected by the fracture wall and other proppant particles, and was accounted for[11] by introducing a retardation factor for the slurry velocity:

$$v_{p,x} = C_{ret}\left(c_p, w\right) v_{sl,x} \tag{12}$$

Fracture energy equation

The energy conservation equation for the fracture is:

$$\frac{\partial}{\partial x}\left(\rho \bar{h} \vec{v}\right)_{sl} w + \frac{\partial}{\partial z}\left(\rho \bar{h} \vec{v}\right)_{sl} w + \frac{\partial\left(\rho \bar{u}\right)_{sl} w}{\partial t} + q_{cond} + q_{leak} \sum_f x_f \rho_f \bar{h}_f + q_{wf} \rho_{wf} \bar{h}_{wf} = 0 \tag{13}$$

where $\bar{h}$ is specific enthalpy and $\bar{u}$ is specific internal energy. We neglect conduction except for heat loss per unit area through the fracture faces, denoted by q_{cond}.

Specific enthalpy and internal energy are dependent on slurry composition and specific enthalpy and internal energy are assumed to be equal. The heat loss per unit area through the fracture faces is formulated as heat loss through a semi-infinite medium:

$$q_{cond} = 2\sqrt{\frac{K_{res}\rho_{res}C_{res}}{\pi\left(t-\tau\right)}}\left(T - T_{res}\right) \tag{14}$$

where K_{res} is reservoir thermal conductivity, ρ_{res} is reservoir density, C_{res} is reservoir heat capacity, and τ is the time that heat loss commences from an area of the fracture face.

Fracture mechanics equations

The fracture width equation is based on the solution for the deflection caused by a load distributed over the boundary of a semi-infinite elastic medium[12]. Because this equation yields a width profile over an infinite plane, we assume that width is zero outside the fracture face. The load is the difference between the fracture pressure and the minimum principal stress, the net pressure, acting on the fracture face, and the above zero width condition requires a net pressure, P_{net}, acting on the area outside the fracture as well:

$$w\left(\vec{r} \in F\right) = \int_{\vec{r} \in F} \frac{2\left(1-\upsilon^2\right)}{\pi E r}\left(P - \sigma_{\min}\right) dA + \int_{\vec{r} \notin F} \frac{2\left(1-\upsilon^2\right)}{\pi E r} P_{net}\, dA \tag{15}$$

$$w\left(\vec{r} \notin F\right) = \int_{\vec{r} \in F} \frac{2\left(1-\upsilon^2\right)}{\pi E r}\left(P_f - \sigma_{\min}\right) dA + \int_{\vec{r} \notin F} \frac{2\left(1-\upsilon^2\right)}{\pi E r} P_{net}\, dA = 0 \tag{16}$$

where E is Young's modulus, ν is Poisson's ratio, σ_{min} is the minimum principal stress, and F is the fracture face area. The factor of two into Equations 15 and 16 arises because the fracture consists of two identical faces.

Continued injection of fluid into a fracture causes the fracture width to increase and the fracture to extend. The criterion for fracture extension is obtained from the stress intensity factor at the fracture tip[13]:

$$K_I = \frac{E}{8(1-\upsilon^2)}\left(\frac{2\pi}{r}\right)^{\frac{1}{2}} w(r) \quad (17)$$

where K_I is the stress intensity factor and r is the inward normal distance from the fracture front. The movement of the fracture front is given as a front velocity, v_{fr}, which has the form[14]:

$$v_{fr} = C_{ext}\left(\frac{K_I - K_{IC}}{K_{IC}}\right)^n \quad (18)$$

where C_{ext} and n are parameters that depend on the medium and K_{IC} is rock toughness, a rock property.

Well mass, flow, and energy equations

In this section, we formulate the mass and energy conservation equations that govern fluid and heat flow in the well. Slurry, injected at the surface, transverses the well before entering the fractures. We represent the well as a series of connected segments that follow the well trajectory. Each well segment has constant outer and inner radii that define an annulus through which slurry flows. We represent slurry flow as flow of a series of slurry segments that travel through the well in a piston-like manner, as illustrated in Figure 1. Each slurry segment is composed of a single fluid and proppant component.

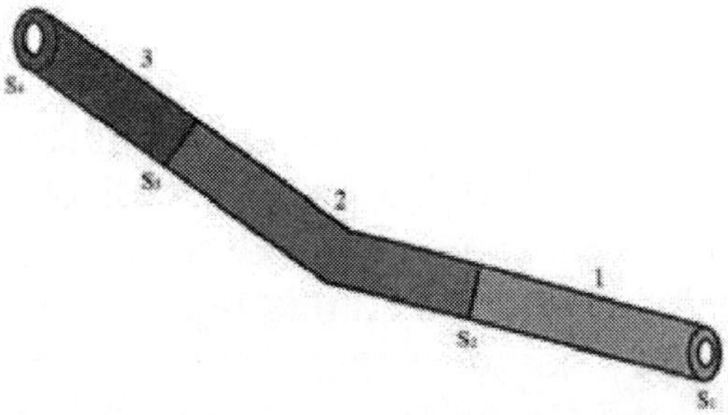

Figure 1. Well consisting of two segments and containing three slurry segments with four boundaries (S_i).

A mass balance for the fluid and proppant components in slurry segment i is:

$$\frac{\partial}{\partial t}\left(\int_{S_{i+1}}^{S_i} A_i \rho_j x_j dS\right) + \int_{S_{i+1}}^{S_i} T_{ws} \rho_j x_j \left(P_w - P_{surr}\right) dS - \delta_{iN} q_{inj} \rho_j x_j = 0, j = f, p \quad (19)$$

where A is the annular area for slurry flow, x denotes component volume fraction, T_{ws} is transmissibility per unit length for slurry loss to the surroundings, P_w is well pressure, P_{surr} is surroundings pressure, and q_{inj} is slurry injection rate, which applies only to slurry segment N (injected into the well at the top) via the Kronecker delta. The transmissibility per unit length for slurry loss to the surroundings is only non-zero along completed intervals.

The pressure along the well has frictional and gravitational components and is obtained from:

$$\frac{\partial P_w}{\partial S} = \left(\rho_f x_f + \rho_p x_p\right)\overline{g} \cdot \hat{S} + \frac{\partial P_{w,fric}}{\partial S} \tag{20}$$

where $\overline{g}$ is the gravitational vector and $\hat{S}$ is the unit vector along the well trajectory. The frictional portion of the pressure gradient, $P_{w,fric}$, is obtained from a correlation that is a function of slurry composition and well geometry, and can include the well-known friction factor.

An energy balance for a slurry segment is:

$$\frac{\partial}{\partial t}\left(\int_{S_{i+1}}^{S_i} A_i\left(\rho_f x_f \overline{u}_f + \rho_p x_p \overline{u}_p\right) dS\right) + \int_{S_{i+1}}^{S_i} T_{ws}\left(\rho_f x_f \overline{h}_f + \rho_p x_p \overline{h}_p\right)\left(P_w - P_s\right) dS - \delta_{iN} q_{inj}\left(\rho_f x_f \overline{h}_f + \rho_p x_p \overline{h}_p\right) + \int_{S_{i+1}}^{S_i} \pi D_o q_{loss} dS = 0 \tag{21}$$

where q_{loss} is the rate of heat loss per unit area from the slurry to the surroundings and D_o is the outer well diameter for fluid flow. Heat loss from the slurry to the surroundings is conceptualized as heat flow through a series of uniform concentric media that include the well tubulars and the surrounding formation.

Stress shadow effect

Fracture width depends on net pressure and the presence of a fracture induces a stress in the nearby formation that is proportional to its net pressure. This induced stress can affect the opening of nearby fractures because it changes the minimum stress magnitude and direction of the stress field nearby. This phenomenon is referred to as the stress shadow effect.

A fracture can be conceptualized as a crack in an infinite elastic medium and the stress field induced by such a crack can be modeled using the Displacement Discontinuity Method[15] (DDM). The DDM is a type of boundary element method. In its two-dimensional version, the boundary of a crack in an infinite elastic medium is represented as a series of connected line segments. Each line segment can undergo shear displacement along its length, as shown in Figure 2a, or normal displacement perpendicular to its length, as shown in Figure 2b. The analytical solution for constant displacement at a line segment is superposed over the crack boundary to yield stresses and other quantities at arbitrary points in the infinite elastic medium. This leads to a series of algebraic equations relating the displacements along the boundary to the boundary conditions there, namely the pressure of the fluid in the crack is equal to the normal stress on the boundary and the shear stress on the boundary is zero:

$$\mathbf{AD} = \mathbf{P} \tag{22}$$

where **A** is the matrix of coefficients that arose from the superposition, **D** is the vector of boundary displacements, and **P** is the vector of boundary fluid pressures.

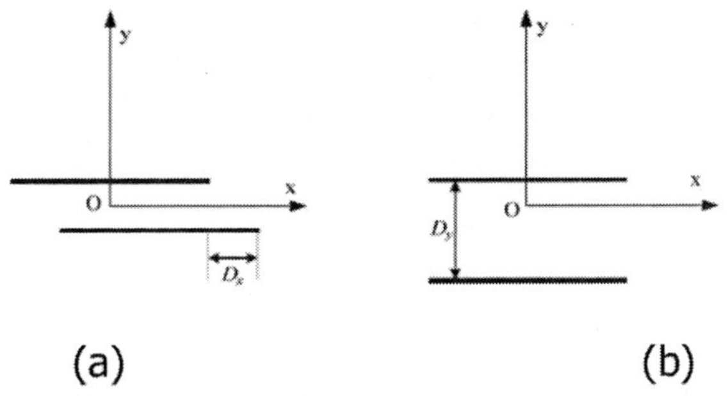

Figure 2. Displacements along crack boundary: a) shear; b) normal.

We utilize the DDM method to calculate the stress shadow effect. The DDM method contains the assumption of an elastic medium with uniform properties. We apply the DDM along two-dimensional cross sections in the x-direction (along the fracture length), because reservoir properties are assumed to vary only with depth and to implement the DDM in a computationally efficient manner. The induced stress tensor at a given point is the sum of the contributions from all the fractures, and the total stress tensor there is the sum of the ambient stresses and the induced ones. The principal stresses and directions are then obtained from the eigenvalues and eigenvectors of the total stress tensor. The fracture width calculations then use the minimum principal stresses.

Governing equation solution

The governing equations for our hydraulic fracturing simulator are solved using the integral finite difference method[16]. We discretize the fractures into a two-dimensional grid that is uniform and Cartesian, shown in Figure 3. Because multiple completion zones may result in multiple fractures, this grid has an additional dimension equal to the number of these fractures. Each grid block has a position expressed in global coordinates and various properties including a normal vector that defines its orientation. Fracture grid blocks are either active or inactive. Before fracture initiation, all grid blocks are inactive. The well pressurizes as slurry is injected into it, and when this pressure reaches a critical value, a fracture is initiated. Then, grid blocks containing completions, shown by the dot in Figure 3, become active, meaning they contain fluid and proppant and have a non-zero fracture width. As the fracture extends, inactive grid blocks that neighbor active ones, containing a "B" in Figure 3, become active. These inactive grid blocks also comprise the second term in Equations 15 and 16.

A normal vector is associated with each active grid block that is orthogonal to the minimum principal stress direction. This vector is calculated when the grid block becomes active. The stress tensor components for the grid block are obtained from the superposition of the ambient stresses and those induced by the stress shadow effect. This normal vector does not change after the grid block becomes active and varying grid block normal vectors result in fractures that are non-planar.

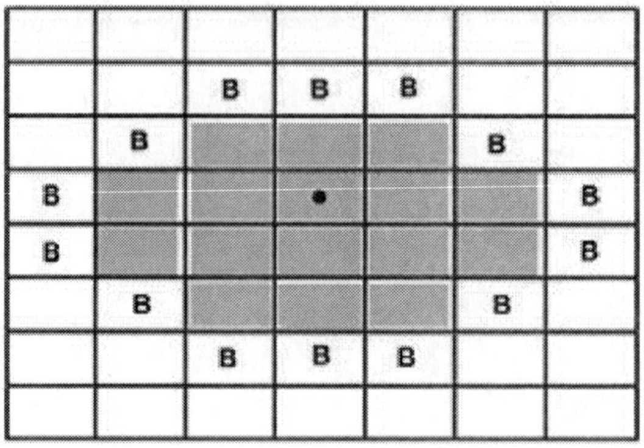

Figure 3. Fracture grid showing active grid blocks (shaded) and inactive (unshaded) ones. Blocks with "B" are adjacent to active ones; the large dot is a block with completions.

The well mass and energy conservation equations are also solved using the integral finite difference method and are integrated over a slurry segment volume. The well pressure equation is integrated from the top of the well (the wellhead) to an arbitrary measured depth, yielding the pressure there.

There is mass and energy exchange between the well and the surroundings (the fracture or the formation), namely fluid flow from the well into the fracture through completions, flow from the well into the formation before fracture initiation, and heat exchange between the well and the formation.

The governing equations are a function of the well and fracture primary variables. The well ones are slurry segment boundaries, fluid volume fractions, temperature, and wellhead pressure. The fracture ones are pressure, width, temperature, proppant component volume fractions, and fluid component volume fractions.

The governing equations for the well and fracture are solved in a sequential manner. First, fracture pressure, width, and all well primary variables except temperature are solved for. Second, fracture fluid and proppant volume fractions are solved for. Finally, fracture and well temperatures are solved for. After each time step, the extension of the fracture, if any, is calculated. Since the finite difference formulation is not fully implicit, we choose the time step size to ensure stability, namely that the Courant number, the ratio of flow through a grid block to its volume, is less than one.

Example problems

We describe three simulations to provide model verification and application examples. The first, radial fracture propagation, is presented for model validation and is compared to both an analytical solution and another hydraulic fracturing simulator. The second demonstrates simulation of the stress shadow effect, and the last one is a simulation using field data that illustrates some of the simulators capabilities, including simulating the stress shadow effect.

Radial fracture propagation

The radial fracture propagation problem is a simulation of fracture propagation in the horizontal plane with no leakoff, gravity, or temperature effects[17]. Slurry consists of one incompressible Newtonian fluid component with 40 cp (0.04 Pa·s) viscosity and no proppant. Young's modulus is 2.5x10^6 psi (17.24 GPa), Poisson's ratio is 0.25, the minimum stress is 5000 psi (34.47 MPa), and slurry is injected at 20 bbl/min (0.053 m^3/s) for 30 min (1800 s) via a point source. Grid blocks are square with length of 32.81 ft (10 m). Figure 4 shows simulated fracture width after 30 minutes of injection (left) along with published results[17] (right) using another hydraulic fracturing simulator. The fracture width profiles from both simulators are very similar. They also compared fracture radius versus time to an analytical solution for radial fractures[5] with no leakoff and obtained good agreement. Figure 5 is the comparison of our results with the analytical solution. Our results track the analytical solution well, with our fracture radius increasing in discrete jumps, a characteristic of our finite difference formulation.

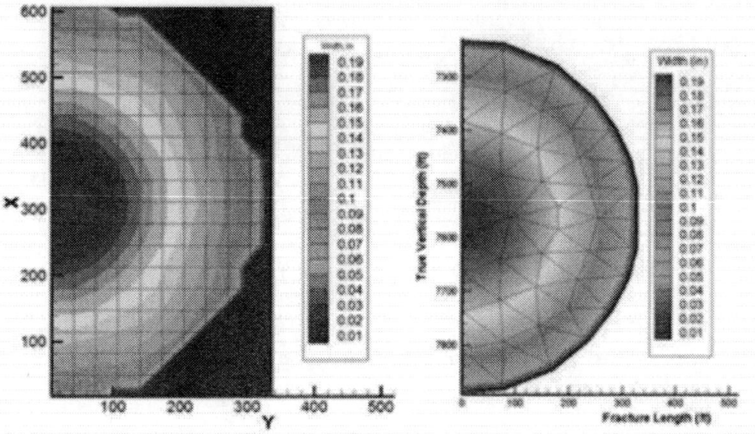

Figure 4. Fracture width after 30 minutes injection; left, our simulator, right, published results[17].

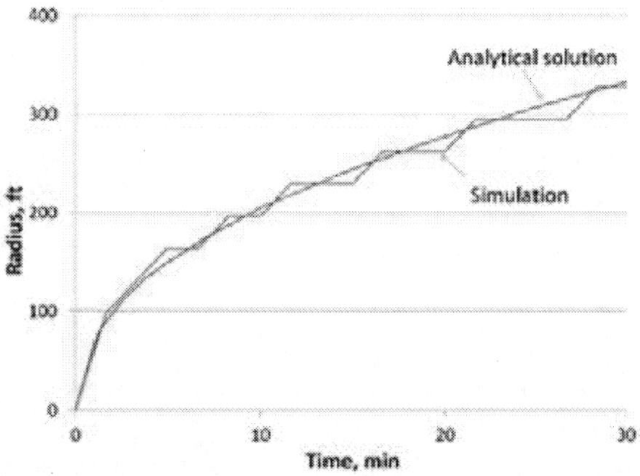

Figure 5. Comparison of fracture radius versus time for our simulator and the analytical solution.

Stress shadow effect simulation

We present a simulation that illustrates the effect of the stress shadow on hydraulic fracturing along a horizontal well. There are six completed intervals along the well that are spaced 10 m apart, and the well is oriented perpendicular to the minimum stress direction.

Rock properties (including porosity, permeability, Young's modulus, Poisson's ratio, tensile strength, and toughness) vary with depth and have one value from 2980 to 3030 m and another one elsewhere. Rock properties are such that fracturing is more amenable in that depth range than elsewhere, and the horizontal well is located in that interval as well, at 3000 m. Figure 6 illustrates the well configuration. There is one fluid component with power-law rheology and one proppant component. Fluid can leak off through the fracture face and there is a frictional pressure gradient along the well. Each stage consists of slurry being injected at $5 \cdot 10^{-2}$ m^3/s for 1200 seconds. Proppant volume fraction in slurry is initially zero (pre-flush), gradually increases to 0.23, and is then zero for the last 120 seconds (post-flush). One such stage was injected into each completed interval.

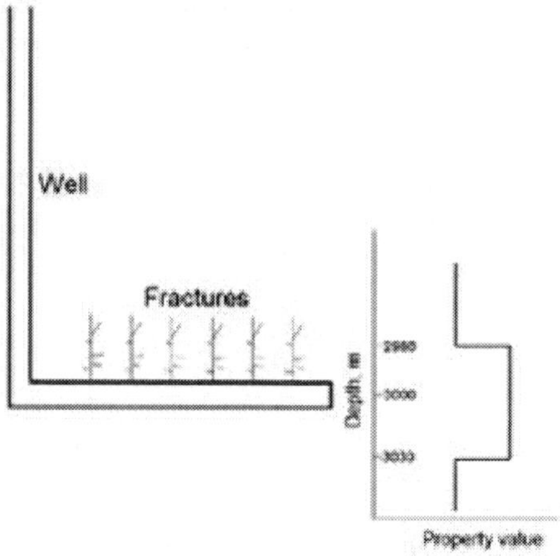

Figure 6. Well configuration for Stress Shadow Effect simulation.

The first completed interval is at the largest measured depth along the well (the toe of the well) and subsequent ones are placed in order of decreasing measured depth. The sequence of stage injection was done in two ways, with both starting with completion 1: 1) in order of decreasing measured depth, equivalent to increasing completed interval number (1,2,3,4,5,6) and 2) odd completion numbers followed by even ones (1,3,5,2,4,6). Figure 7 shows the distance to the fracture plane (the plane the fracture would be in without the stress shadow effect) versus distance along the fracture for the first stage injection sequence.

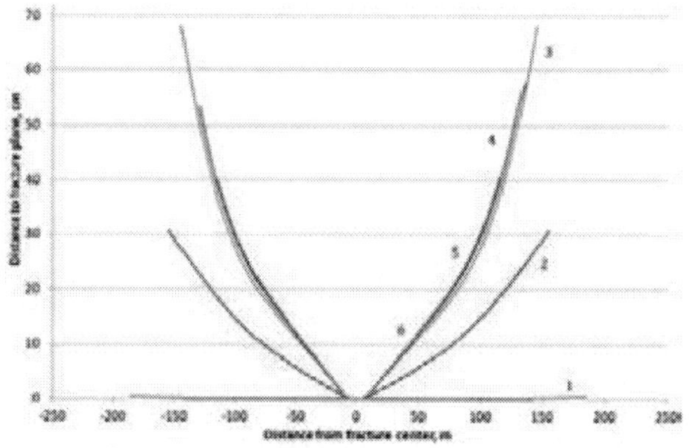

Figure 7. Distance to fracture plane for injection sequence (1,2,3,4,5,6).

The fracture at completion 1 is the longest and has the least deviation; as stage number increases, overall fracture length decreases and deviation increases. This results from increasing minimum stress, which requires higher pressure for fracture propagation. A byproduct of higher pressure is more leakoff and consequently a smaller fracture. Figure 8 shows the distance to the fracture plane for the latter stage injection sequence. Fractures in completions 1, 3, and 5 show the same trend as the first stage sequence. The fracture in completion 2 deviates in the opposite direction because it has two fractures, from completions 3 and 5, on one side and one, from completion 1 on the other side, that contribute to the stress shadow effect. Fractures from completions 4 and 6 continue the trend established by fractures 1, 3, and 5.

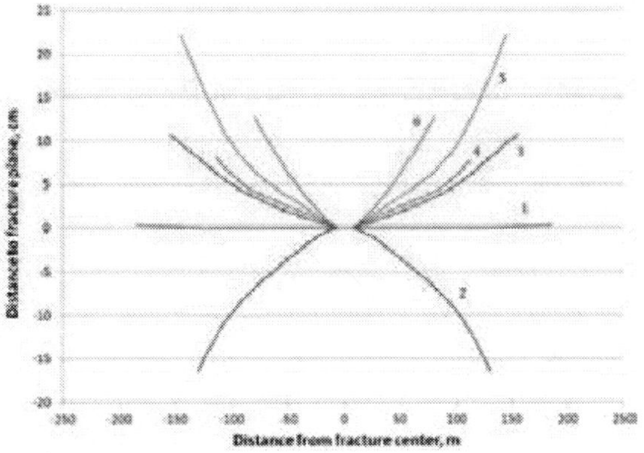

Figure 8. Distance to fracture plane for injection sequence (1,3,5,2,4,6).

Field simulation

We present a simulation using field data. Slurry consists of one fluid component with power law rheology (consistency index of $1.3 \cdot 10^{-1}$ Pa-s^n, exponent of 0.045, and maximum viscosity of $7.4 \cdot 10^{-2}$ Pa) and a proppant component. Fluid is incompressible with density 1000 kg/m^3. The proppant component density is 2600 kg/m^3 and the proppant diameter $6.0 \cdot 10^{-4}$ m. Friction for well slurry flow is obtained from tabular data.

The well consists of six segments whose geometry and trajectory are shown in Table 1. A segment's end and segment properties are given by a row in the table; the segment begins at the depths of the previous row. The well consists mainly of a vertical segment that turns horizontal and is a single tube.

Table 1. Well geometry and trajectory. Azimuth angle of zero is due north.

Measured depth, m	True vertical depth, m	Azimuth angle, °	Outer diameter, m	Inner diameter, m
0	0			
2844	2844	90.0	0.056	0.0
3556	3337	11.0	0.056	0.0
3757	3350	2.2	0.056	0.0

3957	3356	2.4	0.070	0.0
4559	3360	1.1	0.070	0.0
4766	3360	3.1	0.169	0.0

Rock-related properties that are constant with depth are porosity (0.05), Poisson's ratio (0.20), rock density (2720.0 kg/m^3), Biot's coefficient (1.0), and pressure (32.0 MPa). Other such properties have one value in the depth range 3280 to 3360 m and another value elsewhere. Table 2 shows the values for these other properties.

Table 2. Rock properties with step-wise variation. Values are given for depth range of 3280-3360 m and outside that depth range.

Property	Inside depth range	Elsewhere
Permeability, m^2	$1.0 \cdot 10^{-14}$	$1.0 \cdot 10^{-15}$
Young's modulus, GPa	11.6	41.6
Tensile strength, MPa	3.85	6.85
Toughness, MPa	2.0	15.0
Minimum stress, MPa	55.0	77.0

The minimum stress is oriented along the y-direction and the horizontal portion of the well trajectory is roughly due north (from Table 1 data), which coincides with the y-direction.

The slurry injection schedule consists of ten identical stages, each being injected into one of the completed intervals. Injection rate is 0.133 m^3/s. A stage has sixteen slurry segments and the properties of each segment are shown in Table 3.

Table 3. Properties of slurry segments of a stage.

Segment	**Vol., m^3**	**Proppant frac.**	**Segment**	**Vol., m^3**	**Proppant frac.**
1	25.0	0.0	9	29.7	0.092
2	10.0	0.38	10	36.0	0.138
3	25.0	0.0	11	43.3	0.162
4	10.1	0.38	12	48.3	0.185
5	30.0	0.0	13	38.3	0.208
6	10.0	0.38	14	17.3	0.223
7	48.0	0.0	15	9.30	0.238
8	17.7	0.46	16	54.3	0.0

Fracture grid blocks are square with length 10 m. Reservoir compressibility is 10^{-9} Pa^{-1} and reservoir viscosity is 0.0001 Pa-s. The fracture front displacement coefficient and exponent in Equation 18 are 10 m and 1.0, respectively. The maximum proppant fraction is 0.8 and the exponent in Equation 4 (effect of proppant fraction on slurry viscosity) is -4.1. There are ten completed intervals that start at 4580 m measured depth and are evenly spaced at 5 m apart. Completed intervals are numbered starting at one for the largest measured depth and increasing with decreasing measured depth.

Injection starts at the completion at the largest measured depth, the toe of the well, and proceeds in decreasing measured depth. Table 4 shows the overall fracture dimensions, length and height, for the fracture emanating from each completed interval at the end of injection there.

Table 4. Fracture geometry at the end of injection into each completion.

Fracture	Length, m	Height, m	Fracture	Length, m	Height, m
1	250.0	90.0	6	170.0	80.0
2	230.0	80.0	7	170.0	80.0
3	170.0	80.0	8	170.0	80.0
4	170.0	80.0	9	170.0	80.0
5	170.0	80.0	10	170.0	80.0

Fracture heights in Table 4 are fairly uniform, due to the input data constraining fracture height growth to the layer from 3280 to 3360 m. Fracture 1 is the longest, the next two fractures show decreasing length, and the rest of the fractures are similar to Fracture 3. The stress shadow effect increases the minimum stress for Fracture 2 and further for Fracture 3. This increases the leak off rate because the fracture pressure increases in response to increasing minimum stress. The increased leakoff reduces fracture length. Fractures 3 to 10 are similar and have similar minimum stress profiles. The stress shadow from a fracture depends on fracture width. There is still fluid leakoff after injection into a fracture stops. This causes fracture width to decrease, and consequently the stress shadow effect the fracture induces. With increasing time, the contribution to stress shadow for each fracture decreases but the total number of fractures increases. These two effects tend to cancel and result in an overall constant stress shadow effect for larger fracture numbers. Figure 9 shows the width profiles of Fractures 1, 2, and 3, with the decreasing fracture length quite evident.

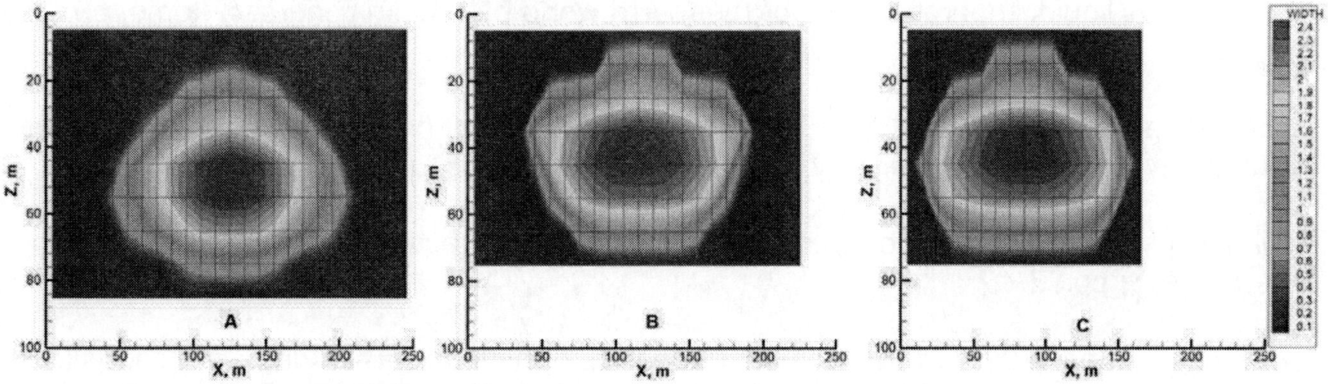

Figure 9. Width profiles for Fracture 1 (A), Fracture 2 (B), and Fracture 3 (C).

Summary and conclusions

We developed a general three-dimensional hydraulic fracturing model that accounts for fluid and proppant flow, and heat transfer in the well and the fracture, and the geomechanics of fracturing including fracture initiation, growth and, propagation, as well as the stress shadow effect.

We ran a simulation to provide model verification, a radial fracture propagation problem. We compared our results to another hydraulic fracturing simulator and to an analytical solution and obtained good agreement. We then ran a simulation that demonstrated the dependence of fracture geometry on the stress shadow effect, and finally a field case that included the stress shadow effect as well. The simulator performed well and we anticipate using in many other applications.

Acknowledgement

This work was supported by EMG Research Center in Colorado School of Mines and CNPC-USA.

References

1. Perkins, T. K. and Kern, L. R. 1961. SPE 89. Widths of hydraulic fractures. J Pet Tech 13 (9):937–49.

2. Sneddon, I. N. 1946. The distribution of stress in the neighbourhood of a crack in an elastic solid,Proc.Royal Society of London, Series A, 187, 229-60.

3. Norgren, R. 1972. Propagation of a vertical hydraulic fracture, SPE Journal, 12(4): 306-314.

4. Khristianovic, S.A. and Zheltov, Y. P. 1955. Formation of vertical fractures by means of highly viscous liquid. In: Proceedings of the fourth world petroleum congress, Rome, pp. 579–86.

5. Geertsma, J. & de Klerk, F. 1969. A rapid method of predicting width and extent of hydraulically induced fractures, J. *Pet. Tech.* 21 (12):1571-1581.

6. Daneshy, A. A. 1973. On the design of vertical hydraulic fractures. SPE 3654, J Pet Tech, January 1973, pp.83–97.

7. Mack M. G. and Warpinski, N. R. 2000. Mechanics of hydraulic fracturing. In: Economides, Nolte, editors. Reservoir stimulation. 3rd ed. Chichester; Wiley:[Chapter 6].

8. Simonson, E. R., Abou-Sayed, A. S., and Clifton, R.J. 1978. SPE 6089. Containment of massive hydraulic fractures. SPE J, 18(1):27–32.

9. Nicodemo, L., Nicolais, L., and Landel, R.F.: Shear rate dependent viscosity suspensions in Newtonain and non-Newtonian liquids, Chemical Engineering Science, 29,729-735, 1974.
10. Schechter, R. S.: Oil Well Stimulation, Prentice Hall, Englewood Cliffs, NJ, 1992.

11. Friehauf, K. E. 2009. Simulation and design of energized hydraulic fractures, Ph.D. thesis, The University of Texas at Austin, 2009.

12. Timoshenko and Goodier: Theory of elasticity, McGraw-Hill Book Company, Inc, New York, NY 1951.

13. Yew, C. H.: Mechanics of hydraulic fracturing, Gulf Publishing Company, Houston, TX, 1997.

14. Mastrojannis, E.N., Keer, L.M., and Mura, T., Growth of planar cracks induced by hydraulic fracturing, International Journal of Numerical Methods in Engineering, 15 (1980) 41–54.

15. Crouch, S.L., 1976. Solution of plane elasticity problems by the displacement discontinuity method. I. Infinite body solution. Int. J. Numer. Methods Eng. 10, 301–343.

16. Narashimhan, T. N. & Witherspoon, P. A.: An integrated finite difference method for analysis of fluid flow in porous media. Water Resour Res 12 (1976), pp. 57–64.

17. Ribeiro, L. H. & Sharma, M. M.: A new three-dimensional, compositional, model for hydraulic fracturing with energized fluids, SPE 159812, presented at SPE Annual Technical Conference and Exhibition, 8-10 October 2012, San Antonio, Texas, USA.

602500 Optimal Proppant Pumping Schedule Design for Hydraulic Fracturing Operation Using a Eulerian-Lagrangian Model

Abstract:

Slickwater hydraulic fracturing is becoming a prevalent approach to economically recover shale hydrocarbon. It is very important to understand the proppant transport behavior during slickwater hydraulic fracturing treatment for effective creation of a desired propped fracture geometry. Many experimental and numerical studies have been conducted to understand the proppant transport process with an aim to achieve improved fracture conductivity [1, 2]. However, the currently available models are either oversimplified or have been performed at limited length scales (i.e., not in field-scale geometries) to avoid high computational requirements [3, 4]. Another limitation is that the currently available hydraulic fracturing simulators are developed using only single-sized proppant particles [5].

Motivated by this, in this work, a computationally efficient three-dimensional multiphase particle-in-cell (MP-PIC) model has been employed to simulate the multi-size proppant transport in a field-scale geometry using the Lagrangian description for the proppant particles, and the Eulerian continuum description for the fracturing fluid [6, 7]. Instead of tracking each particle, groups of particles (called parcel) are tracked, which allows to simulate the proppant transport in field-scale geometries at an affordable computational cost. The MP-PIC model was simulated using the Ada and Terra clusters provided by Texas A&M high performance research computing facility. The MP-PIC method required around 96 hours of calculation of 28 CPUs to simulate proppant transport in the described field-scale fracture geometry. In the literature, available CFD simulations only considered the fracture length of up to 10 m, which is much smaller than field-scale fracture geometries, and fracturing fluid injection times of up to 200 s, which is also not a field-scale injection time [8]. Because of the gain in computational efficiency by the use of the MP-PIC method, we are able to simulate proppant transport in the field-scale fracture geometry; in particular, a fracture of 180 m in half length, 30 m in height, and 0.00762 m in width is considered with 1 hour operation time for hydraulic fracturing. From these simulation results, we found that shale gas production from unconventional reservoirs depends on the propped fracture surface area (PFSA) and average fracture conductivity (AFC) values at the end of hydraulic fracturing. We also found that a smaller size proppant results in a larger PFSA because proppants can penetrate deeper inside the fracture due to less gravitational settling. On the other hand, using large proppant particles results in a high AFC due to their high permeability.

Motivated by these sensitivity studies, we used the MP-PIC model for optimal operation of hydraulic fracturing. The reservoir simulation software from Computer Modeling Group Ltd. (CMG) is used to simulate shale gas production from an unconventional reservoir. To reduce the computational requirement of using the MP-PIC model and CMG within the optimization framework, a reduced order model (ROM) was developed to describe proppant transport during hydraulic fracturing, and a map was developed to predict the cumulative shale gas production volume as a function of PFSA and AFC at the end of hydraulic fracturing. Then, we used the ROM and map in an optimization framework to design an optimal multi-size proppant pumping schedule. Simulation results presented in this work show that the AFC and PFSA values obtained by the proposed pumping schedule were able to maximize the cumulative shale gas production volume from an unconventional reservoir for given fracturing resources.

References

1) Kern, L.; Perkins, T.; Wyant, R. The mechanics of sand movement in fracturing. Journal of Petroleum Technology 1959, 11, 55–57.

2) Patankar, N.A.; Joseph, D.D. Lagrangian numerical simulation of particulate flows. International Journal of Multiphase Flow 2001, 27, 1685–1706.

3) Blyton, C.A.; Gala, D.P.; Sharma, M.M. A Comprehensive Study of Proppant Transport in a Hydraulic Fracture. SPE Annual Technical Conference and Exhibition (SPE 174973), Houston 2015, TX.

4) Tsai, K.; Fonseca, E.; Lake, E.; Degaleesan, S. Advanced computational modeling of proppant settling in water fractures for shale gas production. SPE Journal 2012, 18, 50–56.

5) Siddhamshetty, P.; Liu, S.; Valkó, P.P.; Kwon, J.S. Feedback control of proppant bank heights during hydraulic fracturing for enhanced productivity in shale formations. AIChE Journal 2017, 64, 1638–1650.

6) Mao, S., Shang, Z., Chun, S., Li, J., & Wu, K. 2019. An Efficient Three-Dimensional Multiphase Particle-in-Cell Model for Proppant Transport in the Field Scale. Unconventional Resources Technology Conference. doi:10.15530/urtec-2019-462

7) Mao, S., Siddhamshetty, P., Zhang, Z., Yu, W., Chun, T., Kwon, J., and Wu, K. 2020. Impact of Proppant Pumping Schedule on Well Production for Slickwater Fracturing. Unconventional Resources Technology Conference. doi:10.15530/urtec-2020-2630

8) Zeng, J.; Li, H.; Zhang, D. Numerical simulation of proppant transport in hydraulic fracture with the upscaling CFD-DEM method. Journal of Natural Gas Science and Engineering 2016, 33, 264–277.

610620 Experimental and Numerical Study of Apparent Permeability in the Utica Shale

Abstract:

Shale formations are sedimentary heterogeneous rocks composed of micropores (pores less than 2 nm diameter), mesopores (pores with 2–50 nm diameter), and macropores (pores greater than 50 nm diameter), storing a significant amount of gas and playing an important role in the energy supply. The matrix permeability controls the long-term gas production from shale; therefore, quantifying permeability is essential for predicting gas production. In this study, experimental analyses in tandem with a numerical analysis were conducted to characterize and evaluate the permeability of shale. To determine the permeability of the Utica shale sample, the pressure pulse technique was conducted under varying confining stress while the pore pressure is kept constant. The pore size distribution and the average pore diameter of the Utica shale sample were determined using nitrogen gas adsorption technique. It was found that increasing confining pressure while pore pressure is constant decreases the apparent permeability. The apparent permeability model based on the second-order slip boundary condition was developed considering the poromechanics effects on gas transport mechanisms. The measured pressure profiles perfectly matched with data obtained from numerical simulation, indicating the applicability of the developed model in prediction of permeability in shale gas reservoirs.

598461 Slip Length of Methane Flow Under Shale Reservoir Conditions: Effect of Pore Size and Pressure

Abstract:

Gas recovery from shale reservoirs is drastically different from the conventional reservoirs due to the presence of extensive amount of nano-sized pores. Within nanopores, gas molecule distributions are inhomogeneous and gas slippage on the pore surface plays a crucial role in gas flow. The conventional hydrodynamic equations can no longer describe gas flow in nanopores. Slip length has been widely used to quantify the deviation from the traditional hydrodynamic equations, such as Hagen-Poiseuille (H-P) equation. Determining slip length of a specific fluid can better assess the flow behavior in nanopores as well as multiscale modeling.

Many studies were carried out to clarify the slip length in nanopores in shale, such as experiments, numerical analysis and simulations. Although some qualitative and suggestive conclusions on gas transport have been obtained from previous studies, to our best knowledge, a systematic study about the effects of pore size, temperature and pressure on the slip length of methane transport in nanopores at shale reservoir conditions is still lacking.

Herein, we report a molecular simulation study of methane flow in organic nanopores under shale reservoir conditions (temperature: 300 to 450 K, pressure: 10 to 60 MPa), with pore sizes ranging from 2 to 20 nm. We use the grand-canonical Monte Carlo simulations to determine methane content in nanopores and analyze the density distributions using equilibrium molecular dynamics. The surface adsorption effect is almost negligible under high pressure (60 MPa) conditions on average density, while it plays an important role under low-pressure conditions. Finally, we use nonequilibrium molecular dynamics simulation to study the transport behavior of nano-confined methane molecules. We find that the pore size has a significant effect on the slip length under low pressure (10 MPa) conditions. In contrast, the slip length is almost constant under high-pressure conditions. Under most conditions, the slip length decreases as pressure increases. Such a trend is more obvious in the smaller pores under high-temperature conditions. Our work should provide important insights into the quantification of slip length under shale reservoir conditions.

599931 Hydrotreating Kinetics of Nitrogen-, Oxygen- and Sulfur-Containing Model Compounds for Co-Processing Biocrude Oils with Vacuum Gas Oil

Abstract:

Co-processing of HTL bio-crude oils with petroleum crude oils is a practically feasible way to produce green transportation fuels and dramatically reduce the capital cost. Researchers have performed extensive studies on the hydrotreating kinetics of heteroatoms (N, O, S) removal. However, there is very limited information on the simultaneous hydrodenitrogenation (HDN) and hydrodeoxygenation (HDO) of fatty acid amides which are the major nitrogen content in the biocrudes. It is necessary to develop detailed kinetic model and investigate the impact of co-processing on the hydrotreating chemistry. First of all, we evaluated the activity and product selectivity of a single molecule model compound, N,N-diethyldodecanamide (DEDAD), using a bench-scale fixed-bed, continuous, downflow hydrotreater over a commercial NiMo/Al_2O_3 sulfide catalyst. Especially, we revealed the reaction mechanism of simultaneous HDN and HDO of DEDAD, which was dominated by directly deoxygenation (DO pathway) to produce amine intermediate, instead of C-N bond cleavage through denitrogenation (DN pathway) to yield a nitrogen-containing compound and an oxygen-containing species. In addition, to study the influence of sulfur on simultaneous HDN and HDO, H_2S was added to the DEDAD test, which showed significant inhibition effect on the DO path. Moreover, we have studied the effect of substitution of the amide group on the HDN and HDO chemistry of alkylamides. N-methyldodecanamide (MDAD) was selected as secondary amide which reacted slower compared with tertiary amide DEDAD under the same reaction conditions. Besides, we have also successfully determined the kinetics with respect to HDO of dodecanoic acid (DA) and 4-propylphenol (PP), HDS of dibenzothiophene (DBT), as well as HDN of quinoline (Q) and amines. The commercial NiMo sulfide catalyst showed better HDO and HDS performance than HDN with the highest overall rate constant for hydrotreating of fatty acid.

611886 Biochar from Agricultural Waste Via Microwave-Driven Pyrolysis

Abstract:

In view of obtaining biochar for agriculture usage via pyrolysis, two different heat treatments – microwaves (MWP) and conventional electrical (CEP) heating were investigated and compared. The chosen biomass waste was poplar (Populus alba). MWP pyrolysis was carried out using 2.45 GHz microwaves generated by a semiconductor generator (0-200 W) in a TE01 monomode cavity, while CEP was carried out in a custom-made lab-scale tubular batch reactor electrically heated placed in a commercial NABERTHERM oven. Although the temperature evolution vs. time was not possible in MWP above 240ºC (max. temperature of the fibre optic built-in the experimental set-up), the influence of the microwave power & residence time and the energy/gram of treated biomass were recorded and used for the evaluation. In CEP the influence of pyrolysis temperature (300-800ºC) vs. the characteristics of biochar was evaluated. To determine how heating method affects the properties of biochars and consequently their possible utilization, physical, chemical, thermal, porosity and EDX analysis were carried out for all biochars and for the corresponding feedstock. The experimental study was focused on the comparison of heating method vs. biochar yield distribution, elemental composition, pH, high heating value (HHV), surface area, porosity and morphology. In relation to the microwave heating, the evaluation of the process economics vs. application of the obtained biochar is attempted.

603708 Production and Blending of Biomass-Derived Calcined Coke Using a Rotary Tube Furnace

Abstract:

The aluminum industry relies on calcined petroleum coke as an anode material for smelting of aluminum oxide into aluminum metal. Depending on the starting oxygen content, biomass pyrolysis oils can be distilled and the residues thermally treated into calcined coke. While early studies have shown promise based on elemental and physical properties, their integration with and production from rotary calciners has yet to be realized. Using a rotary tube furnace, we performed continuous calcination of petroleum coke blends with biocoke. We used bio-coke with an oxygen content of ~13 wt%. A solids feeder conveyed the blends at controlled feed rates. Overall, the lab furnace proved capable of converting green petroleum coke into calcined petroleum coke with the properties specified by the industry. We successfully decreased the concentration of problematic trace metals (Ni, V, S, etc.) via blending and co-calcining. The average crystallite height (Lc) decreased from 26.5 to 25.6 Å, which still remains within typical specifications. However, as evidenced through microscopy, some amorphous carbon remained in the product.

CHARACTERIZATION OF FACTORS AFFECTING THE GAS WELL DELIVERABILITY OF THE JIAOSHIBA SHALE GAS FIELD BY MULTIVARIATE STATISTICAL ANALYSIS

***Yizhong Zhang*[1,2], *Maolin Zhang*[1], *Bin Ju*[3], *Jincai Shen*[4], *Lan Ge*[4], *Fanhua Zeng*[2] ***

**Corresponding author. Email address: Fanhua.Zeng@uregina.ca.*

[1] Hubei Cooperative Innovation Center of Unconventional Oil and Gas, Yangtze University, Hubei, Wuhan, China, 430100

[2] University of Regina, Regina, SK, Canada, S4S 0A2

[3] School of Oil and Natural Gas Engineering, Southwest Petroleum University, Chengdu, Sichuan, China, 610500

[4] SINOPEC Chongqing Fuling Shale Gas Exploration and Development Co., Ltd., Chongqing, China, 408014

Abstract

The Longmaxi shale gas in the Jiaoshiba Shale Gas Field is the first and the largest successfully developed shale gas in China. The analysis of key factors of deliverability of gas wells plays a significant role in efficiently developing shale gas reservoirs. However, there are few methods analyzing the key factors of deliverability of shale gas wells by considering both geological characters and engineering parameters. In this study, the reservoir characteristics, drilling and completion data, fracturing data (including the well logging interpretation of horizontal section, the track and layer section that the well bore passed through, and fracturing parameters) and production data of 86 gas wells in the Wufeng-Longmaxi shale, Jiaoshiba Shale Gas Field in Eastern Sichuan Basin, China are statistically analyzed and ordered by multivariate statistical analysis. The gas well deliverability is described by absolute open flow potential and 9 factors (physical parameters of reservoir, including density of rock, porosity, total organic carbon, brittle mineral content of rock and drilling and fracturing engineering quality, including horizontal length of gas well, horizontal length ratio through different stages in the formation, number of fracturing stages, liquid volume, and sand amount) that may affect the deliverability of shale gas wells are considered. The results reveal that the absolute open flow potential has a close relationship with porosity, total organic carbon, brittle mineral content and total liquid volume. The multivariable nonlinear regression models of absolute open flow potential with different key factors considered are built, and the prediction results are reliable by goodness-of-fit test acceptable with real gas well deliverability testing data. The research methods and results in this study can provide a reference basis for reasonable production rate allocation, design, and adjustment of development plan for the shale gas reservoirs.

Keywords: Gas well deliverability, Longmaxi shale gas, Jiaoshiba Shale Gas Field, Multivariate statistical analysis, Multivariable nonlinear regression

1. Introduction

Following the increased demand of petroleum energy and the developed science and technology, the industry nowadays has focused on the unconventional resources. Shale gas reservoirs are one of the most popular kind of unconventional resources with extremely low permeability, tight porous structure, large specific area and high heterogeneity, which have attracted attentions of researchers worldwide [1-3]. Shale gas has self-generated, self-contained, organic-rich, nanoscale pore structure, and other unusual characteristics and it could be considered as a source as well as a reservoir [4-5]. In addition to free gas and small amount of dissolved gas, adsorbed gas is largely existed in the tight formation, which makes it difficult to produce owing to the complexity of the reservoir [6-8]. The Sichuan-Chongqing region in southwest China is one of the fastest shale gas production growing regions in the world. The annual shale gas deliverability of China has reached 2×10^{10} m^3 with the annual shale gas production of 1.5×10^{10} m^3 [9]. As one of the main contributors of natural gas production growth in China, the gas production of Jiaoshiba shale gas reservoir has exceeded the daily production of 1.67×10^{7} m^3/d since 2013 [10]. The large production of commercial shale gas flow of the Longmaxi Formation (e.g. more than 2×10^{5} m^3/d of JY1 well) has made the Jiaoshiba area one of the most important shale gas field in China.

Compared with typical shale gas reservoirs in North America, the shale gas in Sichuan Basin has lower TOC value, lower porosity, lower gas saturation, smaller thickness and higher buried depth, which makes its exploration more complicated [9]. In both the early test stage and current production stage, there are large differences among deliverability of the shale gas wells in the Jiaoshiba area. Therefore, the deliverability evaluation is the basis of reasonable shale gas production and prediction, and the research on deliverability evaluation and its related reservoir and engineering parameters of shale gas wells is particularly important. Previous studies have focused on the impacts of reservoir quality including sedimentary environment [11], structure [12-13], mineralogy and lithology [14-17] and engineering quality including drilling and completion techniques [12, 18] and multi-stage hydraulic fracturing performance [19-20] on the shale gas production. Their results revealed that both the reservoir quality and engineering quality will affect the produced gas rate of the shale gas wells. Most of researches did analysis based on the relationship between the interested variable and its single impacting factor, comprehensive multivariate statistical analysis with consideration of correlation and collinearity among factors needs to be further studied. Moreover, instead of the produced gas rate or gas bearing content, the absolute open flow (AOF) potential may be a better parameter to describe the deliverability of gas well. The AOF potential is a hypothetical gas flow rate that may be produced if the flowing pressure at the bottom of the well (BHFP) were equivalent to the atmospheric pressure, i.e. the maximum flow rate that the gas well may produce [21]. In this study, the well deliverability of Jiaoshiba shale gas is described by the AOF potential based on BHFP and the impacts of reservoir and engineering characteristics on the AOF potential will be comprehensively analyzed. The data of over 86 wells of Fuling Jiaoshiba shale gas field is analyzed and the impacts of the reservoir quality (i.e. density of rock, porosity, total organic carbon, brittle mineral content of rock) and the engineering quality (i.e. horizontal length of gas well, horizontal length ratio through different formation stages, number of fracturing stages, liquid volume, and sand amount) on the AOF are quantified. The most important factors affecting deliverability of the gas wells in the Jiaoshiba area will be obtained by hierarchical clustering analysis method. The regression correlation between the well deliverability and the important parameters will be built. The objective of this study is to find a new method to evaluate the key factors that affect the well deliverability of shale gas reservoir by multivariate statistical analysis with consideration of correlation and collinearity among factors and build multivariable nonlinear regression models based on these key factors, so that the well deliverability of shale gas reservoir could be predicted.

2. Shale gas reservoir characters

The Longmaxi shale gas in the Jiaoshiba area located in the Eastern Sichuan Basin in the southwest region of China is the first and the largest successfully developed shale gas reservoir in China [22]. The shale gas bearing formation is the Upper Ordovician Wufeng Formation to Lower Silurian Longmaxi Formation with a thickness of about 200–300 m, area of 272.44 km^2 and buried depth of less than 3500m. Comprehensively considering the lithology, electrical and physical properties, geochemical and gas-bearing characteristics, the gas bearing shale section in the Upper Ordovician Wufeng Formation to Lower Silurian Longmaxi Formation could be vertically divided into three parts with stable horizontal stratigraphic distributions for each member (i.e. Long1, Long2 and Long3 from bottom to top). Long1 member is rich in fossils (e.g., radiolarian, calthrop, and graptolite) and is composed of black mudstone and argillaceous siltstone. The Long2 member consists of argillaceous siltstone and fine-grained sandstone. The Long3 member is composed of dark-gray mudstone and gray shale with fewer fossils [5, 23].

Previous studies suggested that Long1 member consisted of one 3-ord sequence, and could be divided into three system tracts (i.e. the transgressive system tract (TST), early highstand system tract (EHST), and late highstand system tract (LHST)) [10, 15, 16]. As shown in **Fig.1** [5], this study is targeted in the Long1 member, which is abundant of organic matter with pores and could be divided into 3 stages (i.e. stage1, stage2 and stage3 from bottom to top) based on physical properties with gas bearing increasing from top to bottom. The Long1 member is of good brittleness, and the brittle mineral content gradually increases from top to bottom. The natural gas in the layer belongs to over-mature dry gas, in which methane is the main component without containing hydrogen sulfide, carbon monoxide and hydrogen.

3. Samples

The AOF potential is a hypothetical gas flow rate that may be produced if the flowing pressure at the bottom of the well (BHFP) were equivalent to the atmospheric pressure [21]. The AOF potential may be a better parameter to describe the deliverability of the gas wells as it gives the maximum flow rate that the gas well can produce. As shown in Fig.2, the AOF potential data of and 9 factors that may affect the deliverability of gas wells (i.e. horizontal length, number of fracture stages, total liquid volume, total amount of sand, rock density, porosity, TOC, brittle mineral content of rocks, horizontal length ratio through stage 1 and stage 3 of Long1 member) are collected (data is offered by SINOPEC Chongqing Fuling Shale Gas Exploration and Development Co., Ltd., Chongqing, China). Though **Fig.2** shows certain relationship between AOF potential and single impacting factor, the correlation and collinearity among different factors are not considered. According to the data, the average reservoir pressure of the Long1 member is 37.69MPa with pressure coefficient 1.55. The reservoir temperature is 85.99°C. The pay zone thickness is between 20m-200m and the most favorable thickness is about 38m. The kerogen type of the organic matter is mostly type II_1 to II_2. The total organic carbon (TOC) contents of the shale range from 1%-8% with average value of 5% and it is increasing from top to bottom. The vitrinite reflectance (R_o %) of organic matter ranges from 2.5%-3.7% with average value of 2.9%. The porosity of the shale is between 1.88% to 7.06%, with average value of 4.15% and the permeability of the shale is less than 0.001mD. The horizontal wells in the sample have lateral length between 1500m to 1800m, well spacing between 500m to 600m, and the distance between adjacent clusters between 20m to 25m. All the horizontal wells are multi-stage hydraulic fractured with number of fracturing stages between 15 to 29. The fluid volume per stage is between 1500m^3 to 1800m^3, and proppant volume per stage is between 50m^3-65m^3. The fractured half-lengths range from 300m to 350m, and propped half-lengths range from 250m to 270m. The fracture heights range from 40m to 50m, and main fracture width range from 5mm to 10mm.

4. Methods

4.1. Normality test

When working with multivariable regression problems, some assumptions should be followed. The regression process of independent variables and the target variable are assumed to follow Gauss–Markov theorem (i.e. the ordinary least squares (OLS) estimator has the lowest sampling variance within the class of linear unbiased estimators, if the independent variables have no perfect collinearity, and the errors in the regression model are uncorrelated with equal variances and expectation value of zero) [24]. Thus, the first step of the multivariate statistical analysis is to test whether the sample variables follow normal distribution. The Q–Q (quantile-quantile) plot is applied, which is a probability plot that shows a graphical comparison between the samples and normal distribution by plotting their quantiles against each other [25]. If the points in the Q–Q plot approximately lie on the 45-degree line (i.e. $y=x$), the distribution of the sample being compared are similar enough to normal distribution. If normality test of a variable behaves not well, certain transformation (i.e. logistic transformation, logarithm transformation, index transformation, power transformation, etc.) of the variable and nonlinear regression model should be considered.

4.2. Standardization

Before the key factor analysis, the calculation of correlations between each parameter is essential. Since the units, scales and ranges of the factors are different, the first step is to standardize all the parameters based on the normality test results. The method applied in the study is the Fisher Z-transformation method. The standard score (Z score) for each data point is its difference from sample mean divided by the sample standard deviation.

$$Z = \frac{X - \overline{X}}{s} \tag{1}$$

where

X is the data point.
$\overline{X}$ is the sample mean.
s is the sample standard deviation.

4.3. Correlation analysis

After standardization, the correlations between each factor are calculated. This is to eliminate the repeated consideration of correlation and collinearity in the later multivariable regression. In other words, if two variables are strongly related, only the principle variable between them is considered in later analysis. The sample data has a number of n wells, and each well has a number of k variables including a dependent variable (X_k) which is Z scores for AOF potential and $k-1$ independent variables (X_1, X_2, ..., X_{k-1}) which are Z scores for the factors affecting AOF potential. The sample data could be shown in matrix form as follows

$$X = \begin{bmatrix} x_{1,1} & x_{1,2} & \cdots & x_{1,k} \\ x_{2,1} & x_{2,2} & \cdots & x_{2,k} \\ \cdots & \cdots & & \cdots \\ x_{n,1} & x_{n,2} & \cdots & x_{n,k} \end{bmatrix} \tag{2}$$

Then the correlation matrix for all variables would be

$$R = \begin{bmatrix} r_{11} & r_{12} & \cdots & r_{1k} \\ r_{21} & r_{22} & \cdots & r_{2k} \\ \cdots & \cdots & \cdots & \cdots \\ r_{k1} & r_{k2} & \cdots & r_{kk} \end{bmatrix} \tag{3}$$

$r_{ij}(i,j=1,2,...,k)$ is the correlation coefficient for $x_{ij}(i=1,2,...,k)$ and $x_{ij}(i=1,2,...,k)$, and the calculation equation is

$$r_{ij} = \frac{\sum_{m=1}^{k}(x_{im}-\overline{x_i})(x_{jm}-\overline{x_j})}{\sqrt{\sum_{m=1}^{k}(x_{im}-\overline{x_i})^2 \sum_{l=1}^{k}(x_{jl}-\overline{x_j})^2}} \tag{4}$$

The value of correlation coefficient should be between -1 and 1. If the correlation coefficient is close to 1, the involving two variables have a strong positive relationship, and if it is close to -1, the involving two variables have a strong negative relationship. If the correlation coefficient is close to 0, this suggests that the involving two variables may be independent to each other.

4.4. Hierarchical clustering analysis

As long as correlations among factors are obtained, clustering method could be applied to find the key factors while eliminate the factors with small effect on AOF potential or strong correlation with other factors. Hierarchical clustering analysis (HCA) is a method to find the hierarchy relationship among variables, so that the variables could be divided into clusters [26]. HCA is performed on the normalized sample data set of the 86 well data including AOF potentials and 9 different factors, using centroid linkage clustering (Eq.5) as linkage criteria. The method will try to link the two most related factors in the data set to form a new cluster at each step (**Fig.3** agglomerative hierarchical clustering). The new merged cluster is then considered as a variable in the relation map and agglomerative approach is again used. Finally, a tree-hierarchy structure with linkage lines that join all the factors would be built.

$$D_{uv}^2 = (\overline{X}_u - \overline{X}_v)'(\overline{X}_u - \overline{X}_v) \quad (5)$$

, where $\overline{X}_u$ is the centroid for cluster C_u. $\overline{X}_v$ is the centroid for cluster C_v.

4.5. Multivariable nonlinear regression

Once the main factors that affect the AOF potential are found, other factors are either unessential to AOF potential or have strong correlation to the main factors, which could be eliminated. The next step is to linearize the main factors, including logistic transformation, logarithm transformation, index transformation, power transformation, etc. Then, a set of multivariable linear regressions is carried out to find the optimal regression model. If there are k variables, the linear regression for the variables is

$$y_i = \alpha + \beta_1 X_{i1} + \ldots + \beta_k X_{ik} + \varepsilon_i = \alpha + \sum_{j=1}^{k} \beta_j X_{ij} + \varepsilon_i, i = 1,2,\ldots,n \quad (6)$$

Least squares could be used to find the intercept α and coefficients β_j (j=1,2,...,k), which is to find the optimums for α and β_j (j=1,2,...,k) when the sum of squared residuals, *SSe*, is minimized. The squared residual r_i^2 is defined as the difference between real value and predicted value of the dependent variable.

$$r_i = y_i - \hat{y}_i = y_i - f(x_i, \alpha, \beta) \quad (7)$$

$$SSe = \sum_{i=1}^{n} r_i^2 \quad (8)$$

$\hat{y}_i$ is the estimation value of dependent variable y_i. $a, b_j (j=1,2,\ldots,k)$ are the estimation values of intercept and coefficients $\alpha, \beta_j (j=1,2,\ldots,k)$. Let the partial derivatives with respect to $a, b_j (j=1,2,\ldots,k)$ equals 0.

$$\begin{cases} \dfrac{\partial SS_e}{\partial a} = 0 \\ \dfrac{\partial SS_e}{\partial b_j} = 0 \end{cases} \quad j = 1,2,\ldots,k \quad (9)$$

It could be also proved that $a, b_j (j=1,2,\ldots,k)$ is the unbiased estimation of $\alpha, \beta_j (j=1,2,\ldots,k)$.

5. Results and Discussions

5.1. Normality test

Fig.4 shows the Q–Q (quantile-quantile) plots for the 10 variables of the 86 wells. The results reveal that the points of most variables in the Q–Q plots approximately lie on the 45-degree line, which means they follow normal distribution well. The only variable may not behave with good normality is horizontal length ratio through stage 1 and stage 3 of Long1 member. This variable may need transformation (i.e. logistic transformation, log transformation, logistic transformation, etc.) in the later regression study. Thus, the multivariable nonlinear regression would be needed in this study.

5.2. Standardization and correlation of the variables

The matrix of correlation coefficients of all variables is shown in **Table 1**. **Table 1** suggests that there is no strong linear relation between the AOF potential and single one factor. Relatively, the AOF potential is positively related to the TOC, the brittle mineral content, and the total liquid volume. It is negatively related to rock density. Some factors show strong correlation with each other, and the number of fracture stages and total liquid volume have the highest correlation coefficient of 0.946. The drilling and fracturing engineering factors, including the number of fracturing stages, the length of horizontal well, the total liquid volume and the total sand amount, are highly linear related. Among the geological factors, the rock density is highly negative to TOC, brittle mineral content, and the TOC and the brittle mineral content of rock are highly linear related. The collinearity among some factors is very strong and may affect AOF potential modelling, which should be eliminated in the multivariate statistical analysis of AOF potential.

5.3. Hierarchical clustering analysis

According to **Table 1**, the number of fracturing stages and total liquid volume have the highest correlation coefficient, which is 0.946. These two factors could be firstly clustered together, and the central point of this cluster is calculated by centroid method. Then the correlation analysis is done again after applying the whole cluster as a variable. This progress is continued until all the variables converged to one cluster. The result of hierarchical clustering analysis with linkage lines is shown **Fig.5**, and **Table 2** gives the importance order of the effect of each factor on the AOF potential.

It is suggested in the hierarchical clustering analysis results that the order of the factors that affect AOF potential is: a. porosity, b. TOC, c. brittle mineral content of rock, d. horizontal length ratio through stage 1 and stage 3 of Long1 member, e. number of fracturing stages, f. total liquid volume, g. horizontal length, h. total sand amount and i. rock density. The AOF potential firstly has closest relationship to porosity. It then has closer relationship to the cluster of TOC, brittle mineral content of rock, horizontal length ratio through stage 1 and stage 3 of Long1 member. These four factors belong to the first cluster. The other four factors including number of fracturing stages, total liquid volume, horizontal length and total sand amount belong to the second cluster. Rock density belongs to the third cluster. Thus, the main factors that affect AOF potential are porosity, TOC, brittle mineral content of rock and number of fracturing stages.

5.4. Multivariable nonlinear regression

After the main clusters in the hierarchical clustering analysis are obtained, the key factors that affect the AOF potential are found. The independent variables in the multivariable nonlinear regression models could be chosen according to the cluster results. With consideration of the correlation and collinearity among factors, one factor in each cluster with smallest linkage distance is chosen in the regression model, and four models after testing (**Table 3**) are regressed in this study.

With trials of nonlinear transformation of the involving variables, the linear regression of the transformation is ideal, and all the coefficients have passed the student t test for significance. The four models are shown as follows.

Model 1:

$$Y' = \ln Y, X_1' = \ln X_1, X_2' = (\ln X_2)^8, X_3' = X_3 \tag{10}$$

$$Y' = -13.067 + 1.509X_1' + 0.012X_2' + 0.867X_3' \tag{11}$$

Model 2:

$$Y' = \ln Y, X_4' = \ln X_4, X_2' = (\ln X_2)^8, X_3' = \ln X_3 \tag{12}$$

$$Y' = -1.756 + 1.465X_4' + 0.011X_2' + 1.043X_2' \tag{13}$$

Model 3:

$$Y' = \ln Y, X_4' = \ln X_4, X_5' = X_5^{\frac{1}{100}}, X_2' = (\ln X_2)^8, X_3' = \ln X_3 \tag{14}$$

$$Y' = -1.603 + 1.434X_4' + 0.226X_5' + 0.011X_2' + 0.848X_3' \tag{15}$$

Model 4:

$$Y' = \ln Y, X_1' = \ln X_1^{\frac{1}{10}}, X_3' = X_3^{\frac{1}{100}}, X_2' = (\ln X_2)^8, X_6' = X_6 \tag{16}$$

$$Y' = -12.556 + 5.211X_1' + 0.149X_5' + 0.012X_2' + 0.035X_6 \tag{17}$$

It is shown from **Fig.6** that the total AOF potential of 86 wells is $5420.48\times10^4 m^3/d$ and the total AOF potential of 86 wells predicted by the 4 models range from $5110.24\times10^4 m^3/d$ to $5607.74\times10^4 m^3/d$. The average AOF potential of single well is $63.03\times10^4 m^3/d$ and the average AOF potential of single well predicted by the 4 models range from $59.42\times10^4 m^3/d$ to $65.21\times10^4 m^3/d$. The coefficient of determination (R^2) in the figure is the proportion of the variance in the dependent variable Y that is predictable from the independent variables $X_j (j = 1, 2, ..., k)$. The R^2 values show that all the 4 models could be relatively preferable models to predict the AOF potentials based on different factors. In total, the prediction value of model 1 and model 4 are closer to the real AOF potential value with higher coefficients of determination (R^2) and less prediction errors.

6. Conclusions

In this study, instead of the produced gas rate, the absolute open flow (AOF) potential is analyzed as a better parameter to describe the deliverability of gas well. The reservoir parameters, drilling and completion data, fracturing data (including the well logging interpretation of horizontal section, the track and layer section that the well bore passed through, fracturing scale and monitoring parameters) and sample production data of 86 gas wells in the Wufeng-Longmaxi shale, Jiaoshiba shale gas field in Eastern Sichuan Basin, China are considered in the multivariate statistical analysis. The hierarchical clustering analysis could give the orders of the key factors that affect gas well AOF potential as well as eliminate the collinearity among the factors, which gives the basis of multivariable nonlinear regression. The results suggest that the order of the factors affects AOF potential is: a. porosity, b. TOC, c. brittle mineral content of rock, d. horizontal length ratio through stage 1 and stage 3 of Long1 member, e. number of fracturing stages, f. total liquid volume, g. horizontal length, h. total sand amount and i. rock density. The estimation results of the AOF potential by the multivariable nonlinear regression models based on the key factors from hierarchical clustering analysis have small differences to that of the actual values of the sample with goodness-of-fit test acceptable. The predictions of AOF potential of the models are reliable, and the feasibility of this multivariate statistical analysis is verified in the study of shale gas well deliverability. The research results can provide a reference basis for design of drilling and completion, the reasonable production rate allocation, design and adjustment of development plan of the shale gas reservoirs.

Acknowledgement

The authors would like to acknowledge the support from National Natural Science Foundation of China (NSFC Grant Number: 520004032) and the Thirteen-Five National Science and Technology Major Projects of China: 1. Study on Shale Gas Fracturing and Gas Production Regularity in Wufeng Formation and Longmaxi Formation in Fuling Area (Sub-Issue Number: 2016ZX05060-026) and 2. Research on the efficient development technology and policy of inter-salt shale oil reservoir (Sub-Issue Number: 2016ZX05049005-010).

References

1. Zolfaghari, A., Dehghanpour, H., & Xu, M. (2017b). Water sorption behaviour of gas shales: II. Pore size distribution. *International Journal of Coal Geology, 179*, 187-195.
2. Zhang, Y., Yao, S., Zhang, M., Zhou, X., Mei, H., & Zeng, F. (2018). Prediction of adsorption isotherms of multicomponent gas mixtures in tight porous media by the oil–gas-adsorption three-phase vacancy solution model. Energy & Fuels, 32(12), 12166-12173.
3. Zhang, Y., Zhang, M., Mei, H., & Zeng, F. (2019). Study on salt precipitation induced by formation brine flow and its effect on a high-salinity tight gas reservoir. *Journal of Petroleum Science and Engineering, 183*, 106384.
4. Dong, T., Harris, N. B., Ayranci, K., Twemlow, C. E., & Nassichuk, B. R. (2017). The impact of composition on pore throat size and permeability in high maturity shales: Middle and Upper Devonian Horn River Group, northeastern British Columbia, Canada. *Marine and Petroleum Geology*, 81, 220-236.
5. Wang, C., Zhang, B., Hu, Q., Shu, Z., Sun, M., & Bao, H. (2019). Laminae characteristics and influence on shale gas reservoir quality of lower Silurian Longmaxi Formation in the Jiaoshiba area of the Sichuan Basin, China. *Marine and Petroleum Geology, 109*, 839-851.
6. Cao, T. T., Song, Z. G., Wang, S. B., & Jia, X. (2015). A comparative study of the specific surface area and pore structure of different shales and their kerogens. *Science China Earth Sciences, 58*(4), 510-522.
7. Chalmers, G. R. L., & Bustin, R. M. (2007). The organic matter distribution and methane capacity of the lower cretaceous strata of northeastern British Columbia, Canada. *International Journal of Coal Geology, 70*(1-3), 223-239.
8. Chalmers, G. R. L., & Bustin, R. M. (2008). Lower cretaceous gas shales in northeastern British Columbia, Part II: Evaluation of regional potential gas resources. *Bulletin of*

Canadian Petroleum Geology, 56(1), 22-61.
9. Ma, X., Li, X., Liang, F. , Wan, Y. , & Guo, W. . (2020). Dominating factors on well productivity and development strategies optimization in Weiyuan shale gas play, Sichuan Basin, SW China. *Petroleum Exploration and Development, 47*(3), 594-602.
10. Shu, Y., Lu, Y., Chen, L., Wang, C., & Zhang, B. (2020). Factors influencing shale gas accumulation in the lower Silurian Longmaxi formation between the north and South Jiaoshiba area, Southeast Sichuan Basin, China. *Marine and Petroleum Geology, 111*, 905-917.
11. Wang, Y. , Dong, D. , Li, X. , Huang, J. , Wang, S. , & Wu, W. . (2015). Stratigraphic sequence and sedimentary characteristics of Lower Silurian Longmaxi Formation in the Sichuan Basin and its peripheral areas. Natural Gas Industry B, 2(2–3), 222-232.
12. Hu, M., Huang, W., & Li, J. (2018). Effects of structural characteristics on the productivity of shale gas wells: A case study on the Jiaoshiba Block in the Fuling shale gasfield, Sichuan Basin. *Natural Gas Industry B, 5*(2), 139-147.
13. Zhang, H., Wang, J., & Zhang, H. (2016). Investigation of the main factors during shale-gas production using grey relational analysis. *The Open Petroleum Engineering Journal, 9*(1).
14. Wei, C., Wang, H., Sun, S., Xiao, Y., Zhu, Y., & Qin, G. (2012, January). Potential investigation of shale gas reservoirs, Southern China. In *SPE Canadian Unconventional Resources Conference*. Society of Petroleum Engineers.
15. Wang, C., Zhang, B., Lu, Y., Shu, Z., Lu, Y., Bao, H., ... & Chen, L. (2018). Lithofacies distribution characteristics and its controlling factors of shale in Wufeng Formation-Member 1 of Longmaxi Formation in the Jiaoshiba area. *Petroleum Research, 3*(4), 306-319.
16. Wang, C., Zhang, B. Q., Lu, Y. C., Shu, Z. G., Lu, Y. Q., Bao, H. Y., ... & Chen, L. (2018). Lithofacies distribution characteristics and main development controlling factors of shale in Wufeng Formation-Member 1 of Longmaxi Formation in Jiaoshiba area. *Acta Pet. Sin., 39*(6), 631-644.
17. Kim, G., Lee, H., Chen, Z., Athichanagorn, S., & Shin, H. (2019). Effect of reservoir characteristics on the productivity and production forecasting of the Montney shale gas in Canada. *Journal of Petroleum Science and Engineering, 182*, 106276.
18. Curtis, M. E., Cardott, B. J., Sondergeld, C. H., & Rai, C. S. (2012). Development of organic porosity in the Woodford Shale with increasing thermal maturity. *International Journal of Coal Geology, 103*, 26-31.
19. Cipolla CL, Lolon EP, Erdle JC, Rubin B. Reservoir modeling in shale-gas reservoirs. *SPE reservoir evaluation & engineering*. 2010 Aug 1;13(04):638-53.
20. Zhu W, Ma D, Qi Q, Song Z, Yue M. Multi-zone coupling productivity of horizontal well fracturing with complex fracture networks in shale gas reservoirs. Natural Gas Industry B. 2018 Feb 1;5(1):80-9.
21. Mohammed, K. R., Stigliano, H., Vicente, M. G., & Angulo, R. J. (2010, January). Comparison of Different Deliverability Testing Techniques in Low Permeability Gas Reservoir and Its Impact on the Absolute Open Flow (AOF) Estimation. In *SPE EUROPEC/EAGE Annual Conference and Exhibition*. Society of Petroleum Engineers.
22. Guo, X., Qin, Z., Yang, R., Dong, T., He, S., Hao, F., ... & Liu, K. (2019). Comparison of

pore systems of clay-rich and silica-rich gas shales in the lower Silurian Longmaxi formation from the Jiaoshiba area in the eastern Sichuan Basin, China. *Marine and Petroleum Geology, 101*, 265-280.
23. Yang, R., He, S., Hu, Q., Hu, D., Zhang, S., & Yi, J. (2016). Pore characterization and methane sorption capacity of over-mature organic-rich Wufeng and Longmaxi shales in the southeast Sichuan Basin, China. Marine and Petroleum Geology, 77, 247-261.
24. Henderson, C. R. . (1975). Best linear unbiased estimation and prediction under a selection model. *Biometrics, 31*(2), 423-447.
25. Gnanadesikan, M. B. W. . (1968). Probability plotting methods for the analysis for the analysis of data. *Biometrika, 55*(1), 1-17.
26. Rokach, Lior, and Oded Maimon. (2005). "Clustering methods." Data mining and knowledge discovery handbook. Springer US, 321-352.

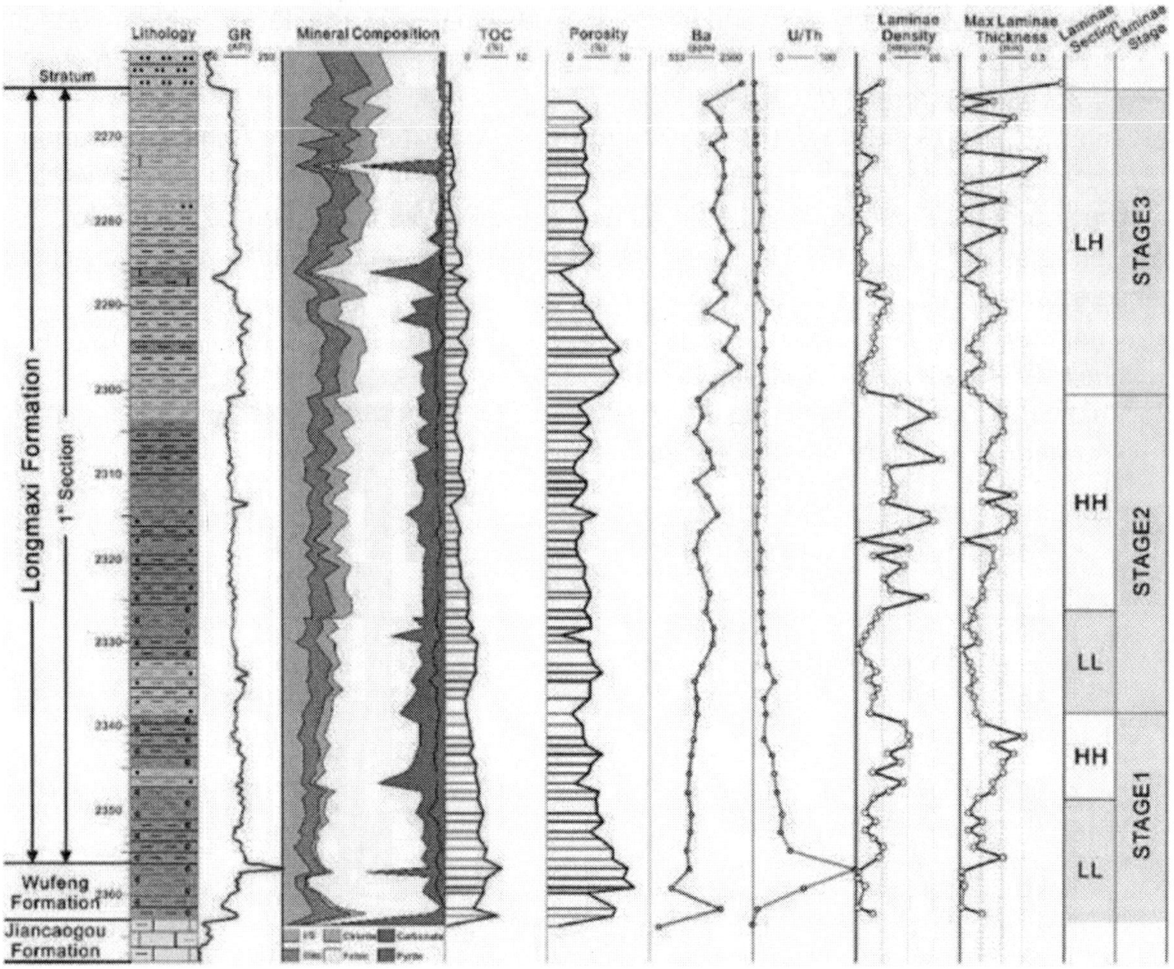

Fig.1 Lithofacies, gamma ray log, mineral content, TOC content, porosity, geochemical proxies, laminae density, and laminae stages of the Long1 member in well A (Wang et al., 2019).

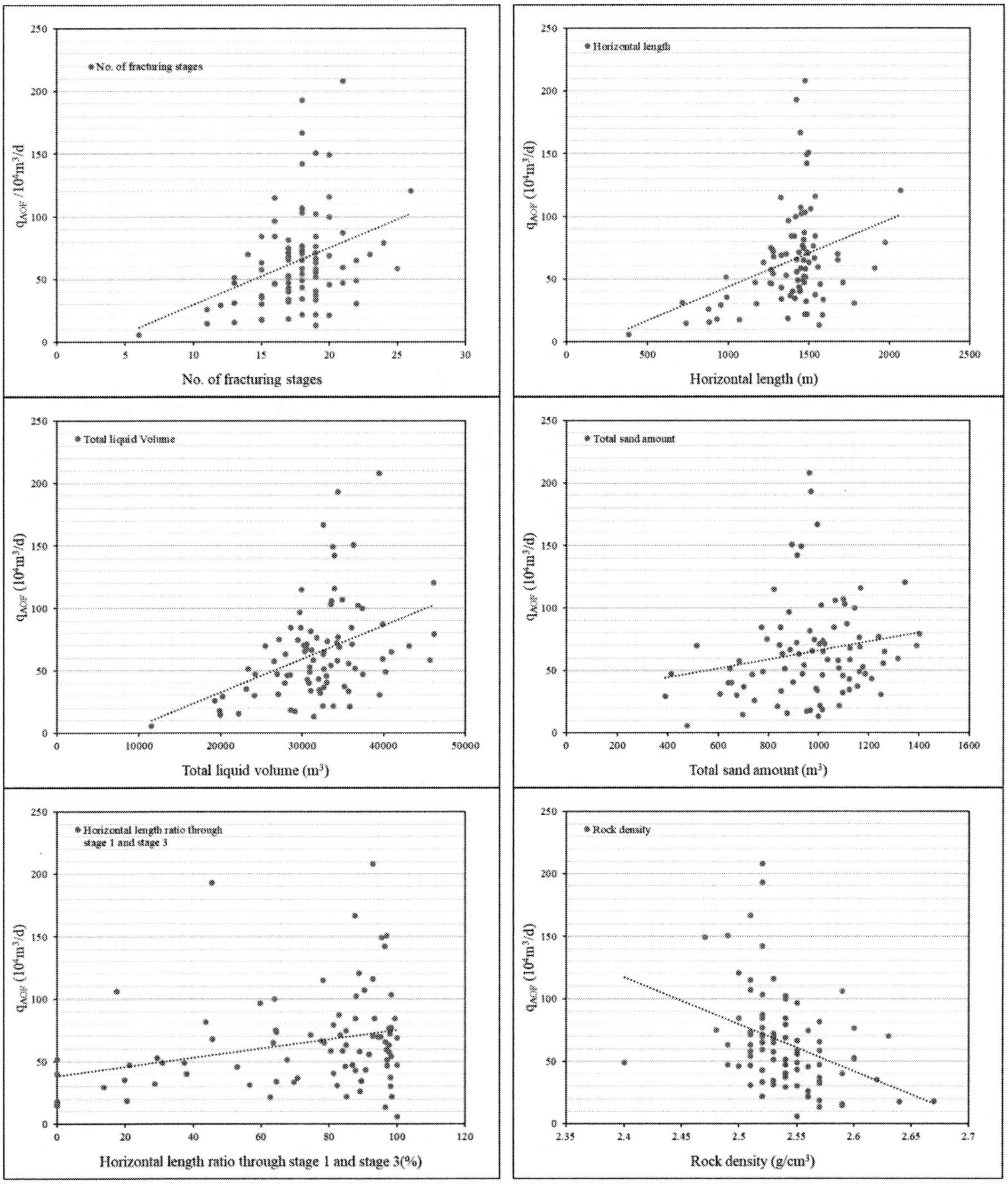
No. of fracturing stages
q_{AOF} /10^4m^3/d
No. of fracturing stages
Horizontal length
q_{AOF} (10^4m^3/d)
Horizontal length (m)
Total liquid Volume
q_{AOF} (10^4m^3/d)
Total liquid volume (m^3)
Total sand amount
q_{AOF} (10^4m^3/d)
Total sand amount (m^3)
Horizontal length ratio through stage 1 and stage 3
q_{AOF} (10^4m^3/d)
Horizontal length ratio through stage 1 and stage 3(%)
Rock density
q_{AOF} (10^4m^3/d)
Rock density (g/cm^3)

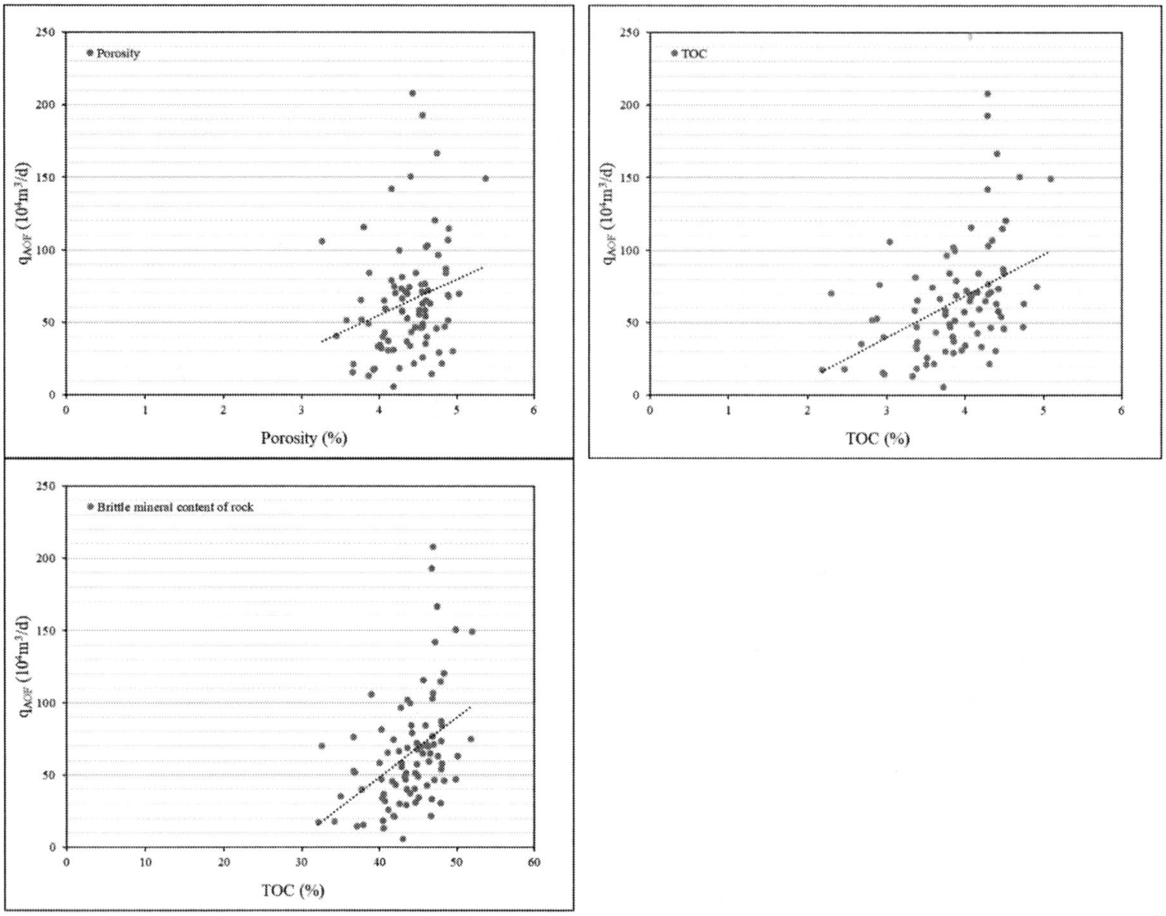

Fig.2 The scatter plots with single linear regression line of sample data of 86 gas wells in Jiaoshiba area with the AOF potential against 9 factors that may affect the deliverability of gas wells (i.e. number of fracture stages, horizontal length of well, total liquid volume, total sand amount, rock density, porosity, TOC, brittle mineral content of rocks, horizontal length ratio through stage 1 and stage 3 of Long1 member)

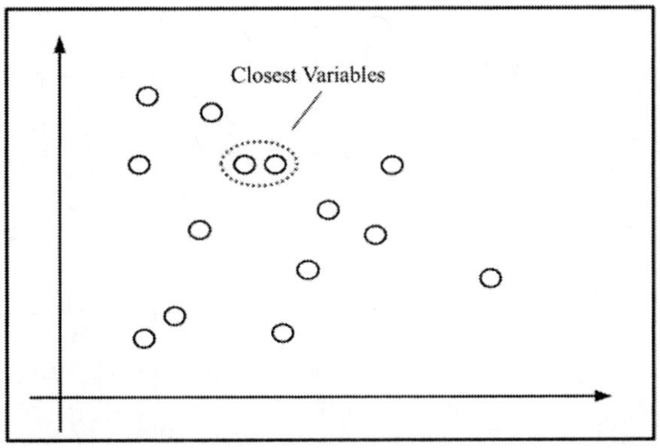

(a)

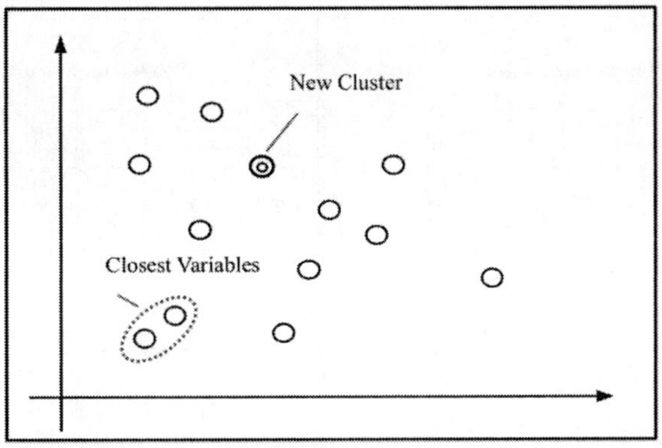

(b)

Fig.3 Hierarchical Clustering Analysis: (a) Clustering the variables by closest relation; (b) The center of new cluster is calculated by centroid method

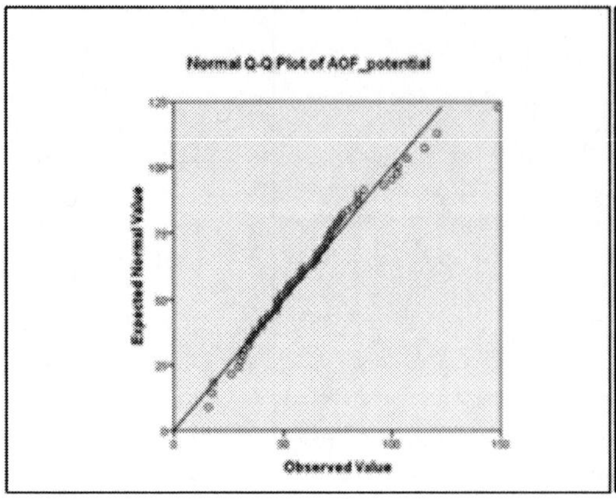
Normal Q-Q Plot of AOF_potential
Expected Normal Value
Observed Value

a.

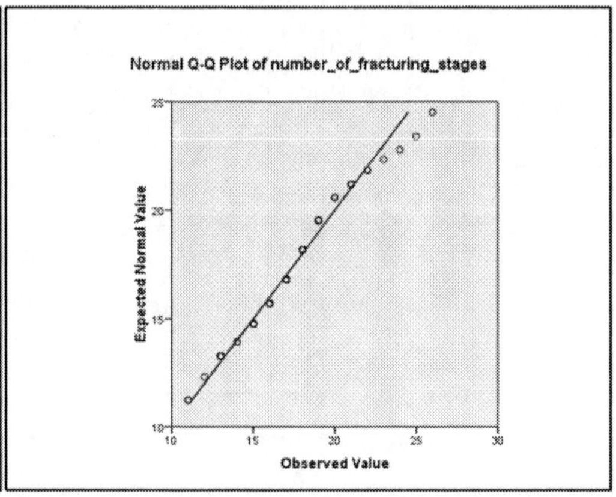
Normal Q-Q Plot of number_of_fracturing_stages
Expected Normal Value
Observed Value

b.

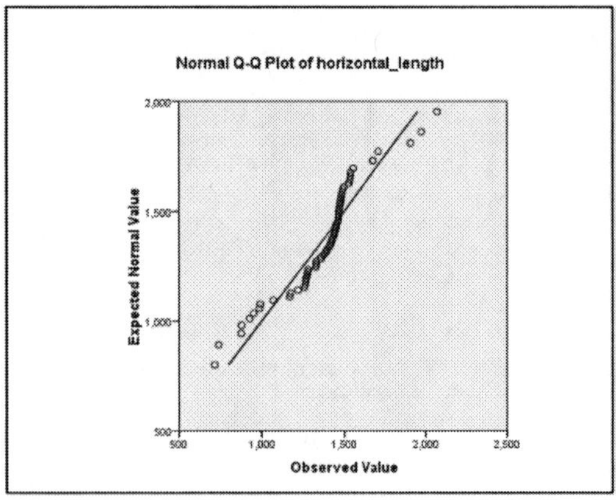
Normal Q-Q Plot of horizontal_length
Expected Normal Value
Observed Value

c.

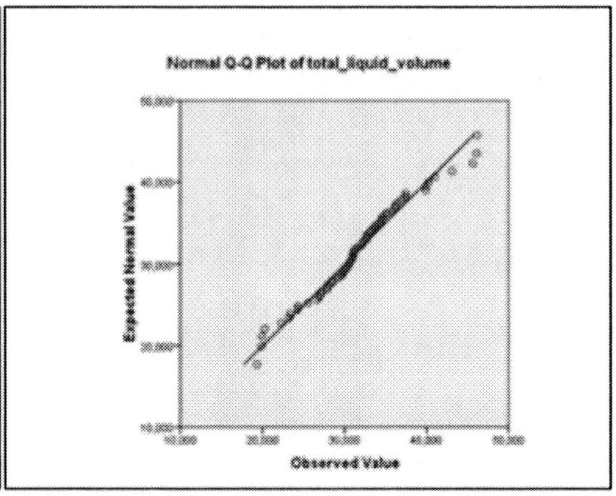
Normal Q-Q Plot of total_liquid_volume
Expected Normal Value
Observed Value

d.

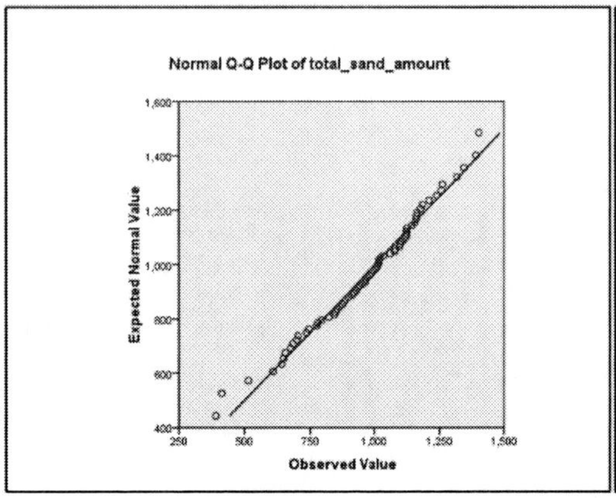
Normal Q-Q Plot of total_sand_amount
Expected Normal Value
Observed Value

e.

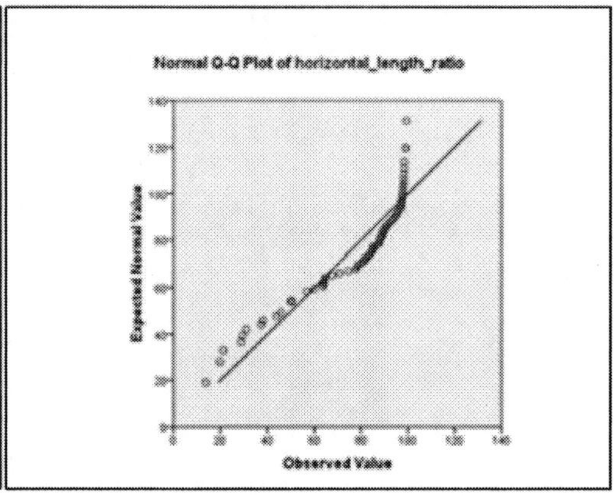
Normal Q-Q Plot of horizontal_length_ratio
Expected Normal Value
Observed Value

f.

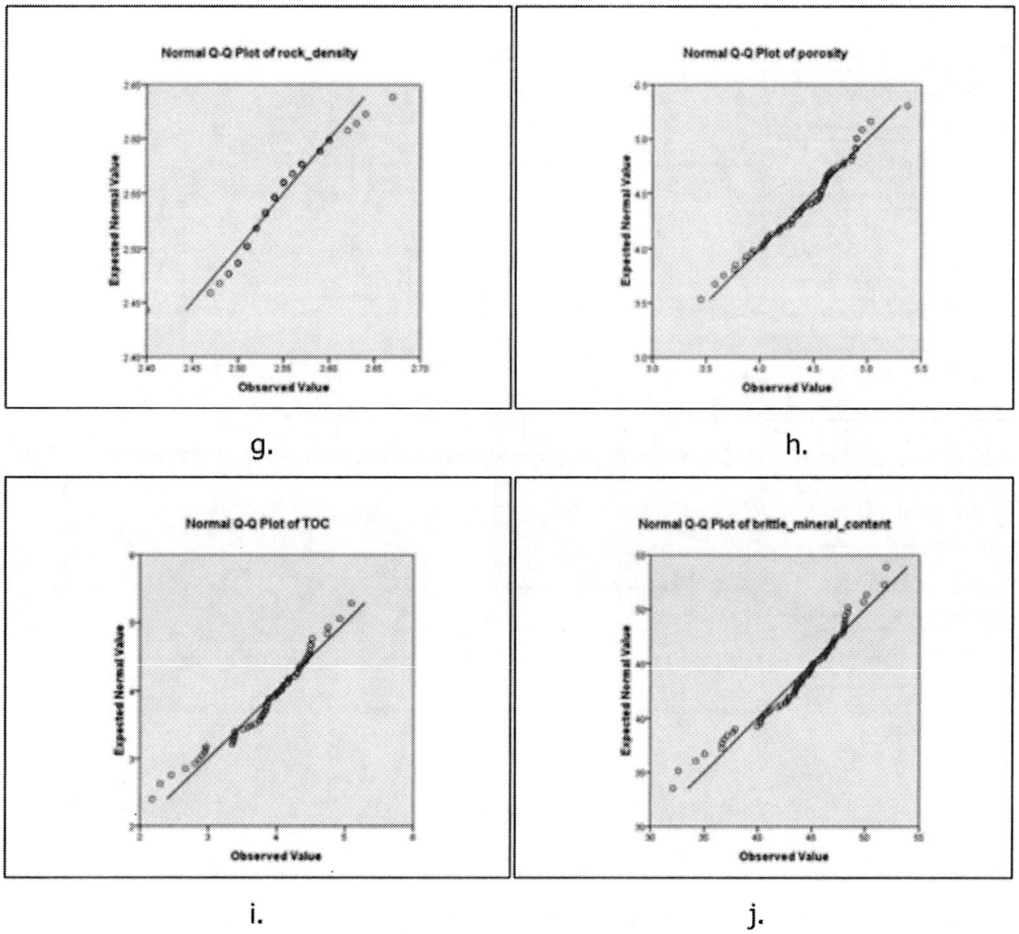

Fig.4 The Q–Q (quantile-quantile) plots for the 10 variables of the 86 wells: a. AOF potential; b. number of fracturing stages; c. horizontal length of well; d. total liquid volume; e. total sand amount; f. horizontal length ratio through stage 1 and stage 3 of Long1 member; g. rock density; h. porosity; i. TOC; j. brittle mineral content of rock

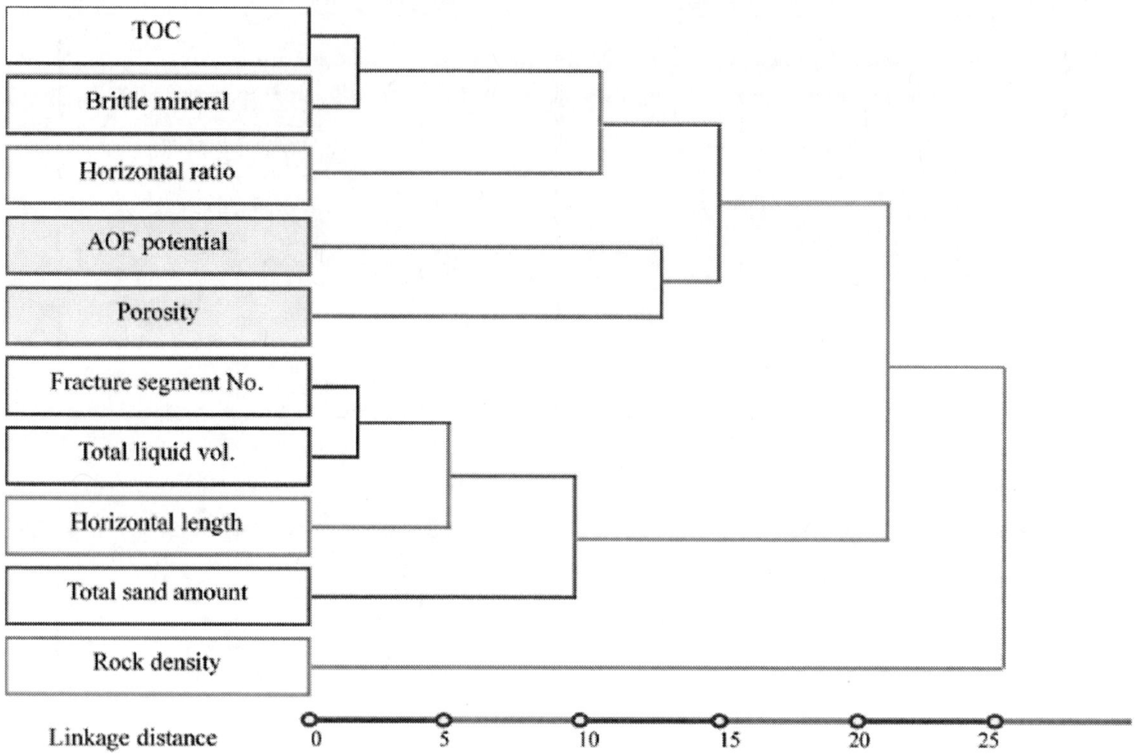

Fig.5 Hierarchical clustering analysis result of the sample data of 86 wells with linkage distance

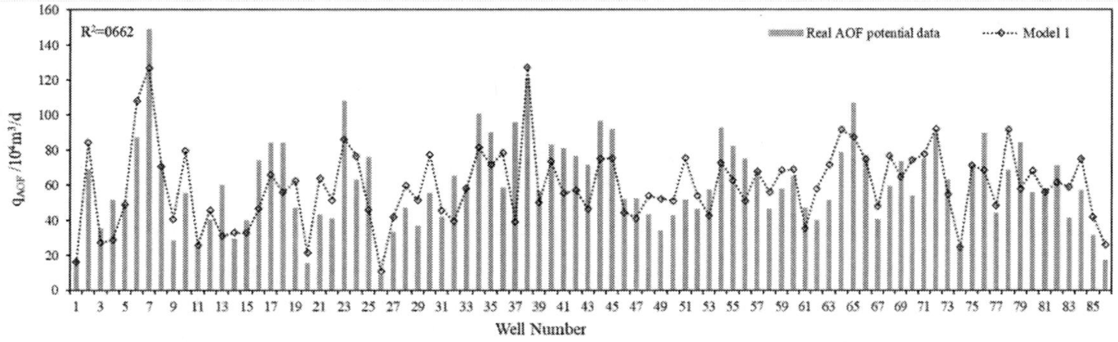

(1)

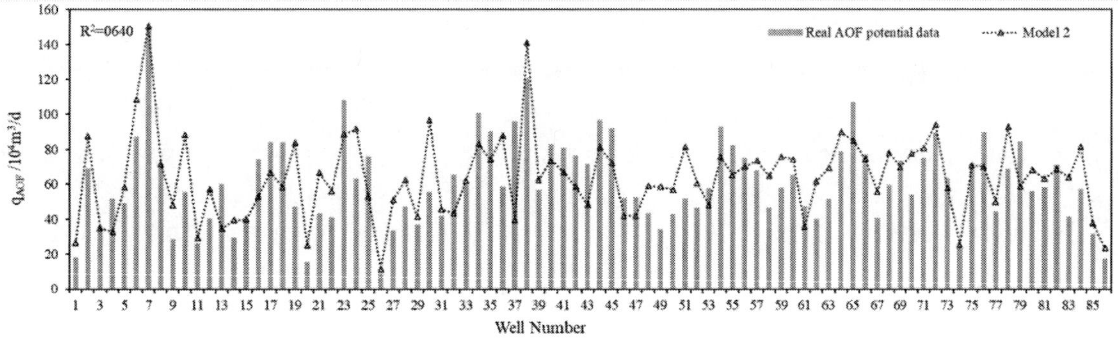

(2)

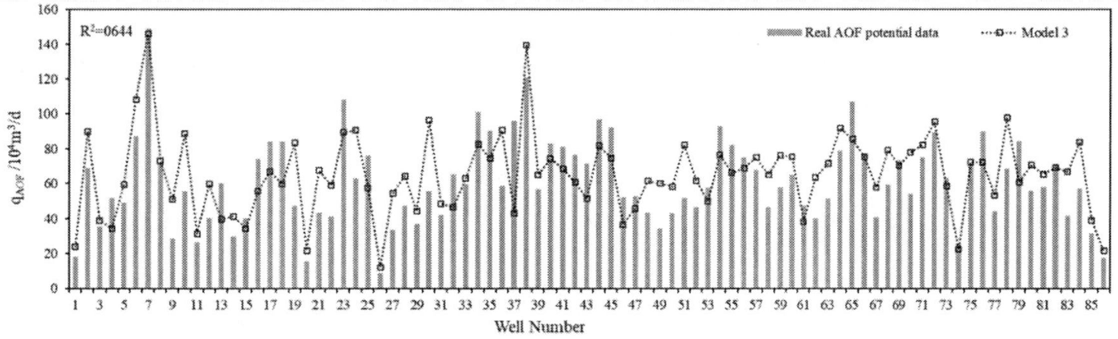

(3)

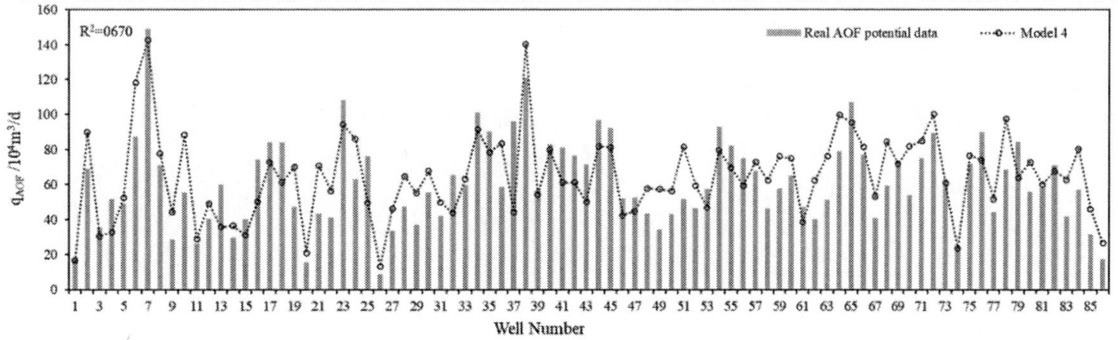

(4)

Fig.6 Comparison of actual and predicted AOF potential by the 4 models for the 86 wells with consideration of different key factors in Fuling Jiaoshiba shale gas field

Table 1 Correlation coefficients matrix of all variables in Jiaoshiba area

Factors	AOF potential	No. of fracturing stages	Horizontal length	Total liquid Volume	Total sand amount	Horizontal length ratio through stage 1 and stage 3	Rock density	Porosity	TOC	Brittle mineral content of rock
AOF potential	1	0.354	0.348	0.405	0.194	0.285	-0.373	0.246	0.433	0.438
No. of fracture stages	0.354	1	0.897	0.946	0.765	0.273	-0.203	-0.013	0.241	0.235
Horizontal length	0.348	0.897	1	0.863	0.633	0.331	-0.241	0.024	0.26	0.252
Total liquid Volume	0.405	0.946	0.863	1	0.736	0.302	-0.248	-0.01	0.274	0.282
Total sand amount	0.194	0.765	0.633	0.736	1	0.22	-0.002	-0.072	0.072	0.07
Ratio through stage 1 and stage 3	0.285	0.273	0.331	0.302	0.22	1	-0.492	0.287	0.568	0.578
Rock density	-0.373	-0.203	-0.241	-0.248	-0.002	-0.492	1	-0.289	-0.908	-0.917
Porosity	0.246	-0.013	0.024	-0.01	-0.072	0.287	-0.289	1	0.419	0.336
TOC	0.433	0.241	0.26	0.274	0.072	0.568	-0.908	0.419	1	0.982
Brittle mineral content of rock	0.438	0.235	0.252	0.282	0.07	0.578	-0.917	0.336	0.982	1

Table 2 Hierarchical clustering analysis result of factors in Jiaoshiba area

Parameter	Hierarchical clustering analysis result
AOF potential	1
Number of fracturing stages	2
Horizontal length	2
Total liquid volume	2
Total sand amount	2
Horizontal length ratio through stage 1 and stage 3	1
Rock density	3
Porosity	1
TOC	1
Brittle mineral content of rock	1

Table 3 Factors chosen in regression model according to hierarchical clustering analysis results

Regression model	Factors considered
1	a. Total liquid volume (X_1) b. Porosity (X_2) c. TOC (X_3)
2	a. Fracture segment No. (X_4) b. Porosity (X_2) c. TOC (X_3)
3	a. Fracture segment No. (X_4) b. Horizontal length ratio through stage 1 and stage 3 (X_5) c. TOC (X_3) d. Porosity (X_2)
4	a. Total liquid volume (X_1) b. Horizontal length ratio through stage 1 and stage 3 (X_5) c. Porosity (X_2) d. Brittle mineral content of rock (X_6)

601617 Solutions of C6F13- and C4F9-Based Polyfluoroacrylates in CO_2; Solubility, Viscosity, and Adsorption Onto Cement

Abstract:

High molecular weight poly(fluorinated acrylates) (PFAs) have drawn interest in the chemical and petroleum engineering fields due to their remarkable solubility in liquid and supercritical carbon dioxide (sCO_2) without the need for any co-solvent. This study does not include any examples of PFA based on monomers with the -C8F17 moiety because the ultimate hydrolysis degradation products of such polymers leads to the formation of PFOA, a compound deemed unsafe by the EPA. Instead, we have synthesized PFA based on fluoroacrylayte monomers containing the -C6F13 group (the functionality used in current commercial fluoroacrylate copolymers). These -C6F13-based polymers would degrade to form perfluorohexanoic acid (PFHxA) that, while significantly safer than PFOA, is also receiving more scrutiny from regulatory agencies in recent days. Therefore, we have also synthesized PFA based on monomers containing the -C4F9 functionality. The -C4F9-PFA ultimate degradation products include perfluorobutanoic acid (PFBA), which is significantly more environmentally benign than the -C6F13-based PFA. Therefore, one of the main objectives of this study was to determine if -C4F9-PFA retained the remarkable CO_2-solubility and CO_2-thickening capability as PFA based on the -C6F13 PFA and previously reported -C8F17-based PFA.

Many samples of PFA based on the -C6F13 and -C4F9-based fluoroacrylate monomers and varying ratios of initiator:monomer were bulk polymerized and then characterized for molecular weight and polydispersity. The -C4F9-PFA was shown to be nearly identical in CO_2-solubility as the -C6F16-PFA. Between 1-4 wt% PFA, the solubility of PFA in CO_2 was nearly independent of PFA molecular weight. Further, the cloud points loci for mixtures of CO_2-C4F9-PFA were nearly identical to the cloud point curves for the CO_2-C6F13-PFA at each temperature, with higher pressure required at higher temperature. This demonstrated that the -C4F9-PFA was just as CO2-soluble as the -C6F13-PFA. Further, both polymers were capable of thickening CO_2 in dilute concentration to a comparable degree (e.g. a 6-fold increase in viscosity with 4 wt% C4F9 PFA, 1.24×10^6 Da PFA in CO_2 at low shear rate).

PFA has previously been considered for use as a "CO_2 thickener" in enhanced oil recovery (EOR) operations. Unlike common chemical engineering applications where low viscosity is beneficial to reduce mass transfer resistance, low CO_2 viscosity is often problematic for EOR because it leads to viscous fingering (mobility control problems) and preferential loss of disproportionate amounts of CO_2 to high permeability "thief zones" that contain little recoverable oil (conformance control problems). However, in 2019 it was shown that PFA dissolved in CO_2 adsorbs so strongly onto sandstone or carbonate rock that the use of PFA in large-scale EOR projects to address mobility control would not be feasible, although near-wellbore conformance control applications (i.e. injecting the CO_2-PFA into isolated thief zones) may be viable.

In the current study, however, we are attempting to exploit the adsorption of PFA on solid surfaces for improved wellbore integrity. We are proposing to inject PFA-CO_2 solutions into small cracks within cement or between steel and cement (i.e. microannular cracks), thereby allowing a portion of the PFA to absorb and greatly reduce the crack permeability or completely seal the crack. Because PFA is a high molecular weight, amorphous, transparent, sticky, stretchy, oil-repellant, water-repellant polymer that is insoluble in water, oil or natural gas, it is a promising sealant for wellbore applications. Even though PFA is inherently expensive, this is a small volume application in which the cost of PFA would not exclude its consideration. Relative to other sealants (cement, resins, aqueous emulsions), PFA-CO_2 solutions are orders of magnitude less viscous, which would allow them deeper access into smaller cracks than other options. The PFA-CO_2 solutions would, however, be an inappropriate selection for the remediation of extremely large cracks or voids in cement.

In order to lay the groundwork for this proposed application, we have also determined the equilibrium adsorption of PFA onto Portland cement particles. This was accomplished by a direct measurement of the mass gain of a known quantity of sieved (22-30 mesh) Portland cement particles that was immersed in a PFA-CO2 fluid of specified original composition (e.g. 4 wt% PFA). The equilibrium concentration of PFA in the liquid CO_2 was determined via mass balance (original PFA placed in the cell – the amount of PFA found to adsorb onto the cement). This is, to the best of our knowledge, the first report of polymeric adsorption levels of solid surfaces immersed in CO_2-polymer solutions

611710 Competitive Sorption of CO2 in Shale Nanopores for Sequestration and Enhanced Gas Recovery Using Molecular Density Functional Theory

Abstract:

Enhanced gas recovery (EGR) by carbon dioxide (CO_2) injection to gas-bearing shale formations has been shown to be feasible and beneficial experimentally from pore scale to field scale. In addition to the improved recovery factor, CO_2 is trapped in the geological formation, which provides an opportunity to store CO_2. When supercritical CO_2 is injected, the preferential adsorption and dissolution of CO_2 in nanopores and organic matter enables the displacement of shale gas. This mechanism has been investigated by molecular simulation in graphite pores, but rarely for permeable pores by simulation or classical density functional theory (DFT). The compositional distribution of components from the organic matter source (kerogen) to nanoscale pores in the organic matter and to large fractures makes the estimation of fluid storage and the multicomponent compositional distributions in shale a challenging problem. We have introduced a density functional theory that accounts for molecular size and shape and kerogen maturity and swelling to predict the partitioning of components between the kerogen matrix and nano-pores in the kerogen [1].

In this work, a molecular thermodynamic model is presented that predicts both the adsorption of hydrocarbons in nanoscale pores and the dissolution of fluids in organic matter in equilibrium with the bulk kerogen phase. The swelling ratios of five kerogens of varying maturities and types in different solvents are well quantified using the new model. This kerogen matrix model is incorporated in interfacial statistical associating fluid theory (iSAFT) DFT to form a nanoporous kerogen composite model. The equilibrium partitioning of mixtures of fluids from the bulk phase to the nanoporous kerogen phase show enrichment of aromatic and cyclic molecules in kerogen, the preferential adsorption/absorption of high molecular weight molecules and carbon dioxide, and the significant amount of hydrocarbon storage both in pore space and kerogen matrix [1, 2].

Further, we apply molecular DFT to study the competitive sorption of CO_2 with shale gas in kerogen at various conditions. The DFT model is verified by grand canonical Monte Carlo (GCMC) simulation in graphite slit pores for pure and binary component systems at different temperatures, pressures, pore sizes and bulk compositions for methane/ethane/ CO_2. [2]. The model is utilized to predict the CO_2 adsorption selectivity in multicomponent systems which include two different shale gases. We show the selectivity of CO_2 decreases as temperature / pressure/ nanopore size/ shale gas average molecular weight increase. Subsequently, we model the CO_2/CH_4 adsorption in pores of more realistic conditions, in which we consider different levels of moisture in the pore and the gas dissolution in organic matrix. Interestingly, the presence of water inside a nanopore greatly reduces the CO2 adsorption but improves the CO_2 selectivity with respect to methane. The presence of organic matter kerogen also increases the CO_2 selectivity. In summary, a novel permeable pore DFT model enables prediction of partitioning and distribution of CO2 and hydrocarbons in shale related to enhanced gas recovery and CO_2 storage. [1, 2].

1. Jinlu Liu and Walter G. Chapman, "Thermodynamic Modeling of the Equilibrium Partitioning of Hydrocarbons in Nanoporous Kerogen Particles," Energy & Fuels, 33, 891-904 (2019).
2. Jinlu Liu, Shun Xi, and Walter G. Chapman, "Competitive Sorption of CO2 with Gas Mixture in Nanoporous Shale for Enhanced Gas Recovery from Density Functional Theory," Langmuir, 35, 8144-8158 (2019).

609261 Spatial and Temporal Evolution of Natural Gas and Natural Gas Liquid Feedstocks in Shale Basins

Abstract:

Technological advancements in horizontal drilling and hydraulic fracturing have led to increasing production of light hydrocarbons from shale resources, including natural gas and natural gas liquids (NGLs). However, some basins that have insufficient gas gathering and pipeline networks (e.g. the Permian Basin in Texas) may flare significant quantities of gas. For example, in the Permian Basin, about 4.6% of gross gas production was estimated to be flared between May 2018 and March 2019 (Zhang et al., 2020). To mitigate flaring issues, hydrocarbon resources can be locally transformed to chemical or energy products. The National Science Foundation's Center for Innovative and Strategic Transformation of Alkane Resources (CISTAR) is developing modular processes to transform natural gas and NGLs to gasoline-like transportation fuels.

Determining the scale of operation of such modular chemical manufacturing requires a comprehensive understanding of the temporal and spatial availability of the hydrocarbon feedstocks. Preliminary work was performed in the Eagle Ford oil and gas producing region in south central Texas. With over 20,000 wells, the Eagle Ford produced 5.4 billion cubic feet natural gas per day and 1.2 million barrels of condensate and oil per day in 2019 (RRC, 2020). From the northwest to the southeast portion in the Eagle Ford Shale, well production varies from oil, to gas with condensate (wet gas), and to dry gas, as shown in Figure 1 (Gherabati et al, 2016). From oil to wet gas to dry gas regions, propane to ethane ratio varies from 1:1 to 1:3 (mass basis). In addition to varying gas compositions, the magnitude of available feedstocks also vary across the region. Ethane available at individual gas plants was estimated by assuming that produced gas from wells is sent to their nearest gas plant. Among 44 gas processing plants in the Eagle Ford, 60% of ethane production are represented by 43 plants with 200,000 – 750,000 kg of ethane per day, with one plant processing at ~3,000,000 kg/day, accounting for 40% of ethane production, shown in Figure 2.

Similar work has been performed in the Permian Basin in Texas. The Permian Basin is one of the largest oil-producing regions in the world and accounts for over 30% of oil production in the United States. Recent study (Zhang et al, 2020) shows that methane emissions in the Permian Basin are the highest among all the U.S. oil/gas producing regions. This presentation will provide detailed characterizations of spatial and temporal variability in light alkane production, and losses due to flaring for a case study region in the Permian Basin.

Collectively, the analyses in the Eagle Ford and the Permian Basin will provide two case studies of spatial and temporal characteristics of feedstock availability for modular light alkane manufacturing.

References

Gherabati, S. A., Browning, J., Male, F., Ikonnikova, S. A., & McDaid, G. (2016). The impact of pressure and fluid property variation on well performance of liquid-rich Eagle Ford shale. *Journal of Natural Gas Science and Engineering*, *33*, 1056-1068.

Railroad Commission of Texas (RRC) – Eagle Ford Shale Information. (2020). Available at: https://www.rrc.state.tx.us/oil-gas/major-oil-and-gas-formations/eagle-ford-shale-information/

Zhang, Y., Gautam, R., Pandey, S., Omara, M., Maasakkers, J. D., Sadavarte, P., ... & Zhang, R.

(2020). Quantifying methane emissions from the largest oil-producing basin in the United States from space. *Science Advances*, *6*(17), eaaz5120.

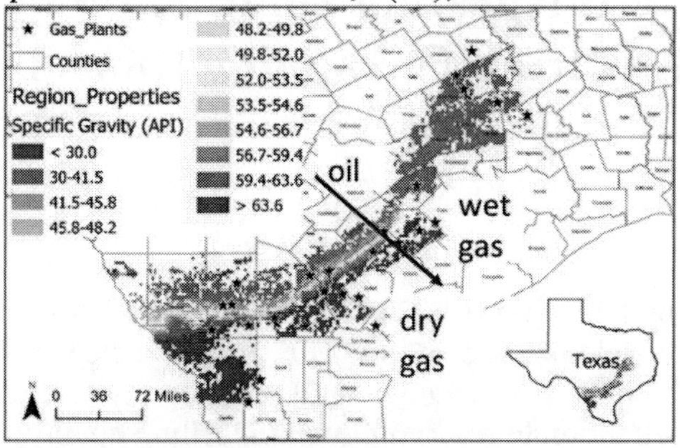

Figure 1. Spatial variabilities in well production in the Eagle Ford Shale.

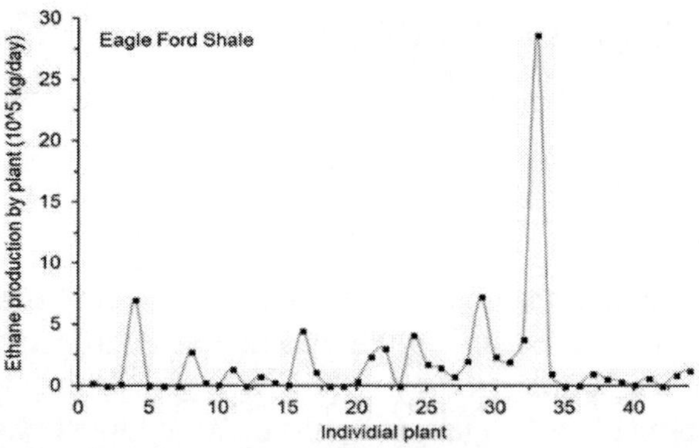

Figure 2. Ethane available at 44 gas processing plants in the Eagle Ford Shale

592706 A New Geology-Engineering Integration Solution for Shale Gas Exploration: A Case Study from Changning Block in China

Abstract:

Rich and quality shale gas reservoirs have been found in the Lower Silurian Longmaxi strata in the southern Sichuan Basin, which was identified as great development potential. However, shale gas reservoir is characterized by ultra-low permeability, which usually has no natural productivity and is called a typical "artificial gas reservoir". We must pay attention to reservoir quality, drilling quality and completion quality in order to obtain the maximum EUR and the highest oil recovery. To tackle this challenge, the integrated Geoengineering solution was deployed in Changning Shale Gas Block, which consists of geological modeling, geomechanics, fracturing simulation and gas reservoir simulation. Geoengineering integration technology is used on the Changning A platform for field applications. The results show that: (1) Carrying out integrated research on geological engineering can realize the organic combination of geology and engineering. The geological model provides technical support for engineering optimization. The engineering practice constantly corrects the geological model and improves the precision; (2) Through the integration of geological engineering, we can create "transparent" shale gas reservoir, which can improve the predictive accuracy of geological and engineering sweet spot mapping; (3) Hydraulic fractures can be fine characterization by using geological engineering integration technology, which is useful for quantitative description and analysis of key parameters such as geometrical shape, net pressure, proppant distribution and conductivity of hydraulic fractures; (4) the refined unstructured-grid (<10m) based 3D numerical simulation model was built for hydraulic fractured horizontal wells in shale gas reservoir, the history match for shale gas wells has been hugely improved: gas production rate was matched to more than 98%; the wellhead pressure was matched to more than 80%; the cumulative liquid production rate was matched to more than 90%. With the implementation of the Geoengineering study, the shale gas production in Changning area has been increasing rapidly in the past 2 years.

593227 New Technology of EOR of Ultra Deep and Complex Condensate Gas Reservoirs in Tarim Basin， China

Abstract:

A number of condensate gas reservoirs (both sandstone and carbonate) with depth of over 5000m have been developed in Tarim Basin, China. Those reservoirs are featured with thick reservoirs pay zones (over 100 meters) with ,strong reservoir heterogeneities in both rock properties and fluid properties. Gas channeling in the process of gas injection development and retrograde condensation in the process of depletion development have seriously affected the recovery of condensate. In order to enhance the condensate recovery in those condensate gas reservoirs, a series of new technologies have been developed during the last decade.

Laboratory experiment analysis and development practice discovered that three major mechanisms controlling the gas migration process in the reservoirs: displacement, gravity and diffusion for cyclic gas injection. It is proposed that gravity differentiation is the main factor affecting the development effect of cyclic gas injection, and the fluid distribution law of underground condensate gas reservoir is re-recognized. For the sandstone condensate gas reservoir developed with large well spacing (over 1000 meters), the displacement and diffusion will be replaced by gravity differentiation gradually in the process of cyclic gas injection, and the fluid will be redistributed vertically according to its density, which makes different phase behaviors. Similarly, the reservoir space of depleted carbonate condensate gas reservoir is mainly dissolution cave and fracture. Under the action of gravity, the reverse condensate oil usually accumulates at the bottom of fracture cave system. On this basis of this understanding, the technology of gravity assisted three-dimensional displacement gas injection for sandstone condensate gas reservoirs and horizontal well placement technology for carbonate condensate gas reservoirs were developed to optimize the well placement. The technologies have been successfully applied to the developed condensate gas field in Tarim Basin, and the recovery rate of condensate oil has been increased by over 19%.

Production capacity evaluation method of condensate gas reservoir considering the influence of stress sensitivity and condensate oil

DU Chunxiao, LIU Xiaohong, ZHANG Fengyi, XU Jing, GENG Zhigang

(Tianjin Branch of CNOOC Ltd., Tianjin 300459)

Abstract: Accurate evaluation of production capacity is an important guarantee for efficient development of condensate gas reservoir. For the deep condensate gas reservoir with low permeability and low pressure difference between the initial formation pressure and the dew point, the influence of reservoir stress sensitivity and condensate oil on the production capacity of the gas reservoir cannot be ignored. Nevertheless, at present, the widely used method to calculate the gas reservoir production capacity is based on single-phase gas, without considering the influence of condensate oil and stress sensitivity on reservoir permeability. On the basis of researching a large number of literatures, combined with laboratory experiments and numerical simulation, the production capacity evaluation model of condensate gas reservoir under the comprehensive influence of stress sensitivity and condensate oil is derived by using the seepage flow mechanics theory, and the research results are applied to the Bohai Z gas reservoir. The results show that the production capacity evaluation model can more accurately reflect the production status of this kind of gas reservoir in different development stages, and can be widely used in the productivity evaluation of similar condensate gas reservoir.

Key words: deep condensate gas reservoir; stress sensitivity; condensate oil; evaluation of production capacity; low permeability

Gas reservoir production capacity evaluation is an important research content of gas reservoir engineering research, which provides key guidance for the design of gas reservoir development plan. For condensate gas reservoir, the fluid flow law and phase change are more complex, and the reverse condensation phenomenon has a serious impact on gas well productivity. According to theoretical and experimental studies, many scholars have confirmed that low permeability reservoirs have strong stress sensitivity. At present, the widely used method to calculate the gas reservoir production capacity is based on single-phase gas, without considering the influence of condensate oil and stress sensitivity on reservoir permeability. Many scholars have carried out research on the influence of reservoir stress sensitivity or retrograde condensation on gas well productivity, but lack of comprehensive consideration of both. Therefore, a new productivity evaluation method of condensate gas reservoir considering stress sensitivity and retrograde condensation is derived by combining laboratory experiments and numerical simulation methods, and is applied to BZ gas reservoir in Bohai oilfield.

1. Influence of stress sensitivity on reservoir permeability

The relationship between rock permeability and effective confining pressure has been studied deeply by many scholars using a large number of experiments. The relationship between permeability and effective confining pressure can be divided into three types: power-law relationship, exponential relationship and polynomial relationship.

BZ gas reservoir in Bohai oilfield is a low permeability gas reservoir. The core stress sensitivity test results of BZ-1 well are shown in Table 1 and figure 1. It can be seen that the dimensionless permeability of core decreases with the increase of effective stress, and BZ gas reservoir has strong stress sensitivity.

Table 1 Data of core stress sensitivity test

Net pressure	Sample 1	Sample 2	Sample 3	Sample 4
MPa	K_n/K_i	K_n/K_i	K_n/K_i	K_n/K_i
3.5	0.748	0.792	0.774	0.781
5.0	0.637	0.670	0.642	0.643
7.0	0.476	0.521	0.503	0.524
9.0	0.391	0.394	0.402	0.414
11.0	0.313	0.329	0.308	0.369
15.0	0.208	0.238	0.239	0.248
20.0	0.152	0.170	0.180	0.170
25.0	0.117	0.121	0.141	0.131
30.0	0.095	0.102	0.122	0.112
35.0	0.081	0.087	0.097	0.083

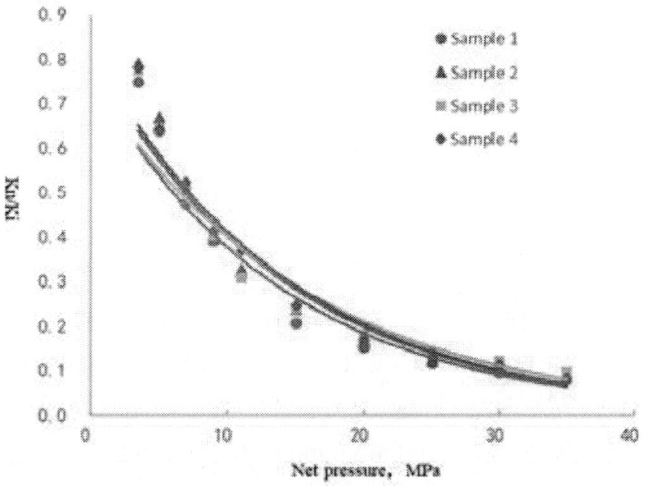

Fig 1 Stress sensitivity curves of the cores

By fitting the relationship between rock permeability and effective confining pressure, it is found that there is a good exponential relationship between them（Table 2）. It can be expressed as：

$$k_n = \mathrm{N}k_i e^{-\alpha(p_i-p)} \quad (1)$$

Table 2 Stress sensitivity index fitting of core permeability

Core number	Exponential fitting expression	R^2
1	$y = 0.7721e^{-0.072x}$	0.9553

2	$y = 0.8186e^{-0.071x}$	0.9595
3	$y = 0.7609e^{-0.064x}$	0.9489
4	$y = 0.8331e^{-0.07x}$	0.9727

According to the principle of seepage mechanics, the square distribution expression of stable seepage pressure of gas reservoir is as follows.

$$p^2 = p_e{}^2 - \frac{p_e{}^2 - p_{wf}{}^2}{\ln\left(\frac{r_e}{r_w}\right)} ln\frac{r_e}{r} \tag{2}$$

Formula (1) is deformed.

$$k_n = Nk_i e^{-\alpha\left\{p_i - \left[p_e{}^2 - \frac{p_e{}^2 - p_{wf}{}^2}{ln\left(\frac{r_e}{r_w}\right)} ln\frac{r_e}{r}\right]^{\frac{1}{2}}\right\}} \tag{3}$$

2 Influence of condensate on gas phase relative permeability

The influence of reverse condensate oil is studied by numerical simulation method. By using single well numerical simulation, the variation law of condensate oil saturation at different distances from well during gas field development is obtained.

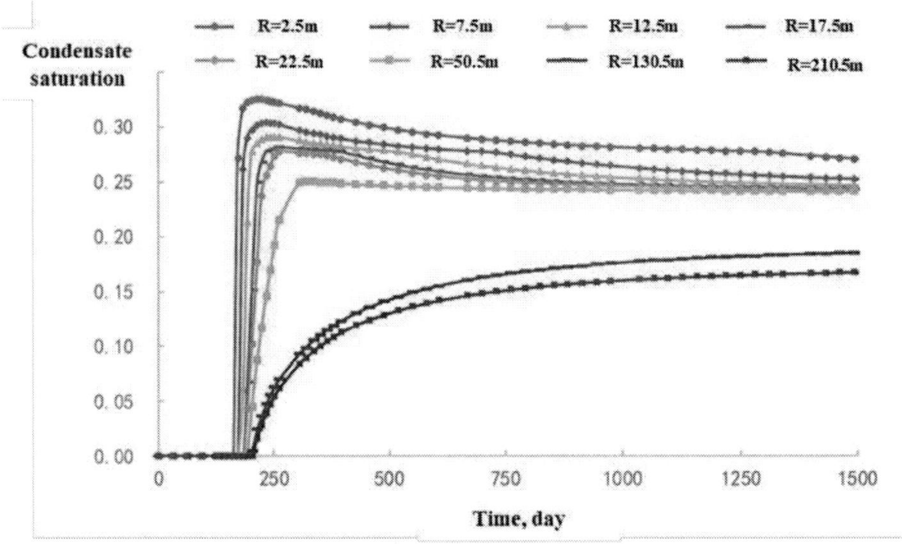

Fig.3 variation law of condensate oil saturation

It can be seen from Figure 3 that the condensate saturation decreases with the increase of radial distance, and tends to be stable after increasing with the extension of production time. The results show that the condensate saturation changes with time in the near well zone, and the distribution function of condensate saturation with radial distance $s_o(\mathrm{r})$ can be obtained at any time.

According to the relative permeability curve, the regression relationship between gas relative permeability and oil saturation can be obtained. The relative permeability curve of BZ gas reservoir in Bohai oilfield is shown in Fig. 4, the regression expression is:

$$k_{rg}(\mathrm{r}) = 4.7367s_o(r)^3 - 2.0497s_o(r)^2 - 2.1398s_o(r) + 1.0015 \tag{4}$$

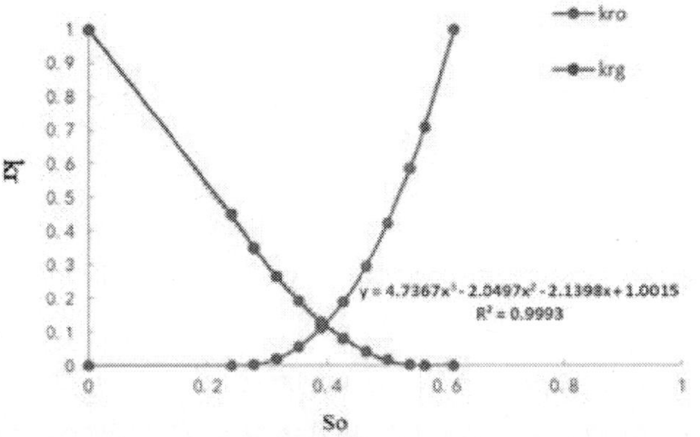

Fig.4 Relative permeability curve of Z gas reservoir

3 Improved gas well productivity evaluation model

The expression of effective permeability is:

$$k_g = k_{\mathrm{n}} \cdot k_{rg} \tag{5}$$

Replace formula (3) and formula (4) into formula (5):

$$k_g = Nk_i[4.7367s_o(r)^3 - 2.0497s_o(r)^2 - 2.1398s_o(r) + 1.0015]e^{-\alpha\{p_i - [p_e{}^2 - \frac{p_e{}^2 - p_{wf}{}^2}{\ln\left(\frac{r_e}{r_w}\right)} ln\frac{r_e}{r}]^{\frac{1}{2}}\}} \tag{6}$$

The condensate saturation beyond a certain distance from the well is lower than the critical flow saturation, the distance is defined as the effective radial distance and expressed as r_o.

The whole stratum can be divided into two regions. The effective permeability can be expressed as:

$$k_g(\mathrm{r}, p_{wf}) = \begin{cases} Nk_a[4.7367s_o(r)^3 - 2.0497s_o(r)^2 - 2.1398s_o(r) + 1.0015]e^{-\alpha\{p_i - [p_e{}^2 - \frac{p_e{}^2 - p_{wf}{}^2}{\ln\left(\frac{r_e}{r_w}\right)} ln\frac{r_e}{r}]^{\frac{1}{2}}\}}, & r_w < r < r_o \\ Nk_a e^{-\alpha\{p_i - [p_e{}^2 - \frac{p_e{}^2 - p_{wf}{}^2}{\ln\left(\frac{r_e}{r_w}\right)} ln\frac{r_e}{r}]^{\frac{1}{2}}\}} & , r_o \le r \le r_e \end{cases} \tag{7}$$

① $r_w < r < r_o$, average permeability can be expressed as:

$$k_1(p_{wf}) = \frac{\int_{r_w}^{r_o} k_g(r, p_{wf}) \cdot 2\pi r dr}{\int_{r_w}^{r_o} 2\pi r dr} \tag{8}$$

② $r_o \le r \le r_e$, average permeability can be expressed as:

$$k_2(p_{wf}) = \frac{\int_{r_o}^{r_e} k_g(r, p_{wf}) \cdot 2\pi r dr}{\int_{r_o}^{r_e} 2\pi r dr} \tag{9}$$

According to the principle of seepage mechanics, the average gas permeability of the formation can be obtained:

$$k(p_{wf}) = \frac{k_1(p_{wf}) \cdot k_2(p_{wf}) \cdot \ln\left(\frac{r_e}{r_w}\right)}{k_2(p_{wf}) \cdot \ln\left(\frac{r_o}{r_w}\right) + k_1(p_{wf}) \cdot \ln\left(\frac{r_e}{r_o}\right)} \tag{10}$$

Based on the above research, the productivity evaluation formula of gas well considering stress sensitivity and condensate oil is obtained.

$$p_e^2 - p_{wf}^2 = \frac{1.273\times10^{-3}T\mu Z}{k(p_{wf})h}\ln\left(\frac{r_e}{r_w}\right) q_{sc} + \frac{2.162\times10^{-10}\gamma_g ZT}{r_w h^2 k(p_{wf})^{1.5}} q_{sc}^2 \quad (11)$$

4 Application

BZ is a rare condensate gas reservoir with complex reservoir and fluid characteristics. The main performance is low pressure difference between the initial formation pressure and the dew point（0.21-7.13MPa）, the maximum amount of condensate is 40%, the reservoir permeability is 1.00~6.80mD， strong stress sensitivity.

The evaluation equation is applied to BZ gas field. The IPR curves of three cases are calculated, under the comprehensive influence of condensate oil and stress sensitivity, only considering the influence of condensate oil or stress sensitivity. The results are compared with those of the traditional binomial formula.

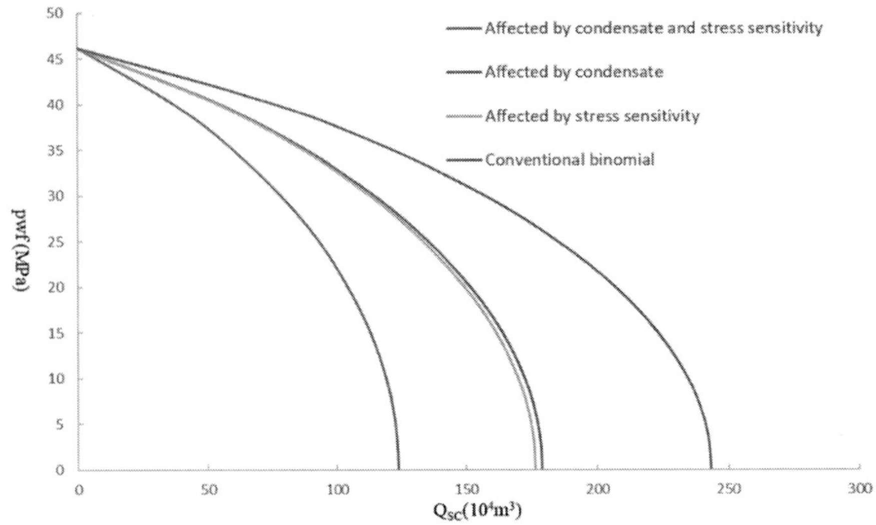

It can be seen that reservoir stress sensitivity and retrograde condensation have great influence on gas reservoir productivity. Under the combined effect of both, the productivity of gas reservoir is reduced by 50%. The IPR curves of gas wells under different formation pressures under the combined influence of stress sensitivity and retrograde condensation are further obtained.

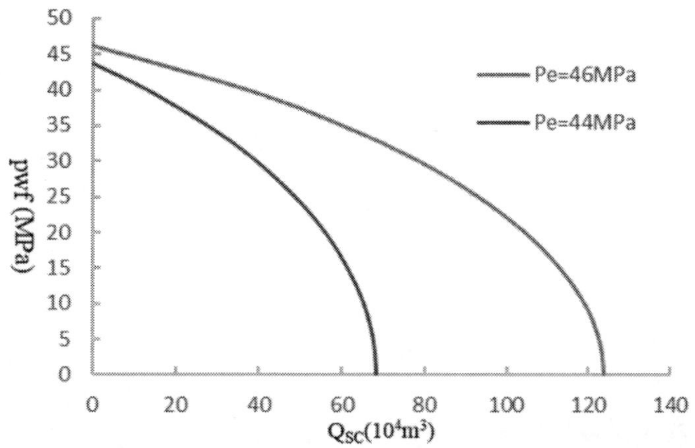

It can be seen that the formation pressure decreases, stress sensitivity and retrograde condensation increase, the productivity of gas wells also decreases obviously. The formation pressure drops by 2MPa, and the open flow rate of gas wells decreases by 57%. Therefore, in order to ensure the gas well productivity, it is necessary to adopt the development mode of gas injection and pressure maintaining, and the gas well production allocation should not be too high, in order to slow down the adverse impact of pressure drop.

6 Conclusion

(1) Stress sensitivity and condensate oil have great influence on the productivity of low permeability condensate gas reservoirs with low pressure difference between the initial formation pressure and the dew point. Most studies lack of comprehensive consideration of the two factors.

(2) Combined with laboratory experiment and numerical simulation, the formation permeability considering the effect of reservoir stress sensitivity and retrograde condensation is derived by using the principle of seepage mechanics, and then the evaluation method which can reflect the real productivity of gas reservoir is obtained.

(3) The research results show the necessity of gas injection to maintain reservoir pressure, and put forward suggestions for reasonable production of gas wells.

Symbol description

k_i-Original permeability, $10^{-3}\mu m^2$

k_n-Absolute permeability, $10^{-3}\mu m^2$

k_g-Effective permeability, $10^{-3}\mu m^2$

k_{rg}-Gas phase relative permeability, $10^{-3}\mu m^2$

p_i-Original formation pressure, MPa

p-Formation pressure, MPa

N-Stress sensitivity coefficient

α-Stress sensitivity index

p_e-Formation pressure at supply radius, MPa

p_{wf}-Bottom hole pressure, MPa

r_e-Supply radius, m

r_w-Wellbore radius, m

s_o-Condensate saturation near wellbore

r-Radial distance from well, m

T-Formation temperature, K

μ-Gas viscosity, mPa·s

Z-Gas deviation factor

h-Thickness of gas reservoir, m

β-Characteristic parameters of pore structure affecting turbulence and inertia resistance

γ_g-Relative density of natural gas

q_{sc}-Gas well production under standard conditions, m^3/d

References

[1] Fussel D.D. Single well Performance for gas condensate reservoir[J]. JPT， 1973， 25(12)： 860-876.

[2] Jones J.R. and Raghavan R. Interpretation of Pressure buildup reservoirs in gas condensate well[J]. SPEFE， 1989.

[3] FAN Xueping, XU Xiangrong. Experimental and mechanism research about permeability damage with the change of stress[J]. PETROLEUM EXPLORATION AND DEVELOPMENT, 2002

[4] XUE Zhan, LIU Pengcheng. A New Method of Correction of Gas Condensate Well Productivity Considering Capillary Number[J]. Journal of Chongqing University of science and Technology (NATURAL SCIENCE EDITION), 2014.

[5] LUO Ruilan, CHENG Linsong. A new method of determining relationship between permeability and effective overburden pressure for low-permeability reservoirs[J]. Journal of China University of Petroleum, 2007.

Numerical simulation study of a complex naturally-fractured gas reservoir with extra-high condensate in Bohai Oilfield

GAO Zhennan, SU Yanchun, GENG Zhigang, LIU Weilin, HUANG Baogang

(CNOOC China Limited Tianjin Branch, Tianjin 300459, China)

Abstract:

The BZ condensate gas field is the largest natural gas field in the history of the Bohai bay basin, with proven reserves of more than 100 billion cubic meters of natural gas and 100 million cubic meters of condensate oil. This gas field has unique features, such as deep burial, fracture development, strong stress sensitivity, small pressure difference between formation pressure and dew point pressure, high condensate content, etc. There is no similar condensate-gas field development cases in the world and proper reservoir numerical simulation study is crucial for efficiently developing such a reservoir.

This paper develops a numerical simulation flowchart for this kind of complex naturally fractured condensate gas reservoirs. First, based on logging, core and well testing data, the reservoir of BZ gas field is divided into three different types of deposits in the vertical direction: glutenite, semi-weathered zone and inner buried hill. The fluid flow among the three different types of deposits is calculated based on the equivalent fluid flow conduction between adjacent grids. Secondly, the correlation of permeability changes with stresses is established based on experimental data and implemented in the reservoir simulation based on the imbedded programming module in the simulator. Thirdly, through PVTi numerical simulation, EOS are tuned based on 7 pseudo-components. Finally, a dual-porosity compositional model is established, and the numerical simulation study is carried out to optimize the development plan. The developed flowchart will be a very useful tool for modeling naturally-fractured condensate gas reservoirs.

The simulation results indicated that: 1) considering the strong stress sensitivity, gas injection can effectively increase the permeability of the reservoir and control the condensate precipitation; 2) considering the structural characteristics of the reservoir, gas injection in the high position can form top-gas drive and expand the swept volume; 3) considering the influence of fracture, injection- production line perpendicular to fracture trend can effectively delay the breakthrough of injected gas and expand the injection capacity; 4) considering the characteristics of condensate, keeping formation pressure higher than dew point pressure can minimize the loss of condensate oil; 5) considering the mechanism of gas injection displacement, slug gas injection can form pressure pulse, expand the sweep volume and improve the recovery of condensate oil. Those findings can also be used to guide the development of other gas condensate reservoirs.

Key words: buried-hill fracture; condensate gas reservoir; stress sensitive characterization; reservoir numerical simulation; cyclic gas injection development

0 Introduction

BZ condensate gas field is the largest natural gas field in Bohai Sea, with proven natural gas reserves of over 100 billion cubic meters and proven condensate reserves of more than 100 million cubic meters. The gas field has the following geological characteristics: buried deep(-4400~-5500m), complex seepage (fracture development), stress sensitivity (permeability loss 89.2%), and the fluid characteristics are as follows: low pressure difference between formation pressure and dew point pressure (1.07MPa), ultra-high condensate (711g / m3), large amount of reverse condensate (32.3%). The efficient development of BZ condensate gas field is a world problem! It is found that there is no similar gas field development case at home and abroad, so it is impossible to accurately evaluate the development effect of this gas field. Therefore, it is urgent to carry out reservoir numerical simulation research to provide support for early development plan and subsequent efficient development[1-3].

1 Research method

According to the reservoir classification, the complex seepage system gas reservoir model is established. Based on the stress sensitive experiment, the variation law of seepage capacity is regressed. Based on the matching of fluid phase, the dual medium component model is established. And the development mode and gas injection development strategy are studied through reservoir numerical simulation.

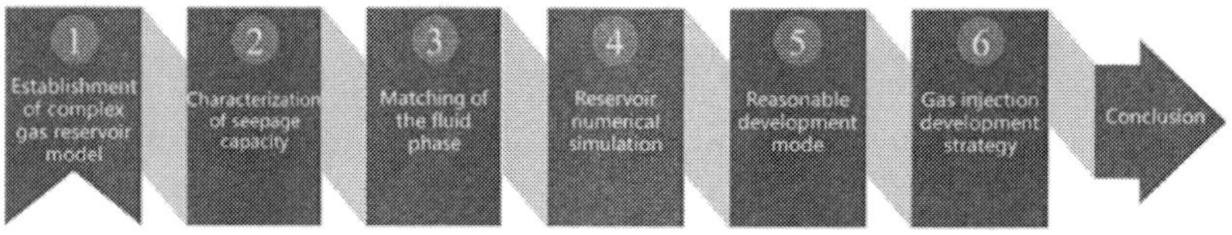

Fig 1. Technology Roadmap

2 Gas reservoir model

2.1 Reservoir zonation

Combined with the analysis of logging, core and testing data, BZ gas field has good reservoir connectivity, and it can be divided into sandy conglomerate zone, weathered zone and buried hill inner zone from top to bottom[4].

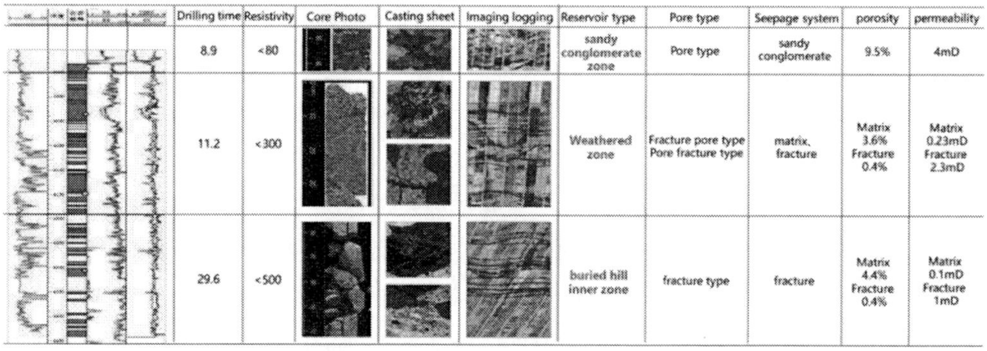

Fig 2. Vertical development characteristics of BZ condensate gas field reservoir

2.2 Establishment of complex gas reservoir model

Establish a structural model based on the understanding of reservoir zoning; and using lithofacies modeling and discrete fracture network modeling to generate matrix and fracture porosity, permeability and interporosity flow coefficient. By zeroing the matrix conductivity inside the buried hill, the coexistence of three types of complex seepage models in the BZ gas field can be realized.

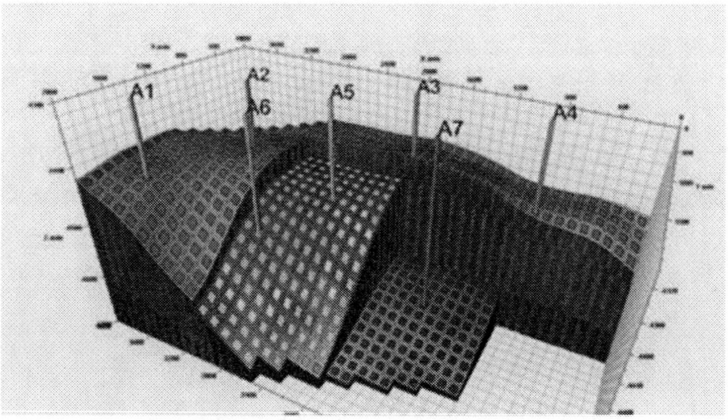

Fig 3. Gas reservoir model of complex seepage system

3 Stress sensitive characterization

3.1 Stress sensitivity analysis

According to the strong stress sensitivity of BZ gas field, the relationship between core permeability residual coefficient and net stress is fitted, and there is a good power function relationship between them; with the increase of net stress (reservoir pressure decreases), the rate of permeability loss gradually decreases[5-11]. In early stage, the net stress increased to 10MPa, Kr reduced by 60%; in mid stage, the net stress increased to 20MPa, Kr reduced by 20%; when reach the abandonment pressure(the net stress increased to 30MPa), the remaining permeability is about 10%.

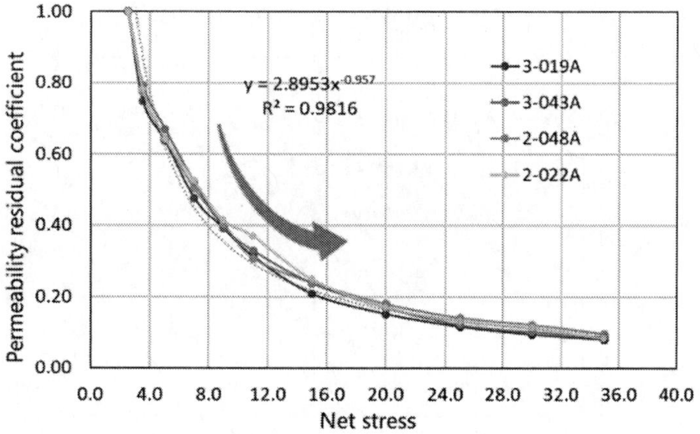

Fig 4. Permeability versus stress curve

$$K_r = 2.8953 \ (P - P_0)^{-0.957}$$

K_r: Ratio of permeability to initial permeability, zero dimension;

P: Reservoir pressure at any time, MPa;

P_0: Initial formation pressure, MPa。

3.2 Stress sensitive characterization

According to the regression relationship, the characterization code of reservoir physical properties with time is compiled, and the sensitivity analysis is carried out by reservoir numerical simulation.

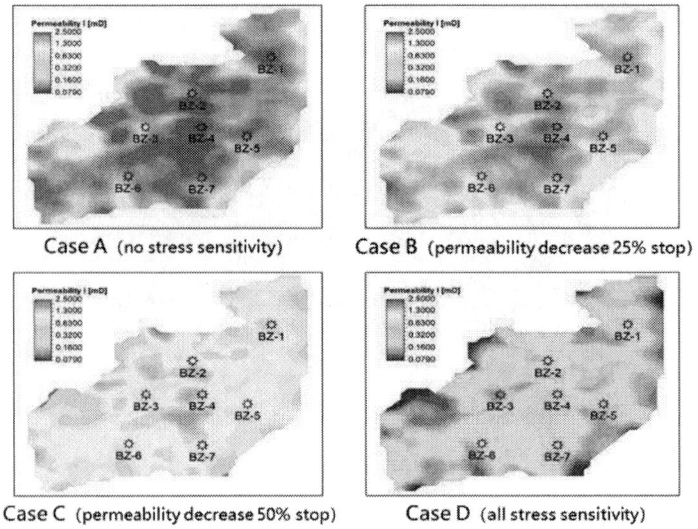

Fig 5. Comparison of permeability of different stress sensitive cases

● 4 Characterization of fluid characteristics

4.1 Fluid phase analysis

The BZ gas field is a condensate gas reservoir with high saturation and ultra-high condensate content through the study of fluid phase behavior characteristics, and the maximum condensate oil saturation is as high as 41%. When the pressure drops about 15MPa, the maximum saturation can be reached. Condensate oil precipitation seriously pollutes the reservoir, increases gas phase seepage resistance, and reduces gas well productivity and recovery[12-13].

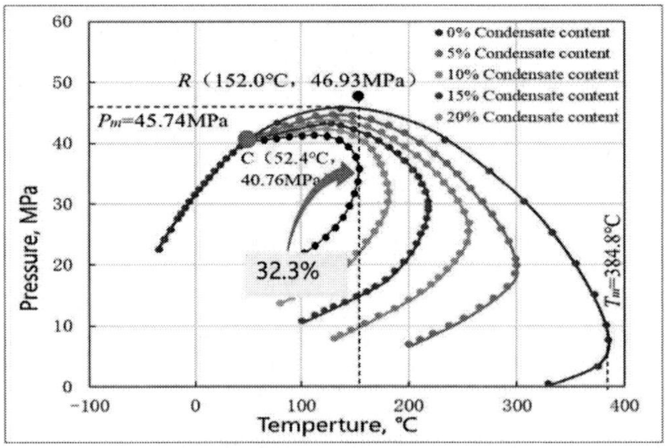

Fig 6. Phase diagram of fluid temperature and pressure in BZ gas field

4.2 Fluid phase matching

Considering the accuracy and speed of numerical simulation, 4 cases of component splitting combinations were set according to the principle of similarity combination. Combined with the calculation time and results of Numerical simulation of component model, the 7-component splitting plan is determined as the best plan.

11 components: C1、N2、C2、CO2、C3、C4、C5、 C6、C7-C10、C15+C22、C23+

7 components: C1+N2、C2+CO2、C3-C4、C5-C6、C7-C10、C15+C22、C23+

5 components: N2、CO2、C1、C2- C6、C7+

3 components: C1+N2、C2-C6+CO2、C7+

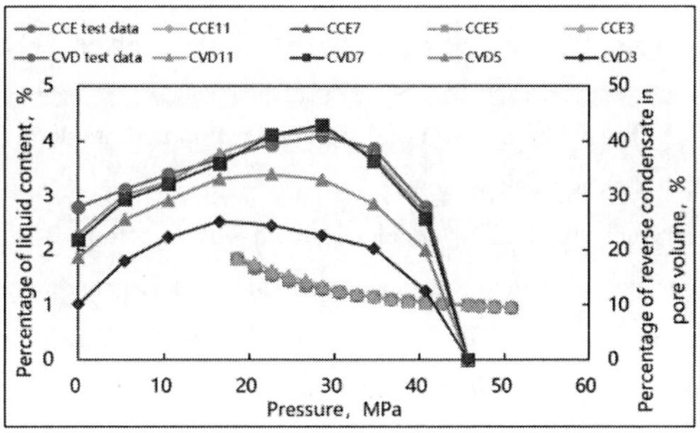

Fig 7. Comparison of experimental fitting accuracy

5 Study on gas reservoir numerical simulation

5.1 Optimization of development mode

As oil and gas fluids are produced, the formation pressure decreases, the physical properties become worse, the seepage ability decreases, and the recovery of gas field decreases. In depletion exploitation mode, the formation pressure drops rapidly, the condensate oil quickly reaches the maximum condensate velocity, and lost in the formation, so the condensate oil recovery rate is low. In gas injection development mode, the formation pressure is maintained through gas injection, and the retrograde condensate rate is controlled, and the condensate oil recovery is high. Considering reservoir stress sensitivity and retrograde condensate characteristics, from the perspective of improving oil and gas recovery and reservoir protection, it is recommended that the BZ condensate gas field development adopts first cycle gas injection and then depletion development[14-17].

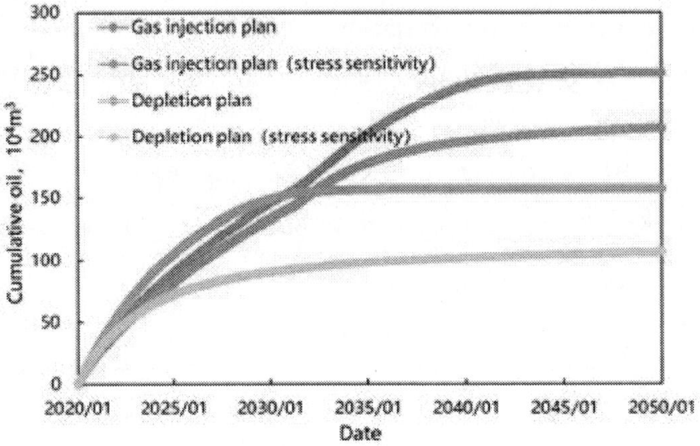

Fig 8. Comparison of different development mode

5.2 Cycle gas injection development strategy

Condensate gas reservoirs are usually developed with gas injection. The main purpose is to supplement formation energy, reduce condensate oil precipitation, and improve condensate oil recovery. When evaluating the effect of development, gas injection sweep volume and formation pressure maintenance level are two very important factors.

5.2.1 Optimization of injection and production

During gas injection development, the gas injection well position is directly related to the dry gas sweep volume. The structural and the fracture trend are two important factors.

A. Structural factor

The sandy conglomerate zone in structural high area of the BZ gas field has good physical properties. Priority is given to production in the early stage, and gas injection at the low area. In later stage, the condensate oil is enriched in the structural low area by gravity, and the top gas injection forms secondary gas and top drive. Vertical displacement, effectively expanding the gas sweep volume.

Tab 1. Development effect of structural factor

Plan	Cumulative gas 10^8m^3	Gas recovery %	Cumulative oil 10^4m^3	Oil recovery %
Inject in low area 10 years	36	53	181.3	26.6
Inject in low area 5 years,And in high area 5 years	36.6	53.3	192.4	28.3
Inject in high area 10 years	36.1	53.1	189.2	27.8

B. Fracture trend factor

When the injection-production line is parallel to the fracture trend, it is easy to form gas channeling, which reduce oil and gas recovery. When the injection-production line is vertical to the fracture trend, it can maximize the sweep volume, and effectively maintain formation energy and the seepage ability, and inhibit condensate oil to separate out.

Tab 2. Development effect of fracture trend factor

Plan	Cumulative gas 10^8m^3	Gas recovery %	Cumulative oil 10^4m^3	Oil recovery %
The injection-production line isparallel to the fracture	35.1	51.1	168.6	25.1
The injection-production line isperpendicular to the fracture	36.6	53.3	192.4	28.3
The injection-production line isoblique to the fracture	36	53	186.1	27.3

5.2.2 Reasonable pressure maintenance level

The most important thing in the development of condensate gas reservoirs is control the formation pressure to slow down the precipitation of condensate oil. The timing of gas injection and the rate of reinjection are two key factors.

A. Timing of gas injection

The timing of gas injection determines the degree of condensate oil loss, seepage capacity, and condensate precipitation rate during gas injection development. Since the difference between the formation pressure and the dew point pressure of the BZ gas field is very small, gas injection at the beginning of production can effectively delay the retrograde condensation time and reduce pollution, which is the optimal solution.

Tab 3. Development effect of timing of gas injection

Plan	Cumulative gas 10^8m^3	Gas recovery %	Cumulative oil 10^4m^3	Oil recovery %
Formation pressure 30MPa (4 years later)	26.4	48	127	23.4
Formation pressure 40MPa (2 years later)	26.3	47.9	132	24.4
Formation pressure 50MPa (0years later)	26.3	47.9	135.9	25.1

B. Re-injection rate

The higher the gas re-injection rate, the more stable the formation pressure remains. Before the formation pressure is higher than the dew point pressure, the 100% re-injection rate can always maintain the formation pressure higher than the dew point pressure, minimize the loss of condensate oil, and increase condensate oil recovery rate.

Tab 4. Development effect of Re-injection rate

Plan	Cumulative gas 10^8m^3	Gas recovery %	Cumulative oil 10^4m^3	Oil recovery %
100% re-injection rate	36.2	52.7	198.5	29.2
80% re-injection rate	36.4	53	196.1	28.8
50% re-injection rate	36.6	53.3	192.4	28.3

5.2.3 Optimization of gas injection mode

Considering the mechanism of gas injection and the demand for gas supply in the Beijing-Tianjin-Hebei region in winter, a comparative study of 3 plans: cyclic gas injection, slug gas injection, and Inject gas and soak. Circulating gas injection is prone to gas channeling, thereby reducing the condensate oil and gas recovery rate. The plan of inject gas and soak delays the gas channeling time on the basis of maintaining formation pressure, so the recovery rate is higher. In slug gas injection plan, the pressure pulse formed by the gas injection and gas production cycle can effectively expand the swept volume. And when injection stopped, the condensate in matrix will flow into the fracture due to the gravity. Therefore, the slug gas injection plan has the highest oil and gas recovery.

Tab 5. Development effect of Optimization of gas injection mode

Plan	Cumulative gas 10^8m^3	Gas recovery %	Cumulative oil 10^4m^3	Oil recovery %
Cyclic gas injection	35.3	50.2	188.4	26.8
Slug gas injection	37	53.9	198.4	29.1
Inject gas and soak	36.5	52.3	194.7	28.2

6 Conclusion

(1) The BZ gas field are divided into sandy conglomerate zone, weathered zone and Buried hill inner zone. The gas reservoir model of complex seepage system is established by the zeroing matrix conductivity inside the buried hill.

(2) The BZ gas field has strong stress sensitivity, and Nearly 90% loss of seepage capacity under abandonment pressure. Therefore, it is necessary to carry out numerical simulation research through time-varying reservoir physical properties.

(3) BZ gas field is a gas field with high saturation and high condensate content. Through PVTi matching and numerical simulation sensitivity analysis, the 7-component splitting plan is determined as the best plan.

(4) The numerical simulation study of the BZ gas field shows that: firstly inject gas in low area of the structure and then in the high area, inject gas at the start of production, the injection rate is 1.0, and seasonal slug gas injection is the optimal plan. This research provides strong support for

ODP program preparation and subsequent efficient development.

Reference:

[1] Fang Na, Jiang Guanghong, Li Yunpeng, et al. Development characteristics and potential tapping direction of buried-hill fractured oil reservoirs[J]. Special Oil & Gas Reservoirs, 2017,24(3):90-94.

[2] Xu Chunqiang, Zhang Zhen, Zhang Xintao, et al. Tectonic evolution and hydrocarbon accumulation model of western sag in western Bohai sea[J]. Journal of Northeast Petroleum University, 2018, 42（1）: 68-76.

[3] Zhang Feifei. Characterization and evaluation of pore-throat structures in ultra-low permeability and tight sandstones and their impacts on petrophysical properties: a case study of the lower Cretaceous in northern Songliao Basin[J]. Journal of Northeast Petroleum University, 2018,42(5):54-62.

[4] Tong Kaijun, Zhao Chunming, Lv Zuobin, et al. Petroleum exploration and development[J]. Petroleum Exploration and Development, 2012,39(1):56-63.

[5] Lin Guangrong, Liu Xuegang, Lu Yan, et al. Impact of stress change on percolation pattern of two-phase in low permeability reservoir[J]. Special Oil & Gas Reservoirs, 2008,15(1):43-45.

[6] Yang Bin, Hou Dali, You Gang, et al. Research and application of stress sensitive experiment on low permeable reservior[J]. Reservoir Evaluation and Development, 2012,5(2):41-45.

[7] Cao nai, Lei Gang. Stress sensitivity of tight reservoirs during pressure loading and unloading process[J]. Petroleum Exploration and Development, 2019,46(1):132-138.

[8] Jiang Ruizhong, Huang Lei, Cui Yongzheng, et al. Transit pressure analysis of a horizontal well in fractured composite reservoirs with stress-sensitivity effect[J]. Journal of Northeast Petroleum University,2018,42（3）: 84-91.

[9] Fang Yang. Research on relationship between deliverability of gas reservoir and the stress-sensitiveness of the reservoir[J]. Well Testing,2009,18（6）:18-21.

[10] Fu Lanqing. The effect of fractures on stress sensitivity and seepage characteristics in sandstone reservoirs[J]. Journal of Yangtze University（Natural Science Edition）,2016,13（20）: 14-18.

[11] Zhang Rui, Ning Zhengfu, Zhang Haishan, et al. New insights and discussions on stress sensitivity of fractured tight reservoir[J]. Natural Gas Geoscience, 2016,27(5):918-923.

[12] Wang Lezhi, Qi Zhilin, Jiang Yiwei. Study on retrograde condensate characteristics of deep condensate gas reservoirs with low permeability [J]. Natural Gas Technology, 2007, 1（1）: 52-54.

[13] Xiao Lixian, Du Jianfen, Guo Ping, et al. Pollution caused by retrograde condensate in condensate gas reservoir[J]. Fault-block Oil & Gas Field, 2009, 16（4）: 102-104.

[14] Cheng Liang, Zhang Xiaodong, Wang Juan, et al. Study on periodic gas injection mechanism of condensate gas reservoirs with high to high-super content of condensate oil in central Tarim by numerical simulation[J]. Journal of Xi'an shiyou University (Natural Science Edition),2016,31(2): 57-64.

[15] Wang Jing, Liu Huiqing, Liu Renjing, et al. Numerical simulation for low-permeability and extra-low permeability reservoirs with considering starting pressure and stress sensitivity effects[J]. Chinese Journal of Rock Mechanics and Engineering, 2013,32(2):3317-3327.

[16] Li Yong, Zhang Jing, Wu Xueyong, et al. A new reservoir simulation approach for fractured gas-condensate reservoirs[J]. Petroleum Exploration and Development, 2010,37(5):592-595.

[17] Jiang Hai，Yang Zhaozhong，Hu Yuehua，et al. Optimization of the cyclic gas injection parameters in condensate gas reservoir[J]. Journal of Xi'an shiyou University (Natural Science Edition),2009,24（2）:54-56.

609157 Feasibility of Demand Response in the Chemical Industry

Abstract:

Abstract

In the field of power generation, the aim is to increase the part of renewable energy. This brings an increase in fluctuations in power supply, which leads to a higher need for adaptability, meaning more flexibility in production and consumption of electricity. Load changes in energy-intensive industries as seen in the chemical industry would be a way to provide this. In the joint project ChemEFlex the feasibility of demand response in Chlor-Alkali-Electrolysis (CAE) as well as in other electrochemical processes of the chemical industry is analyzed. In this work, two processes with different levels of development were studied by evaluating suitable operating parameters and their tolerable limits.

The commercialized process for the electrohydrodimerization of Acrylonitrile (ACN) to Adiponitrile (ADN) was the first case considered suitable for further studies (Fig. 1). Adiponitrile is an intermediate in the production of Nylon. Worldwide the production rate of ADN is around 1,31 Mio. t/a, of which 300.000 t are made through cathodic hydrodimerization. This makes the so-called Monsanto-Process the most important electroorganical synthesis. An advantage of this process is, that it does not need a membrane, which makes it very attractive for demand response because this bottleneck in flexibility options has not to be considered.

The second case is the piloted process of the electrohydrodimerization of Formaldehyde to Ethylene glycol (Fig. 2). Ethylene glycol nowadays is almost exclusively made from direct oxidation of Ethylene, followed by hydrolysis of Ethylene oxide. Even though this process has a poor selectivity and is high in energy consumption, the use of mineral oil results in a very low price. Still, with increasing prices for crude oil and the additional profit from the control energy market, the electrochemical process could be a good alternative. Looking at the high production rates for Ethylene glycol, this could be the third-largest electrochemical process after Aluminum production and CAE.

Experimental section

Two experimental plants were built in a laboratory scale [1], [2] to test the application of demand-side management. The changes in selected parameters like aging of electrodes, product quality, etc. are documented and evaluated while various flexibility profiles (load changes in seconds or minutes, adapted to the requirements of the control energy market) were applied.

Experiments were carried out in Microflowcells by ElectroCell AB (Sweden). For the Monsanto-Process the effects on an undivided cell were studied, whereas for the synthesis of Ethylene glycol, a Nafion 324 membrane was used for separation.

References

[1] N. L. Weinberg, D. J. Mazur, J. Appl. Electrochem., 21 (1991) 895 – 901.

[2] K. Scott, B. Hayati, Chem. Eng. Sci. 45, 8 (1990) 2341–2347.

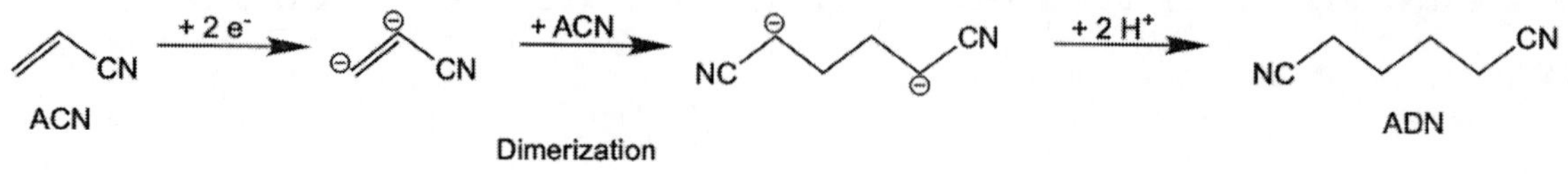

Fig. 1: Electrohydrodimerization of Acrylonitrile to Adiponitrile.

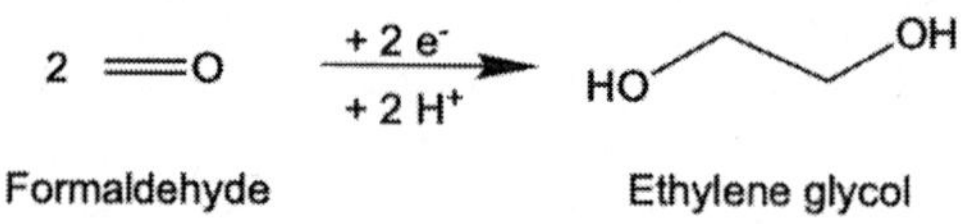

Fig. 2: Electrohydrodimerization of Formaldehyde to Ethylene glycol.

611749 Insights into the Electrochemical Conversion of Biomass Derivatives to Fuels and Chemicals

Abstract:

A number of biomass processing routes yield aqueous solutions of highly-functionalized small molecules. The removal of oxygen-rich functional groups (hydroxyl, carboxyl, ketone, aldehyde) and unsaturated bonds from these species is crucial for upgrading to fuels. Additional types of transformations (partial reductions or oxidations) are also important enabling technologies, as the displacement of fossil fuels will entail a viable biomass-derived chemicals industry.

Electrochemical upgrading routes pose several potential advantages for processing aqueous bio-derived feedstocks. Rearrangements of hydrogen and oxygen are accessible at ambient temperatures and pressures, often with novel selectivity profiles, as elementary reaction steps that are mediated by electron transfer can be accelerated relative to neutral atom-transfer steps. Deoxygenated products can often be separated naturally, and utilizing electric potential as a driving force can be advantageous when large sensible heats associated with solvents hamper the efficiency of using heat.

Our work has focused on electrochemical conversion of two groups of biomass-derived intermediates. We will show recent progress on (i) partial oxidation of furans (mainly derived from hydrolysis) to commodity chemicals and (ii) reduction of C_4-C_6 carboxylic acids (derived by a novel fermentation process) to light alkanes and alcohols. We have performed a number of kinetic studies, including isotopic labeling with differential electrochemical mass spectrometry (DEMS), as well as *in-situ* spectroscopic (ATR-FTIR) characterization of these chemistries. Reactivity trends with respect to conditions (pH, potential, electrolyte ion and solvent) as well as over a series of electrocatalytic materials will be discussed. The work supports a broader goal to establish principles for selective tuning of functional group transformations in the electrochemical environment.

599241 Electrochemical Synthesis of Adiponitrile with a Rotating Rod Electrode

Abstract:

Adiponitrile (ADN) is an important raw material for the production of nylon 66. In industry, ADN can be produced by coupling acrylonitrile (AN) at the cathode with a constant current method. Mass transfer of reactant plays an important role in the reaction. In this study, a rotating rod electrode was applied to simulate the electrolyte flows which are commonly in industry. A higher rotation rate enhanced the formation of ADN and trimer and suppressed the propionitrile (PN) formation. When the rotation rate was higher than 2000 RPM, the oil droplets of the upper organic phase were vigorously drawn into the lower aqueous solution, thereby the proportion of AN in the water phase was kept at a high level and the generation of hydrogen was suppressed, resulting a high selectivity and production of ADN. At the rotation rate of 2000 RPM, the velocity at the electrode surface was about 0.83 m/sec, close to the range for the flow systems in industry. The effects of electrolyte composition on the product composition will be also reported. The conversion, current efficiency and selectivity were discussed by exploring the effects of the concentration ratio of AN and ADN in the electrolyte on the electrosynthesis of ADN.

The experimental results showed that with a fixed total amount of AN and ADN, more oligomers, i.e. ADN and trimer, were produced than PN initially. Along the reaction time, AN concentration decreased, oligomers production rate slowed down and PN production rate increased. A higher AN ratio in the electrolyte gave more initial oligomers than lower AN ratio did. If AN ratio was too low, hydrogen evolution became the main reaction.

When AN concentration was fixed at high values, concentration ratio of AN to ADN did not affect the production rate of ADN, but higher ratio of AN to ADN resulted in higher AN conversion and trimer production rates. Adding an appropriate amount of ADN into an initial electrolyte could increase the selectivity of ADN. When AN concentration was fixed at low values, the addition of ADN were not favor for the oligomerization, but resulted in an increase in the productions of PN and hydrogen. Experiments also showed that the addition of ADN could reduce the volatilization of AN. When AN was fixed at high concentrations, current efficiency always remained higher than 90%, which meant that high AN concentration could suppress hydrogen generation.

609358 Impact of Fouling on Kinetics of the Electrochemical Hydrogenation and Hydrogenolysis of Furfural in Acidic Media over Copper

Abstract:

Electrochemical hydrogenation and hydrogenolysis (ECH) can be used for upgrading of biomass-derived species, and as opposed to thermocatalytic process, has the benefits of producing adsorbed hydrogen for reaction in-situ and can operate at room temperature and pressure. Thermocatalytic methods require externally supplied hydrogen gas for the reduction reactions, such as hydrogenolysis and hydrogenation. This external hydrogen often comes from non-sustainable processes that produce large amounts of CO_2. Furfural (FF) is a biomass-derived species, that will form furfuryl alcohol (FA) via hydrogenation, and 2-methylfuran (MF) via hydrogenolysis. FA and MF are value added products used as a binder and a fuel additive, respectively. An important aspect to study for FF ECH is the kinetics for the process. Side reactions decrease the yield of desired products and can foul the electrode, which has negative impacts onto the desired reaction kinetics.

In this work the effect of electrode fouling on the kinetics of FF ECH to MF and FA was investigated though bulk electrolysis experiments run using fouled and fresh electrodes. Our findings on homogeneous side reactions are presented as well. Experiments were done in acidic media (0.5 M H_2SO_4) over Cu catalysts and run between -500 and -800 mV vs RHE with initial concentrations of FF between 10 and 200 mM in a temperature range of 15°C to 45°C. The amount of fouling was characterized by XPS. Our results showed that high concentrations of substrate and longer experiments lead to fouling of the electrode, which has negative impact onto desired reaction rates. Additionally, a temperature increase from 15°C to 45°C was found to decrease the mass balance closure from 88% to 70%. When the temperature changed from 35°C to 45°C between independent experiments, a 40% drop in FA yield was found after 30 minutes as well as a discolored electrolyte.

609601 Advances in the Use of Electrochemical Oxidative Dehydrogenation to Convert Natural Gas Liquids to Value Added Chemicals

Abstract:

The U.S. has seen tremendous growth of natural gas liquids (NGLs) supply over recent years, due to the development of unconventional resources. The processing of natural gas into pipeline-quality dry natural gas is complex/costly and involves several processes to remove oil, water, acid gases, and NGLs. NGL recovery is a capital and energy intensive process utilizing cryogenic distillation requiring stages of compression/expansion with high degrees of heat integration. Natural gas processors reject C_2H_6 at the separation facility sending it to the natural gas pipeline for sale. Although C_2H_6 rejection is useful in eliminating the costs associated with cryogenic separation, it has limited applicability and results in overall revenue loss as ethane typically sells at a small premium in comparison to natural gas.

Ohio University (OHIO), via U.S. Department of Energy's National Energy Technology Laboratory (NETL) DE-FE0031709 award, is developing a modular technology that simultaneously separates NGLs from shale gas and converts these NGLs to olefin building blocks. The modularity of the technology provides turn-key integration either at the well or further downstream as a distributed chemicals platform for stranded shale gas. Specifically, OHIO is developing an oxidative dehydrogenation (ODH) technology based on a solid oxide fuel cell platform, with a solid electrolyte (membrane) that separates the fuel from oxygen and heterogeneous electro-catalysts that facilitate olefin generation. The reactions below describe the anodic and cathodic reactions based on ethane (other NGLs would follow a similar mechanism) as well as overall ODH and thermal cracking reactions and associated thermodynamics.

Anode: $C_2H_6 + O^{2-} \rightarrow C_2H_4 + H_2O + 2e^-$

Cathode: $1/2O_2 + 2e^- \rightarrow O^{2-}$

Overall ODH: $C_2H_6 + 1/2O_2 \rightarrow C_2H_4 + H_2O$ $\Delta H^\circ_{reaction} = -103.7kJ/mol$

Thermal Cracking: $C_2H_6 \rightarrow C_2H_4 + H_2$ $\Delta H^\circ_{reaction} = 143.9kJ/mol$

This specific approach provides significant advantages in comparison to similar oxidative dehydrogenation technologies including scalability and electricity as a process by-product, while offering higher thermal efficiency than thermal cracking – the state-of-the-art dehydrogenation technique at scale.

Prior development efforts at OHIO have shown electrode choice and fabrication technique significantly affect the SOFC's performance as screen-printed perovskites with rare-earth and alkali-earth chemistries were initially tested. The alkali-earth chemistries showed promising alkane conversion and olefin yield but would require additional improvements for industrially relevant figures of merit. Recent development efforts have prioritized catalyst infiltration as an electrode fabrication technique, thus leveraging enhanced catalytic surface area, lower sintering temperatures and improved triple-phase boundary sites. The results of this fabrication technique including materials characterization and catalyst performance for electrogenerative ODH will be presented at the meeting.

611376 Low-Cost 3D Ni-Based Electrocatalysts for HER and Oer for Fuel Cells

Abstract:

Low-Cost 3D Ni-based Electrocatalysts for HER and OER for Fuel Cells

Matthew DiBiase[1], Sean P. Rogers[1], Thomas Gascoigne[1], Jiangtian Li[2], Rongzhong Jiang[2], F. John Burpo[1], Deryn Chu[2], Enoch A. Nagelli[1*]

[1]Department of Chemistry & Life Science, Chemical Engineering Program, United States Military Academy, West Point, New York 10996

[2]U.S. Army Combat Capabilities Development Command, Army Research Laboratory

2800 Powder Mill Rd., Adelphi, MD 20783-1107

The implementation of porous electrodes in electrochemical applications requires fundamental understanding of the factors controlling electronic conduction through the Faradaic and non-Faradaic charge transfer processes. Moreover, recent advances in nanotechnology and materials science offers a selection of low-cost non-precious metal porous electrodes such as nickel for electrochemical applications. Conventional methods of electrolysis to produce H_2 and O_2 gas necessary for proton exchange membrane (PEM) fuel cells rely on expensive catalysts. Alternative catalysts such as three-dimensional (3D) nickel foams with high surface area and surface to volume ratio are promising materials to reduce kinetic losses and enhance mass transfer rates, respectively. In our study, we investigate the impact of surface morphology and elemental composition of Ni-S and Ni-Fe based foams from hydrothermal synthesis and electrodeposition techniques on catalysis for electrochemical anodic and cathodic water splitting reactions (hydrogen evolution reaction/HER and oxygen evolution reaction/OER). Electrochemical performance of the Ni-S and Ni-Fe based 3D foams was evaluated using linear sweep voltammetry to determine the minimum overpotentials of the catalyst necessary to drive HER and OER. Scanning electron microscopy (SEM), energy dispersive x-ray spectroscopy (EDX), and x-ray diffractometry (XRD) were used to characterize surface morphology, surface elemental composition, and crystallinity of the 3D Ni foams, respectively.

KEYWORDS: Fuel Cells, 3D Foams, Electrocatalysis, Oxygen Evolution Reaction, Hydrogen Evolution Reaction, Electrochemical Engineering

CONTACT: Enoch A. Nagelli, Email: enoch.nagelli@westpoint.edu

609451 Advanced Mass Transport Properties of 3D-Printed Porous Periodic Lattice Graphene Flow-through Electrodes

Abstract:

Since heterogeneous electrochemical reactions under conditions relevant to commercial applications depend on mass transport, liquid type reactors gain efficiency from forcing solutions through 3D porous flow-through electrodes with high internal surface area and enhanced mixing properties. We consider electrochemical reactor design based on maximizing mass transport performance and explore electrodes with periodic lattices enabled by additive manufacturing fabrication tools. Controllable periodic pore structure electrodes solve the problem of uneven velocity distribution (channeling) known to decrease mass transport rates in disordered porous foams and felts and enable engineering of flow patterns for improving mass transport. In this report, by means of numerical simulation and experiment, we study flow rate dependent mass transport of periodic porous lattice electrodes prepared by direct ink write method as simple cubic and face centered cubic structures with 0.8 mm lattice parameter using 0.4 mm nozzle. Volumetric mass transport coefficients were assessed from limiting currents measured for simple one electron transfer reaction in ferrocyanide solutions as a function of flow rates. Results show that while staying at laminar regime ($Re < 10$), the power-law exponent of mass transfer coefficient as a function of flow velocity increases from about 0.4 to 0.6 after threshold velocity is reached. Typically, sharp increases in the power-law exponent are observed due to changing from laminar to turbulent regimes at much higher Re numbers ($Re > 1000$). Numerical simulations reveal that scaling factor increase at much lower Re numbers in this work was possible due to controlled geometry of printed electrode allowing generation of secondary flow patterns promoting enhanced local mass transport as flow switches from viscous to inertial. Electrodes that yield high mass transport coefficients at low flow velocity are advantageous due to the pumping power inefficiencies present at high velocities. Further improvements in the mass transport properties can be achieved by 3D-printing lattices with modified design parameters (e.g. pore size, filament diameter, lattice structure, etc.). The results are discussed in the context of theory and previous literature.

600757 Porous Metallic Membranes As High-Performance Gas Diffusion Electrodes for CO_2 Reduction in Zero-Gap Electrolyzers

Abstract:

Combining the electrochemical reduction of CO_2 (eCO_2RR) with carbon-neutral energy sources creates groundbreaking possibilities for using CO_2 as a raw material for the sustainable production of fuels and chemicals. In this context, recent technoeconomic analyses have emphasized the importance of achieving high current densities (i.e., hundreds of mA cm^{-2}) for the eCO_2RR to be practically viable.[1]

Cell designs that integrate selective eCO_2RR catalysts into a membrane electrode assembly (MEA) are a promising approach to reduce ohmic losses and achieve high energy efficiency at elevated current densities.[2,3] In this contribution, we show that silver membranes—commonly used as porous filtration media in biotechnological applications—sustain high rates of CO_2 reduction to CO (> 200 mA cm^{-2}) when used as GDEs in a zero-gap configuration.[4] These metallic electrodes are simple, highly stable, and cost-competitive vis-à-vis carbon-based GDEs. However, CO_2 crossover across the anion-exchange membrane (AEM) results in poor reactant utilization and in the early onset of mass transfer limitations, highlighting the importance of carrying out a full mass balance to adequately characterize eCO_2RR devices.[4] Furthermore, we will discuss how these porous silver membranes can be used as templates for the synthesis of freestanding copper GDEs, and how different electrode morphologies and compositions can be targeted via the synthesis procedure. In tandem with a judicious selection of anolyte and AEM, these GDEs show promising performance for the conversion of CO_2 into multi-carbon products at high current densities.

[1] M. Jouny, W. Luc, F. Jiao, *Ind. Eng. Chem. Res.* **2018**, 57, 2165-2177

[2] T. Burdyny, W.A. Smith, *Energy Environ Sci.* **2019**, 12, 1442-1453

[3] D. Higgins, C. Hahn, C. Xiang, T.F. Jaramillo, A.Z. Weber, *ACS Energy Lett.* **2019**, 4, 317-324

[4] G.O. Larrazábal, P. Strøm-Hansen, J.P. Heli, K. Zeiter, K.T. Therkildsen, I. Chorkendorff, B. Seger, *ACS Appl. Mater. Interfaces* **2019**, 11, 41281-41288

618254 Sponsored Technology Workshop - The Use of Explosion Cladding to Improve the Life and Reduce the Capital Expense of Process Equipment - NobelClad

Abstract:

Process equipment used in the Oil and Gas, Chemical and Petrochemical industries are often exposed to aggressive environments. Conventional steels in these environments are subject to failure by crevice corrosion, pitting corrosion or stress corrosion cracking. Stainless steels and exotic corrosion resistant alloys are increasing in demand to reduce mechanical and chemical corrosion-induced failures. Unfortunately, the use of many of these alloys are cost prohibitive and do not offer the pressure rating of carbon steel. One solution, explosion cladding of dissimilar metals, has been proven to reduce the capital cost of the heat exchanger while retaining the mechanical and corrosion properties of both the base metal and the cladded surface metal.

During this presentation we'll explain the science of explosion cladding and how to ensure a sound bond. We will share various applications that leverage clad for cost savings in heat exchangers and pressure vessels, and how the welding of internals through direct attach allows for a superior and cost effective way to weld attachments and internals.

613161 Water Is the Oxygen Source for Methanol Produced in Partial Oxidation of Methane in a Flow Reactor over Cu-SSZ-13

Abstract:

The direct oxidation of methane to methanol represents a grand challenge problem. Copper-containing zeolites have attracted attention recently due to their high selectivity to methanol under moderate reaction conditions. The partial oxidation of methane to methanol can be achieved via chemical looping or continuous processes using molecular oxygen as an oxidant and water as an extracting solvent. It has been long considered that water is only necessary for extraction of surface bound methanol while being kinetically inconsequential. However, the current work demonstrates that water can be used as a "soft" oxidant for methane oxidation in a continuous partial oxidation of methane reaction. The continuous partial oxidation of methane to methanol over Cu-SSZ-13 with water in the absence of molecular oxygen showed that active copper site motifs under steady-state conditions can in fact be formed and maintained without apparent deactivation for over 30 hours. The operando Fourier transform infrared spectroscopy (FTIR) measurement technique was applied to confirm the source of the oxygen in the methanol obtained. The isotopically labelled $CH_3{}^{18}OH$ and regular $CH_3{}^{16}OH$ have two distinct C-O stretching modes at 1007 cm^{-1} and 1033 cm^{-1}, respectively. The evidence that the water vapor present in the continuous partial oxidation of methane is the main source of the oxygen atom in the methanol synthesis, was obtained for reactions ran with 18O-labelled water. The time-resolved IR spectra of C-O stretching mode has shown formation of 1007 cm^{-1} peak and it remained constant for the duration of the reaction. This result is a very strong proof that the oxygen in the methanol under long time steady-state conditions in fact comes from the water fed into the system.

609198 Novel Approach for the Delamination of the Zeolite MCM-22

Abstract:

Zeolites are well known for their unique microporous structures that allow for highly shape-selective catalysis and separations. One of the challenges with many zeolites is a high diffusional barrier that leads to low utilization of interior micropore volumes and faster deactivation. This can be overcome by including mesopores in the crystal structure, which shortens the diffusive path length and significantly increases transport. These can be created in a variety of ways including templating, desilication, dealumination, and delamination. However, most of these techniques involve either costly reagents, extended growth/reaction times, harsh chemicals, or some combination thereof.

To improve catalyst production at scale, it would be beneficial to develop easier and less costly methods for preparing hierarchical zeolites. One particularly interesting candidate is the delaminated zeolite ITQ-2. This hierarchical zeolite is traditionally prepared by synthesizing a layered precursor (MCM-22), which is then swollen using surfactants, kept under reflux for several hours, and sonicated in a pH monitored solution. This results in thin sheets of zeolite, only a few unit cells thick, which then stack against each other in a random arrangement creating mesopores in the delaminated ITQ-2 structure. While this method has been shown to be effective at lab scale, commercial production of delaminated zeolites would be improved with shorter processing time and fewer chemicals. In this work, we present a simplified method for producing delaminated MCM-22 without harsh post-synthesis treatments. These materials demonstrate comparable physical properties to traditionally delaminated ITQ-2 without significant loss of acid sites. They also show improved activity for the cracking of propylphenol to phenol and propylene. In addition, the rate of deactivation by coking is reduced because the diffusional path remains available longer as coke deposits accumulate. By further developing this technology, it may be possible to create other delaminated zeolites without the use of expensive treatments.

Catalysis Science & Technology

ROYAL SOCIETY OF CHEMISTRY

PAPER

Cite this: Catal. Sci. Technol., 2020, 10, 5069

Ethane dehydrogenation over Cr/ZSM-5: characterization of active sites through probe molecule adsorption FTIR†

Noah W. Felvey, [a] Michael J. Meloni, [a] Coleman X. Kronawitter [*a] and Ron C. Runnebaum [*ab]

Received 18th May 2020, Accepted 5th July 2020

DOI: 10.1039/d0cy01022g

rsc.li/catalysis

Dispersed Cr species supported on zeolite ZSM-5 were investigated in the context of catalytic ethane dehydrogenation. Probe molecule FTIR and XANES were used to characterize the nature of Cr sites across a range of Cr loadings and in the presence of various extra-framework cations on ZSM-5. These characteristics were related to catalytic performance in ethane dehydrogenation toward the goal of developing design principles for Cr/zeolite catalysts for improved alkane conversions. Diffuse reflectance infrared Fourier transform spectroscopy (FTIR-DRIFTS) with nitric oxide (NO) and carbon monoxide (CO) probe molecules was determined to be an effective tool for distinguishing various types of Cr species present after exposure to reducing conditions typical of alkane dehydrogenation catalysis. It was found that 0.5 wt% Cr/H-ZSM-5 is a stable, ethylene-selective catalyst whose Cr sites are reduced at reaction temperature (650 °C) in the presence of either ethane or N_2. Zeolite framework Al sites were determined to be critical for the stabilization of active Cr sites during ethane dehydrogenation – in the absence of available Al sites, Cr sites quickly deactivated. It was also found that greater Cr/Al molar ratios yielded a greater abundance of electron-rich Cr^{2+} species after reduction. This trend correlated with increased ethane dehydrogenation activity and decreased stability in time-on-stream studies. Both high activity and enhanced stability were observed for Cr/Ca-ZSM-5. Taken together, these results indicate that ethane dehydrogenation over Cr/ZSM-5 is optimized when Cr preferentially situates adjacent to isolated framework Al sites.

1. Introduction

Zeolite-supported metal catalysts with well-defined structure present a unique opportunity to develop highly active and selective catalysts while simultaneously establishing a more fundamental understanding of reaction sites. The zeolite surface provides precise binding sites for supported metals, resulting in active sites that are amenable to more targeted characterization. Under reaction conditions, zeolite micropores can stabilize small metal and metal oxide clusters that would not normally be stable on conventional (bulk) metal oxide supports.[1,2] The structure and stability of dispersed metals on zeolites are impacted profoundly by the proximity to one another of tetrahedral aluminum (Al) sites in the zeolite framework as well as the distribution of these Al sites among possible crystallographic locations.[3–5] The relationship between Al siting and supported metal properties was made clear for cobalt-exchanged zeolites developed for selective catalytic reduction of NO_x (SCR),[5] and it is a central theme in the more recent literature on copper-exchanged zeolites for SCR[2] and selective partial methane oxidation.[6,7] Leveraging control of Al siting in a wide variety of zeolite frameworks is a promising strategy for development of unique and productive active-site structures using Earth-abundant metals. While zeolite-supported metals such as Co,[5] Cu,[2,6,7] Ni,[8] Zn,[9–11] and Ga[12,13] are often studied, a deeper understanding of Cr in zeolite-based catalysts has yet to develop.

Supported-Cr catalysts are applied industrially for catalytic dehydrogenation of light alkanes[14] (typically with Cr/Al_2O_3), a technology crucial to the efficient utilization of natural gas feedstocks. Active sites are known to be mononuclear or oligonuclear Cr species, which are highly sensitive to their coordination to the support.[15–17] Process limitations associated with their industrial utilization include non-selective side reactions, coke formation, and in the case of ethane dehydrogenation, low catalyst activity.[14]

In this context, the hydrothermal stability and controllable acidity of zeotype supports create opportunities to design new

[a] Department of Chemical Engineering, University of California, Davis, CA, 95616 USA. E-mail: ckrona@ucdavis.edu, rcrunnebaum@ucdavis.edu
[b] Department of Viticulture & Enology, University of California, Davis, CA, 95616 USA

† Electronic supplementary information (ESI) available. See DOI: 10.1039/d0cy01022g

Cr-based catalysts. In fact, Cr/zeolite catalysts have been shown to be active for a number of reactions,[18–22] but determination of the binding sites of Cr within zeolite structures remains unresolved. One notable exception is the insertion of Cr into tetrahedral vacancy defects, or silanol nests, in all-silica zeotypes.[22–24] Cr located at these sites resemble the mononuclear Cr species formed on amorphous SiO_2 but show enhanced stability under certain reaction conditions. In contrast to silanol nests, Al sites in zeolites display a variety of configurations, providing the potential for a diversity of supported Cr structures, including multinuclear metal oxide clusters. A critical evaluation of the Cr structures present in zeolites is lacking; it remains unclear whether Al binding sites can be exploited to stabilize active and selective Cr structures.

Probe molecule spectroscopy is a powerful tool used to provide information about the electronic state and coordinative availability of dispersed metal sites. Earlier studies of Cr supported on zeolite ZSM-5 (Si/Al = 27) successfully observed multiple types of Cr by FTIR using nitric oxide (NO) as a probe molecule.[25,26] It was also shown that carbon monoxide (CO) was an ineffective probe at room temperature, as it was not able to coordinate these Cr cations. The two most abundant Cr species displayed differing reduction behaviors and responses to NO adsorption after reduction, implying that these Cr cations resided at distinct binding sites on the ZSM-5 surface. More recently, comparison of experimental Raman bands with results from density functional theory (DFT) calculations of calcined Cr/ZSM-5 allowed identification of two types of Cr differing in the number of Al sites with which they interact.[21] These two types of Cr displayed distinct reduction behavior, consistent with previous reports. At low Cr loading (0.8 wt%) these catalysts were active for methane conversion, but additional amounts of Cr did not increase activity. In this case, the state of Cr after reduction was not determined, as the distinct Raman signals associated with Cr-oxo species were not present after reduction.

In the present study, we combine FTIR probe molecule spectroscopy with relevant catalytic reactor measurements in order to relate the state of reduced Cr species in Cr/ZSM-5 to the catalytic activity and stability for ethane dehydrogenation. In contrast to trends observed for methane conversion, we find that ethane dehydrogenation activity per Cr site increases substantially with Cr loading. At the highest loadings examined, high activity per Cr is accompanied by decreased stability, which provides additional insight into the nature of Cr present. Our results enable a first-pass evaluation of the potential to design effective Cr-based dehydrogenation catalysts through use of zeolite supports.

2. Experimental methods

2.1 Catalyst preparation

Support preparation and ion exchange. NH_4-ZSM-5 (Si/Al = 15, Zeolyst CBV-3024E) as received was converted to the protonic form by calcination under 50 mL min^{-1} flow of synthetic air in a 19 mm OD quartz tube fitted with valves to exclude ambient air, which was placed into a horizontal tubular furnace. Temperature was ramped at a rate of 2 °C min^{-1} to 500 °C, with 1 h isothermal holds at 120 °C, 200 °C, and 300 °C, and held at 500 °C for 3 h. Obtained H-ZSM-5 was transferred without exposure to ambient air into an argon-filled glovebox where it was stored.

Na-ZSM-5 was prepared by ion exchange of H-ZSM-5 with sodium nitrate ($NaNO_3$). H-ZSM-5 was stirred at room temperature for 3 days in an aqueous solution of 1 M $NaNO_3$ (80 mL g^{-1} zeolite), which was periodically adjusted to pH 7 with 1 M sodium hydroxide (NaOH). The solids were subsequently separated by centrifugation and washed four times with 50 mL ultra-pure water per g zeolite. After drying for at least 3 h in an oven at 120 °C, the sample was crushed to a powder form and calcined under the identical conditions as used to prepare H-ZSM-5. The calcined sample was stored in the glovebox prior to use.

Ca-ZSM-5 was prepared by ion exchange with calcium nitrate ($Ca(NO_3)_2$). H-ZSM-5 was stirred for 2 h in an aqueous solution of 1 M $Ca(NO_3)_2$ in a closed glass vessel placed in an oil bath at 80 °C. After 2 h, the vessel was cooled in an ice bath and centrifuged. Recovered solids were resuspended in a fresh 1 M $Ca(NO_3)_2$ solution and exchange was repeated at 80 °C for 2 h. This process was repeated once more for a total of three exchanges. The cooled solution was then separated by centrifugation and washed four times with 50 mL ultra-pure water per g zeolite. The sample was dried and calcined using the same procedure as that for Na-ZSM-5.

Deposition of $Cr(acac)_3$. Preparation of Cr/H-ZSM-5, Cr/Na-ZSM-5, and Cr/Ca-ZSM-5 followed the same procedure. All steps of the preparation were performed without exposing the samples to ambient air. In the glovebox, zeolite (up to 0.5 g) was added with an appropriate amount of chromium(III) acetylacetonate ($Cr(acac)_3$) to a 2 mL glass ampoule. The ampoule was transferred to a Schlenk line, where it was evacuated and flame-sealed at approximately 80 mtorr pressure. The sealed ampoule, containing the light purple mixture, was heated in the calcination furnace at 5 °C min^{-1} to 200 °C and held for 2 h. Then temperature was ramped at 5 °C min^{-1} to 300 °C and held for 2 h. Sample color changed to either green or brown/gray during vacuum heat treatment. After vacuum heat treatment, the sample was transferred to the glovebox, where it was removed from the ampoule and placed into the quartz tube calcination reactor. The sample in the calcination tube was transferred to the furnace, where it was heated under 50 mL min^{-1} flow of air at 5 °C min^{-1} to 300 °C and 2 °C min^{-1} from 300 °C to 500 °C, at which temperature the furnace was held for 3 h. After cooling, the calcination tube was transferred back to the glovebox, where samples were stored prior to use. X-ray diffractograms verify crystallinity of the MFI structure before and after Cr deposition (Fig. S1†).

2.2 Diffuse reflectance Fourier-transform infrared spectroscopy (DRIFTS)

Powder samples were prepared for DRIFTS by loading pure sample (7–9 mg) into a ceramic sample cup and pressing gently to achieve a flat surface. The loaded sample cup was placed into a Pike Technologies DiffusIR Heated Chamber inside the glovebox, allowing air-free transfer to the spectrometer. DRIFTS measurements were performed using a Bruker Tensor II FTIR Spectrometer equipped with a Pike Technologies DiffusIR MidIR Accessory. Spectra were collected using 256 scans and 2 cm^{-1} resolution. For spectra of samples under nitrogen atmosphere, a background spectrum of dried potassium bromide powder (KBr) under N_2 was used, and data are reported as Kubelka–Munk spectra. For spectra of samples with adsorbed probe molecules, a background spectrum of the sample was collected just prior to introduction of the probe gas, and data are reported as log of inverse reflectance. More detail is given in the ESI.†

Sample reductions were performed in the DRIFTS heated chamber. The chamber was ramped at 10 °C min^{-1} to the desired temperature under 30 mL min^{-1} N_2 flow. At the desired temperature, composition of the gas flow was changed and held for 30 min, keeping total flow rate at 30 mL min^{-1}. After 30 min treatment, the heated chamber was purged with 30 mL min^{-1} N_2 for another 30 min before cooling the chamber under N_2 flow. Probe molecules were dosed at 30 °C. Details regarding probe molecule dosing are included in the ESI.†

2.3 Catalytic activity

Ethane dehydrogenation was carried out in a quartz down-flow, packed-bed reactor with 4 mm ID. Sample powders were sieved between mesh sizes 60–40 before use. Reactor effluent was measured online by an Agilent 7890a gas chromatography (GC) unit equipped with FID and TCD detectors. Details about the reactor and GC are included in the ESI.†

In a typical reaction experiment, the reactor was heated in a tubular furnace at 5 °C min^{-1} to 650 °C under 48 mL min^{-1} N_2 flow. Holding at 650 °C, flow was switched to bypass the reactor in order to initialize gas flow rates and composition for ethane dehydrogenation. Unless otherwise specified, the flow rates used were 44 mL min^{-1} N_2 and 4 mL min^{-1} C_2H_6. Space velocity was adjusted by changing the mass of catalyst in the bed. After initial flow rates stabilized, flow was reintroduced to the reactor and the reaction began. The reactor spent about 20 min under static N_2 at 650 °C before flow of C_2H_6 began. Catalyst bed pressure was 3 psig.

2.4 X-ray absorption near edge structure (XANES) spectroscopy

XANES fluorescence spectra were collected at Beamline 4-3 at the Stanford Synchrotron Radiation Lightsource at SLAC National Laboratory. Samples were loaded into a stainless-steel flow cell equipped with cartridge heaters and Kapton windows. Reported XANES spectra were energy-calibrated and normalized.

2.5 Thermogravimetric analysis-mass spectrometry (TGA-MS)

TGA-MS experiments were performed using a Netzsch STA 449 F3 Jupiter thermogravimetric analysis unit with QMS 403 D Aeolos mass spectrometer. Samples (20–30 mg) were loaded into alumina crucibles and placed into the TGA under ambient air. For measurement of carbon deposition, in order to remove adsorbed moisture, samples were held for one hour at 120 °C under flow of synthetic air. Then temperature was ramped 5 °C min^{-1} to 700 °C. Between 150 °C and 500 °C, CO_2 was detected by MS, and the corresponding mass loss was calculated. To estimate the Brønsted acid site density of H-ZSM-5, temperature-programmed desorption (TPD) of 2-propanamine was performed. 2-Propanamine was injected into a stream of N_2 flowing over the sample at 30 °C. Temperature was ramped at 5 °C min^{-1} to 150 °C, where it was held for 30 min to allow desorption of physisorbed 2-propanamine and moisture. Next, temperature was ramped at 5 °C min^{-1} to 527 °C and held for 30 min. Results from TPD of 2-propanamine are included in the ESI† (Fig. S2).

2.6 Physisorption

Nitrogen physisorption measurements were performed using a Micromeritics 3Flex physisorption/chemisorption Instrument. Nitrogen adsorption isotherms were measured at −196 °C. Surface areas were determined by the BET method, and micropore volume was determined by the *t*-plot method. Results of physisorption measurements are included in the ESI† (Fig. S3 and Table S1).

3. Results and discussion

3.1 Al sites in ZSM-5 enable stable dehydrogenation activity

The premise of this investigation is that the activity of zeolite-supported Cr depends strongly on the coordination of Cr to the support. We first distinguish between two categories of binding sites for Cr: (1) charge-balancing bridging hydroxyls (Al–OH–Si), referred to here as "Al sites", where Al is located in a tetrahedral position of the zeolite framework and (2) silanols (Si–OH) that exist at defects in the zeolite lattice and external surfaces of zeolite crystallites. To begin to assess the impact of binding site on the activity of Cr, we prepared two 0.5 wt% Cr/ZSM-5 samples, one using the H-form of ZSM-5 and the other using the Na-exchanged form of the same ZSM-5.

Fig. 1 contrasts 0.5 wt% Cr/H-ZSM-5 and Cr/Na-ZSM-5 samples. The initial activity, expressed as turnover frequency (TOF), of Cr/H-ZSM-5 for ethane dehydrogenation at 650 °C was nearly twice that of Cr/Na-ZSM-5 (Fig. 1a). Cr/H-ZSM-5 also displayed stable activity while Cr/Na-ZSM-5 lost more than 70% of its initial activity during three hours under reaction conditions (Fig. 1b). Initial ethane conversion was below 3% in both experiments. The DRIFTS spectrum of Na-ZSM-5 (Fig. S4†) indicates the disappearance of the peak at 3608 cm^{-1}, caused by the replacement by Na^+ of protons at Al sites. It was expected that the presence of Na^+ cations would block Cr from binding

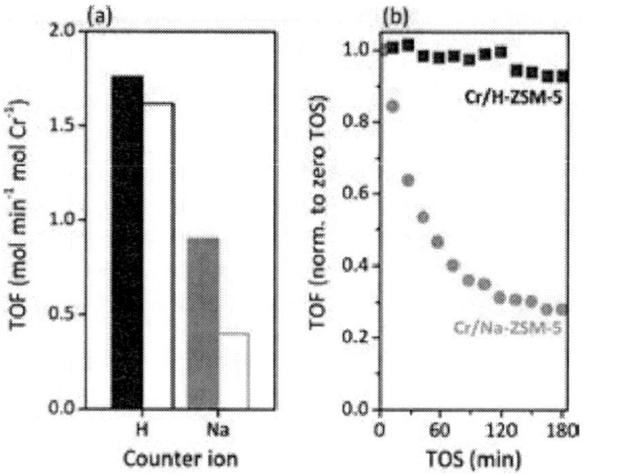

Fig. 1 (a) Rate of ethane conversion at 650 °C at zero time on stream (filled bars) and after one hour on stream (empty bars) for 0.5 wt% Cr/H-ZSM-5 and 0.5 wt% Cr/Na-ZSM-5 catalysts. (b) Normalized rate of ethane conversion *versus* time-on-stream for same samples and reaction conditions as Fig. 1a.

at associated Al sites, forcing Cr to bind at silanol sites in Cr/Na-ZSM-5. For Cr/H-ZSM-5, where hydroxyls at Al sites were available to anchor Cr, a stable and more active dehydrogenation catalyst resulted. Results from a control experiment using bare H-ZSM-5 were used to estimate that the contribution of Brønsted acidity to the measured ethane conversion accounts for only one sixth of the total activity of 0.5 wt% Cr/H-ZSM-5, indicating that the measured activity is representative of the activity of supported Cr species.

At comparable ethane conversion (Table S2†), the initial ethylene selectivity of Cr/Na-ZSM-5 was higher (99.6%) than that of Cr/H-ZSM-5 (93.8%), consistent with the expectation that side reactions (*e.g.* oligomerization, cracking) occur at Brønsted acid sites in Cr/H-ZSM-5. The small amount of deposited carbon (less than 1 wt%) on Cr/H-ZSM-5 after 3 h on stream measured by TGA-MS (Fig. S9†) did not cause extensive deactivation. An even smaller amount of coke was observed for Cr/Na-ZSM-5 (Fig. S8†), but extensive deactivation was evident in the time-on-stream data. This result implies that Cr aggregation likely plays a major role in the deactivation of Cr/Na-ZSM-5.

To gain insight into the state of Cr in these samples, XANES and NO adsorption DRIFTS measurements were recorded after reduction in CO at 300 °C. Reduction with CO at 300–350 °C is a common treatment for activation of supported Cr catalysts, and the use of these reduction conditions enables direct comparison of the present results to literature, which documents the formation of mainly Cr^{2+} in Cr/SiO_2 catalysts[27] and a mixture of Cr^{2+} and Cr^{3+} in Cr/Al_2O_3 catalysts.[17] No consensus regarding the oxidation states of Cr in reduced Cr/H-ZSM-5 has been established.

XANES spectra of fresh Cr/H-ZSM-5 and Cr/Na-ZSM-5 display intense pre-edge peaks at 5993 eV, indicating that Cr was present as dispersed, tetrahedral Cr^{6+} species after calcination (Fig. S5†).[27,28] After reduction in CO at 300 °C, XANES spectra display edge energies at *ca.* 6002 and pre-edge features at 5988, 5990, and 5995 eV, characteristic of supported Cr^{2+} (Fig. 2a).[27,28] These results indicate the average oxidation state of Cr was close to 2+ in both samples after reduction.

DRIFTS spectra of NO adsorption onto CO-reduced Cr/H-ZSM-5 and Cr/Na-ZSM-5 reveal clear differences (Fig. 2b). NO adsorption onto Cr/Na-ZSM-5 reduced in CO at 300 °C yielded broad peaks centered at 1747 cm^{-1} and *ca.* 1880 cm^{-1}, similar to those often observed for Cr^{2+}/SiO_2 treated under the same conditions. In Cr^{2+}/SiO_2 samples,[23,27,29–31] peaks often observed at 1743–1755 cm^{-1} and 1856–1880 cm^{-1} have been shown to arise from antisymmetric and symmetric N–O stretching vibrations of Cr^{2+} dinitrosyl species, respectively. The similarity of the NO spectrum of Cr/Na-ZSM-5 to that of Cr/SiO_2 leads us to conclude that Cr in Cr/Na-ZSM-5 was located mainly at silanol sites after CO reduction at 300 °C.

It is possible for Cr to bind at either external silanols or internal silanols at framework vacancies (silanol nests). DRIFTS spectra of Na-ZSM-5 and Cr/Na-ZSM-5 (Fig. S4†) show that the intensity of the O–H stretching band due to external silanols (3744 cm^{-1}) decreased considerably after Cr loading, whereas the broad band due to internal silanols (3530 cm^{-1}) displayed no obvious change.[24] In a previous study, when Cr was located mainly at silanol nests in dealuminated zeolite beta, a Cr-mononitrosyl characterized by a sharp N–O stretching peak at 1794 cm^{-1} was observed.[24] This peak was not observed for Cr/Na-ZSM-5 in the present study. These results are consistent with the location of Cr at external silanol sites, but the coexistence of Cr at silanol nests cannot be conclusively ruled out.

NO adsorption onto Cr/H-ZSM-5 after CO reduction at 300 °C (Fig. 2b) resulted in a clearly different spectrum from adsorption onto Cr/Na-ZSM-5. Peak intensities are higher, and the dominant peaks are blue-shifted more than 30 cm^{-1} with respect to Cr/Na-ZSM-5. It is also apparent from the line shape that the spectrum associated with Cr/H-ZSM-5 is comprised of several overlapping, sharper peaks, suggesting

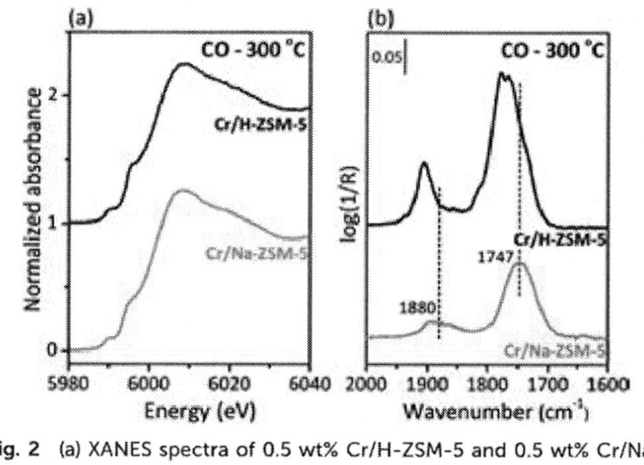

Fig. 2 (a) XANES spectra of 0.5 wt% Cr/H-ZSM-5 and 0.5 wt% Cr/Na-ZSM-5 after 30 min reduction in CO at 300 °C. Samples remained under CO flow at 300 °C while spectra were collected. (b) FTIR-DRIFTS spectra of NO adsorbed at 30 °C on 0.5 wt% Cr/H-ZSM-5 and 0.5 wt% Cr/Na-ZSM-5 samples after reduction in CO at 300 °C.

that multiple Cr species with distinct structures were present in Cr/H-ZSM-5. In contrast, the broad peaks in the Cr/Na-ZSM-5 spectrum imply a higher degree of heterogeneity of Cr structures due to aggregation of Cr or a distribution of surface binding environments occupied by dispersed Cr species.

DRIFTS spectra of NO adsorption after reduction in H_2 at 650 °C, conditions more similar to those present in the reactor during ethane dehydrogenation, indicate an even greater contrast between Cr/H-ZSM-5 and Cr/Na-ZSM-5 (Fig. 3). In Cr/H-ZSM-5, several sharp peaks appeared that were not present after reduction at 300 °C. The availability of Al binding sites for Cr in Cr/H-ZSM-5 resulted in the formation of multiple, distinct Cr species that were not present in Cr/Na-ZSM-5. These observations are consistent with the expectations and motivations of this study – that Al sites in zeolites provide a mechanism to control Cr speciation. Peak assignments are discussed in more detail in the next section.

These experiments, and the ability to control Cr siting, were uniquely enabled through anhydrous Cr deposition. Although aqueous techniques can be employed to produce highly dispersed supported-metal species, aqueous methods for Cr dispersion onto zeolites have complications. The extent of ion exchange of high charge (typically 3+), hydrated Cr cations for one or more singly-charged zeolite charge-balancing protons is not easily controlled. Speciation of hydrated Cr cations to structures with higher nuclearity depends on pH, which varies with solution concentration and surface potential of the support.[32] For this study, we controlled both Cr loading and the zeolite surface potential (*e.g.* by exchanging zeolite H^+ for Na^+), while attempting to maintain high Cr dispersity. In aqueous conditions, Cr may exchange with Na^+ at Al sites. In contrast, the anhydrous approach facilitated anchoring of Cr selectively at silanol sites in Cr/Na-ZSM-5. The sample preparation method used here, discussed in detail in the ESI,† was adapted from techniques which were previously shown to be effective for grafting the molecular Cr precursor $Cr(acac)_3$ onto amorphous silica.[33–35]

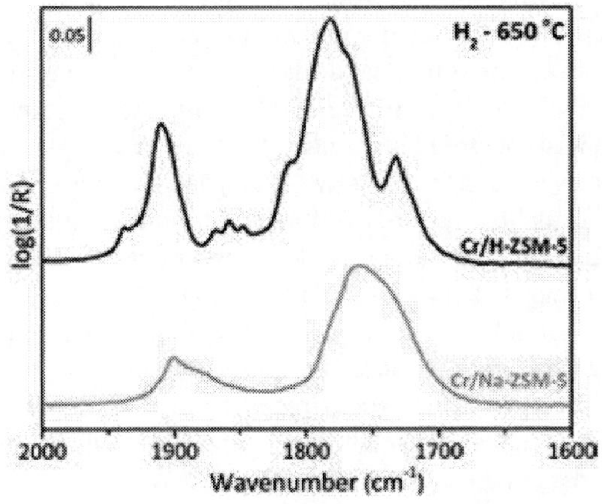

Fig. 3 FTIR-DRIFTS spectra of NO adsorbed at 30 °C on 0.5 wt% Cr/H-ZSM-5 (black) and 0.5 wt% Cr/Na-ZSM-5 (green) after reduction at 650 °C in H_2.

Furthermore, it was necessary to use complementary techniques for the above characterization. The oxidation state of Cr in Cr/H-ZSM-5 is difficult to determine on the basis of NO adsorption with DRIFTS alone, especially considering the limited number of reference spectra available in the existing literature. The location of Cr on the zeolite surface, however, is not readily apparent in the XANES spectra. The variety of Cr speciation and relative uniformity of supported Cr are distinguished more clearly using DRIFTS with NO. The results discussed so far demonstrate the potential to control the speciation of Cr using Al sites in ZSM-5 and highlight the complexity of the Cr/H-ZSM-5 material system.

3.2 Further evaluation of Cr sites in Cr/H-ZSM-5 by DRIFTS

Reduction of Cr/H-ZSM-5. Additional insights were expected to be gained by evaluating the state of Cr after reduction in conditions close to those used for ethane dehydrogenation. To further explore the state of Cr after reduction at higher temperatures, we carried out NO adsorption experiments after reduction at 650 °C in different environments. Notably, Cr was reduced even under inert gas flow (N_2) and yielded similar NO adsorption results (Fig. 4) to the sample treated at the same temperature in H_2 (Fig. 3). Furthermore, exposure at 650 °C to a 10% ethane (C_2H_6) gas flow for 30 minutes yielded similar results (Fig. 4).

Auto-reduction of Cr under inert gas has been previously observed at 500 °C by *in-situ* Raman spectroscopy of Cr/H-ZSM-5 samples prepared by aqueous impregnation.[21] In that study, Wachs and coworkers identified two types of Cr-oxo species assigned to Cr at Al sites in H-ZSM-5. The Raman band for one of these species disappeared under Ar flow at 500 °C, implying reduction of this type of Cr, while the other band was eliminated only upon exposure of the sample to methane. Similarly, the Cr/H-ZSM-5 samples in this study likely contained multiple types of Cr, but exposure to a hydrocarbon feed did not induce further reduction of Cr. Thus, it seems Cr in these samples was fully reduced under inert flow at 650 °C. Either the Cr species previously observed

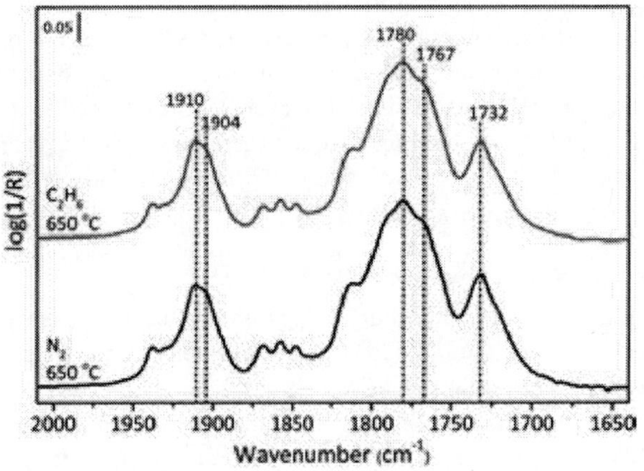

Fig. 4 FTIR-DRIFTS spectra of NO adsorbed at 30 °C on 0.5 wt% Cr/H-ZSM-5 after treatment at 650 °C in N_2 (black) or 10% C_2H_6 in N_2 (purple).

to be reduced only in the presence of methane above 500 °C was not present in the sample evaluated here, or it was present and reduced under inert flow at temperatures between 500 °C and 650 °C.

Possible forms of Cr in ZSM-5. In silicon-rich zeolites, three types of binding sites exist for bare, divalent, extra-framework cations.[5,36–39] In ZSM-5, the α, β, and γ sites are composed of Al–O–Si–O–Si–O–Al sequences present in six-membered rings, where the two negatively charged Al sites balance the 2+ charge of the extra-framework cation. In Co^{2+}-exchanged ZSM-5, it was determined that α sites are located in the walls of the straight channels, β sites are located in the intersections of the straight and sinusoidal channels, and γ sites are located in the sinusoidal channels.[37] Local environments and reactivities of extra-framework cations have been shown to depend on the sites at which cations are located.[5,36,40] For the zeolite used in this study (Zeolyst 3024E, Si/Al = 15), the fraction of Al present in these paired Al sites was estimated by Bell and coworkers[41] to be 0.39. A 3 : 1 ratio of β sites to α sites was found, along with a negligible amount of γ sites.

After calcination, dispersed Cr on amorphous SiO_2 is mostly present in the 6+ oxidation state as $[Cr({=}O)_2]^{2+}$ chromate species bound to support surface oxygens.[27] Reduction in CO above 300 °C converts these species mostly to bare, divalent Cr^{2+} cations. Either of these forms of Cr may exist at paired Al sites in ZSM-5. Cr^{3+} may also be charge-balanced by a paired Al site if it bears a single negatively-charged extra-framework ligand, such as hydroxyl in $[Cr–OH]^{2+}$. In addition, it is possible for extra-framework Cr to bind at single Al sites. Oxidized forms of these may include Cr^{5+} as $[Cr({=}O)_2]^+$ and Cr^{6+} as $[Cr({=}O)_2(OH)]^+$. During reduction, these could hypothetically convert to Cr^{3+} as $[Cr{=}O]^+$ and Cr^{2+} as $[Cr–OH]^+$, respectively. Multinuclear Cr species are possible as well, which may take the form of Cr_xO_y balanced by one or more Al sites. Other Cr species are also possible, but the structures listed here are thought to be most likely to exist and provide a framework for discussing the speciation of extra-framework Cr in ZSM-5.

The Raman bands reported in ref. 21 were compared with results from density functional theory (DFT) calculations, allowing the authors to propose structures of the two types of Cr species observed, which differ by the oxidation state of Cr and the number of Al sites that each Cr cation occupies. The structure that was reduced under inert flow was proposed to be $[Cr({=}O)_2]^{2+}$ on two Al sites or $[Cr({=}O)_2(OH)]^+$ on one Al site, and the other structure was proposed to be $[Cr({=}O)]^+$ on one Al site. We use these assignments as a basis for interpreting our DRIFTS data.

Based on the number of peaks in the NO adsorption spectra in Fig. 3 and 4, it seems there were multiple Cr species present after reduction in N_2, H_2, or C_2H_6. The population of Cr was most likely split between isolated Al sites and paired Al (α and β) sites. Additional adsorption experiments were performed in order to gain information about Cr present in reduced catalysts. Although these experiments alone were not intended to definitively identify the structures and binding sites of Cr, the study was designed as a preliminary assessment of the properties of reduced Cr species in ZSM-5 and their respective relevance to catalysis.

Cr nitrosyls on Cr/H-ZSM-5. Using DRIFTS with NO adsorption, two types of Cr-dinitrosyl species in Cr-exchanged H-ZSM-5 were distinguished by Hadjiivanov and coworkers in ref. 25. The first type, which were predominant after reduction below 500 °C, formed dinitrosyls with antisymmetric/symmetric N–O stretching bands at 1782 cm^{-1}/1910 cm^{-1}. These were assigned as Cr^{3+}-dinitrosyl species. The second type, which became more abundant with reduction temperatures above 500 °C, formed dinitrosyls observed at 1768 cm^{-1}/1902 cm^{-1}. These were assigned as Cr^{2+}-dinitrosyls. During evacuation, the second type of dinitrosyl decomposed, forming mononitrosyl species with an N–O stretching band at 1783 cm^{-1}. No Cr-nitrosyls were formed before reduction.

These Cr-nitrosyl species were present in the Cr/H-ZSM-5 prepared in this study as well. After reduction in N_2 flow at 650 °C, dinitrosyl antisymmetric/symmetric modes at 1780 cm^{-1}/1911 cm^{-1} and 1767 cm^{-1}/1904 cm^{-1} and a mononitrosyl N–O stretching mode at 1789 cm^{-1} were observed under static NO pressure at 30 °C (Fig. 5a, gray curve). As the DRIFTS cell was purged with N_2, weakly bound NO desorbed (Fig. 5a, red curves). As described in ref. 25, peaks at 1767 cm^{-1}/1904 cm^{-1} decreased simultaneously with an increase of intensity at 1789 cm^{-1}, consistent with conversion to a mononitrosyl. A slight gain of intensity at 1780 cm^{-1}/1911 cm^{-1} suggests a small fraction of these dinitrosyls formed by desorption of NO from Cr species with more than two nitrosyl ligands. In addition to the peaks observed in ref. 25, minor peaks at 1732 cm^{-1}, 1814 cm^{-1}, 1848 cm^{-1}, 1858 cm^{-1}, 1869 cm^{-1}, 1930 cm^{-1}, and 1938 cm^{-1} decreased in intensity during brief N_2 flow. Previously, addition of more than two NO ligands into the coordination sphere of Cr in Cr/SiO_2 resulted in shifting of dinitrosyl bands and the appearance of additional IR bands.[23] It was proposed that a small fraction of tetranitrosyl species formed at Cr sites with excess electron density, giving rise to peaks blue-shifted from the dinitrosyl symmetric stretching frequency (above 1875 cm^{-1}) and red-shifted from the dinitrosyl antisymmetric stretching frequency (below 1740 cm^{-1}). In the Cr/H-ZSM-5 investigated here, peaks at high and low frequency ends of the spectrum may be due to Cr species with more than two NO ligands. Another possibility is that peaks blue-shifted from the dominant dinitrosyl peaks result from a small amount of Cr present in higher oxidation states. Peaks at 1814 cm^{-1}, 1930 cm^{-1}, and 1938 cm^{-1} could be due to a small amount of Cr^{3+} remaining after reduction, in line with peaks observed previously for Cr/Al_2O_3.[42] Alternatively, Cr^{3+}/Cr^{4+} could result from oxidation of Cr^{2+} by NO or N_2O. In fact, a small peak at 2249 cm^{-1} due to N_2O was observed during NO adsorption (Fig. S11†).

Co-adsorption of NO and CO. To obtain additional information on the oxidation state of Cr, the sample was exposed to CO, which is a sensitive probe of electronic state

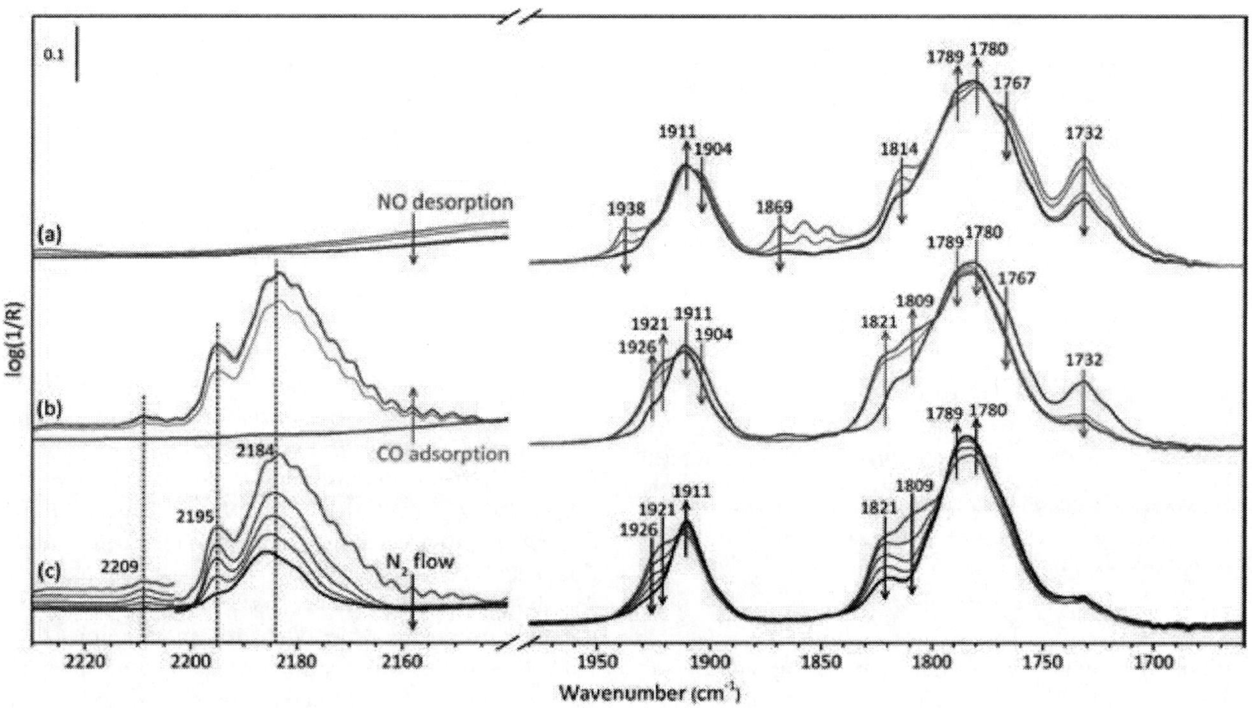

Fig. 5 (a) FTIR-DRIFTS spectra of NO adsorbed at 30 °C on 0.5 wt% Cr/H-ZSM-5 after treatment at 650 °C in N_2 (gray) and subsequent spectra taken during N_2 flow (red). (b) FTIR-DRIFTS spectra taken after 10 minutes of NO desorption under N_2 flow (red) and during subsequent CO adsorption (blue). (c) FTIR-DRIFTS spectra taken after 10 minutes of CO adsorption (blue) and during CO desorption under N_2 flow (black). Data above 2203 cm^{-1} offset in (c) for clarity.

of transition metals. Although it was previously reported that CO did not form stable carbonyl complexes with Cr cations in reduced Cr-H-ZSM-5 at room temperature,[26] we report here that Cr-carbonyl species formed on Cr cations already containing one or more NO ligands.

Following NO adsorption (Fig. 5a, gray curve), weakly bound NO desorbed under N_2 flow until the spectra were fairly stable over time (Fig. 5a and b, red curves). Next, CO was introduced to the cell (see ESI† for methodology). Peaks at 1732 cm^{-1}, 1767 cm^{-1}/1904 cm^{-1}, 1780 cm^{-1}/1911 cm^{-1}, and 1789 cm^{-1} lost intensity, while new peaks developed at higher frequencies (Fig. 5b and c, blue curves). This was due to addition of CO to Cr-dinitrosyl species, as observed in Cr/SiO_2 and Cr/silicalite samples.[23,31] Blue-shifting of dinitrosyl peaks is explained by decreased electron density at Cr after addition of CO. Simultaneously, carbonyl peaks appeared in the region 2170 cm^{-1} to 2210 cm^{-1}, typical of supported-Cr catalysts.[27]

As gas-phase CO was purged from the DRIFTS cell by N_2 flow, Cr-carbonyl peaks became more visible (Fig. 5c). The major peak at 2184 cm^{-1} and shoulder at 2179 cm^{-1} can be assigned to the symmetric and antisymmetric stretching vibrations of Cr^{2+} dicarbonyls.[27] Interestingly, carbonyl peaks at 2184 cm^{-1} and 2179 cm^{-1}, which are considered a fingerprint of Cr^{2+} dicarbonyls in Cr/SiO_2 Phillips catalysts, were not observed for Cr/Na-ZSM-5 under similar conditions (Fig. S13†). Cr/Na-ZSM-5 displayed only one broad carbonyl band centered at 2174 cm^{-1}. A small shoulder at 2171 cm^{-1} was also present for Cr/H-ZSM-5. Peaks around 2170 cm^{-1} were assigned in previous reports to carbonyls co-adsorbed with NO to Cr^{2+} sites in Cr/SiO_2.[27,43] Presumably, electron donation from NO to Cr causes red-shifting of carbonyl bands from those formed in the absence of NO.

Finally, two carbonyl peaks with unusually small FWHM (<10 cm^{-1}) were observed at 2195 cm^{-1} and 2209 cm^{-1}. The sharpness of these peaks suggests they are due to Cr carbonyl species with uniform structures, likely located at specific Al sites in the zeolite.[44] Cr carbonyls around 2195 cm^{-1} can result from Cr^{2+} or Cr^{3+} cations,[17] but definitive assignment is difficult due the confounding impacts of the zeolite support and co-adsorbed nitrosyls in the ligand sphere of Cr. The 2209 cm^{-1} peak is likely due to the presence of a small amount of Cr^{3+} or Cr^{4+}. A carbonyl band at 2204–2206 cm^{-1} was previously assigned to adsorption of CO on Cr^{4+}(=O) sites in Cr/SiO_2 formed by oxidation of Cr^{2+} by N_2O.[43,45]

Upon desorption of CO (Fig. 5c, black curves), the intensity of nitrosyl bands associated with Cr species having both NO and CO ligands decreased, while bands of Cr nitrosyl species at 1780 cm^{-1}/1911 cm^{-1} and 1789 cm^{-1} were restored. The overlap of peaks in Fig. 5c makes distinguishing individual peaks difficult, but close examination of difference spectra facilitates assignment of groups of peaks to individual Cr species. Fig. 6 displays difference spectra obtained by subtraction of spectra collected at increasing times during CO desorption. Sharp loss peaks at 1809 cm^{-1} and 1921 cm^{-1} dominate during the first minute of N_2 flow, suggesting that a mixed NO/CO Cr complex lost at least one

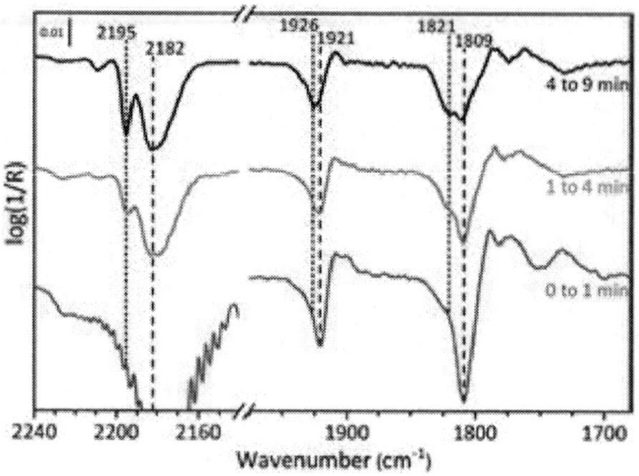

Fig. 6 FTIR-DRIFTS difference spectra of NO and CO adsorbed at 30 °C on 0.5 wt% Cr/H-ZSM-5 after treatment at 650 °C in N_2. Spectra were taken during increasing time of N_2 flow and were processed by subtracting the earlier time spectrum from the later time spectrum, as labelled on the figure.

weakly bound carbonyl ligand. Absorbance of gas-phase CO obscured the carbonyl region in this spectrum but was no longer a problem at later times. Between one and four minutes of CO desorption, the nitrosyl region was still dominated by loss peaks at 1809 cm^{-1} and 1921 cm^{-1}, and in the carbonyl region by a loss peak at 2182 cm^{-1}. From four to nine minutes, loss peaks at 1821 cm^{-1}, 1926 cm^{-1}, and 2195 cm^{-1} grew more prominent. This allows tentative assignment of two Cr NO/CO species: $Cr^{2+/3+}(NO)_2(CO)$ with $\nu_{NO,as}/\nu_{NO,s}/\nu_{CO}$ at 1821/1926/2195 cm^{-1}, and $Cr^{2+}(NO)_2(CO)$ with $\nu_{NO,as}/\nu_{NO,s}/\nu_{CO}$ at 1809/1921/2182 cm^{-1}. The carbonyl loss peak at 2182 cm^{-1} is indicative of a Cr^{2+} monocarbonyl, whereas peaks at 2184/2179 cm^{-1} in Fig. 5c are characteristic of Cr^{2+} dicarbonyls.[27] Thus it is possible that dicarbonyl species $Cr^{2+}(NO)_2(CO)_2$ or $Cr^{2+}(NO)(CO)_2$ were formed as well. These results show that several highly coordinatively unsaturated Cr species existed after reduction in N_2 at 650 °C.

It is evident that further characterization is necessary to definitively claim the surface binding configuration of Cr in reduced Cr/H-ZSM-5. Rather than attempt to extract more information from the single sample, we prepared a set of Cr/H-ZSM-5 catalysts with increasing Cr loading in order to observe changes in the capacity of ZSM-5 to stabilize dispersed Cr active sites. It was expected that at higher loadings, Cr would anchor at additional locations on the zeolite surface and would be more likely to form multinuclear Cr species. Meaningful trends in ethane dehydrogenation performance would indicate broadly which forms of Cr are most relevant for dehydrogenation, and how a zeolite support may be utilized to design improved dehydrogenation catalysts.

3.3 Capacity of ZSM-5 as a support for Cr

On the basis of interpretation of Raman spectra of Cr/H-ZSM-5 (Si/Al = 15, Zeolyst CBV 3024E), it was observed in ref. 21 that at Cr loading above 0.5 wt% (0.10 Cr/Al), Cr tended to bind at silanol sites after calcination. Particularly when there were fewer Al binding sites for each Cr (high Cr/Al), reaction conditions were thought to force Cr from Al to silanol sites where they were less active for methane conversion. Similarly, activity and stability of ethane conversion reported herein were higher for Cr located at Al sites compared with Cr at silanols (Fig. 1). Motivated by these observations, we prepared Cr/H-ZSM-5 samples using the same zeolite product (Zeolyst CBV 3024E) varying Cr loading from 0.27 to 1.6 wt%, corresponding to 0.05 to 0.30 Cr/Al (molar ratio). It has been reported[41] that the fraction of Al present in pairs in this zeolite was 0.39 – therefore, a maximum of *ca.* 0.20 Cr/Al could be stabilized at Al sites if a pair of Al sites is required to stabilize each Cr cation. The range of Cr/Al in this set of samples was chosen in order to observe differences among compositions below and above this 0.20 Cr/Al threshold.

Ethane DH performance as a function of Cr loading. Catalysis results for ethane dehydrogenation at 650 °C by Cr/H-ZSM-5 samples with varying Cr loading are shown in Fig. 7. All reactions were performed at ethane conversion below 8% (see ESI† for details). Initial rates of ethane conversion, normalized per mole of Cr, were essentially constant at low Cr loadings (Fig. 7a). The slight downward trend at low Cr/Al is thought to be caused by added ethane conversion from Brønsted acid sites in the zeolite, which becomes more relevant with decreasing Cr loading (increasing Brønsted acid/Cr). Based on the ethane conversion measured for bare H-ZSM-5, it is estimated that for 0.10 Cr/Al, 18% of the measured ethane conversion may be attributed to activity of the zeolite support. This number decreases with increasing Cr loading. For 0.30 Cr/Al, only 2% of total activity is estimated to be due to activity of the zeolite support. If a simple correction is made, subtracting the conversion measured for an equivalent mass of bare H-ZSM-5, the trend at Cr/Al $\leq$ 0.12 becomes relatively flat (Fig. S15†). Above Cr/Al = 0.12, rates of ethane conversion per mole Cr increase monotonically, contrary to what was found for methane conversion.[21] Doubling Cr/Al from 0.15 to 0.30 results in more than doubling of the initial rate of ethane conversion from 2.3 to 4.8 mol min^{-1} mol^{-1} Cr. Apparent activation energies for 0.10 Cr/Al and 0.17 Cr/Al samples are fairly similar (Fig. 7c), indicating that similar active sites may have been responsible for most of the activity in these samples.

Ethylene selectivity also increased with Cr/Al, while selectivities to side products methane, propane, and propylene decreased (Fig. 7d–f). Most likely, methane and C_3 side products formed largely by oligomerization and cracking on Brønsted acid sites. When ethane dehydrogenation was performed over H-ZSM-5, 21.4% selectivity to methane and 10.8% selectivity to C_3 products were measured at 0.72% ethane conversion (Table S2†). With increasing Cr loading, the activity of ethylene-selective Cr sites was increasingly dominant with respect to catalysis at Brønsted acid sites, resulting in higher ethylene selectivity. Because Cr can bind at Al sites,

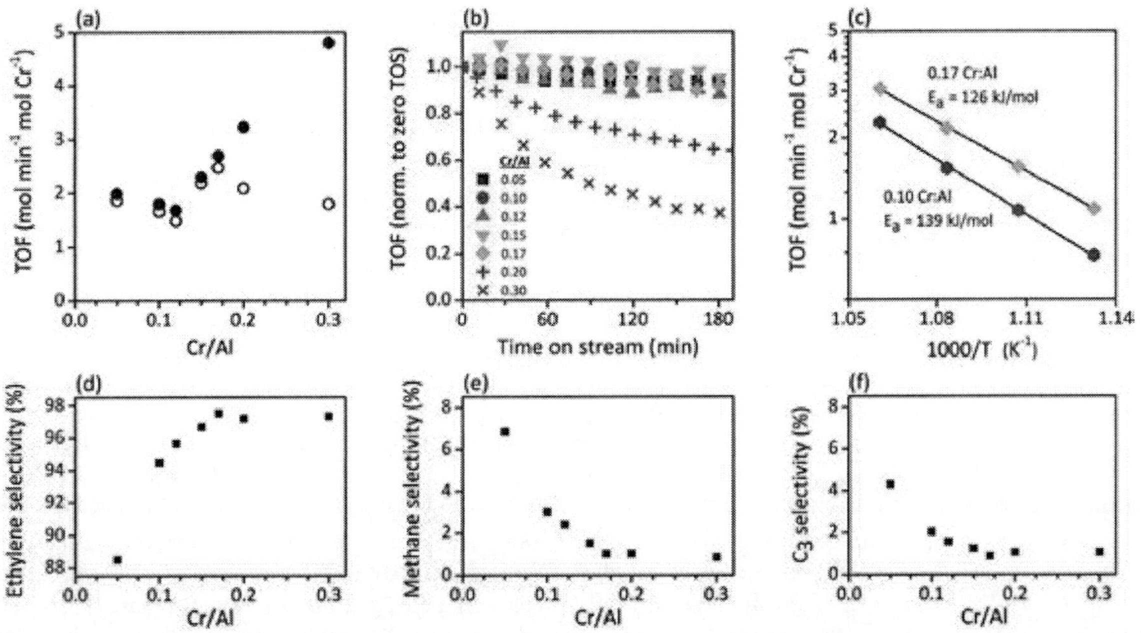

Fig. 7 Ethane dehydrogenation performance of Cr/H-ZSM-5 catalysts with increasing Cr loading (0.10 Cr/Al corresponds to 0.54 wt% Cr). Reaction conditions: T = 650 °C; P = 3 psig; flow rates: 4 sccm ethane, 44 sccm N_2; catalyst amount: 8–45 mg. (a) Initial turnover frequency (TOF) of ethane conversion (filled circles) and TOF after 3 h (empty circles) as a function of Cr/Al molar ratio. (b) TOF of ethane conversion, normalized to the initial value, as a function of time on stream. (c) TOF of ethane conversion *versus* inverse temperature. Apparent activation energies labelled on the plot were calculated by fitting data to the Arrhenius equation. Ethane conversion was measured at 670, 650, 630, and 610 °C. (d)–(f) Initial selectivity to (d) ethylene, (e) methane, and (f) propane and propylene combined *versus* Cr/Al.

increasing Cr/Al likely also decreased the concentration of Brønsted acid sites, further suppressing side reactions.

Differences in stability of ethane dehydrogenation activity on stream were observed as well (Fig. 7b). For Cr/Al ≤ 0.17, more than 85% of initial activity was retained after three hours on stream. Increasing Cr/Al above 0.17, faster deactivation was observed, with the 0.30 Cr/Al sample losing over 60% of its initial activity over 3 h. With increasing Cr/Al, the rate of ethane conversion per mole Cr after 3 h on stream passed through a maximum at 0.17 Cr/Al. Carbon deposition (coking) and aggregation of Cr are thought to be the main contributors to deactivation observed in these experiments. N_2 physisorption measurements of the 0.10 Cr/Al sample showed no loss of micropore volume after reaction (Fig. S3 and Table S1†). After 3 h operating at 4% ethane conversion, the 0.17 Cr/Al sample contained about 1.2% mass of carbon, as measured by TGA-MS (Fig. S10†). This amount of carbon deposition did not, however, result in substantial deactivation. Furthermore, catalytic tests of the 0.30 Cr/Al sample performed at different initial ethane conversions (7.5% and 2.8%) resulted in nearly identical amounts of deactivation (Fig. S16†), suggesting that increasing partial pressure of products (*i.e.* ethylene) did not influence the rate of deactivation. Together, these results suggest that, although coke can form at Cr sites or Brønsted acid sites, coking was not the primary cause of deactivation. Instead, when substantial deactivation occurred, aggregation of Cr was likely the cause.

Paired Al sites are expected to be the most stable binding sites for Cr, as Cr generally requires at least two negatively charged support oxygens.[32,46] With increasing Cr/Al, if paired Al sites become populated, additional Cr is more likely to either: i) populate isolated Al sites, ii) form multinuclear Cr species, or iii) populate silanol sites. To assess whether any of these possibilities were responsible for the observed trends in dehydrogenation activity with increasing Cr/Al, samples with various Cr loadings were characterized by DRIFTS.

DRIFTS of Cr/H-ZSM-5 with varying Cr loading. DRIFTS spectra in the hydroxyl region of H-ZSM-5 and Cr/H-ZSM-5 samples with 0.10 Cr/Al, 0.17 Cr/Al, and 0.30 Cr/Al after 1 h

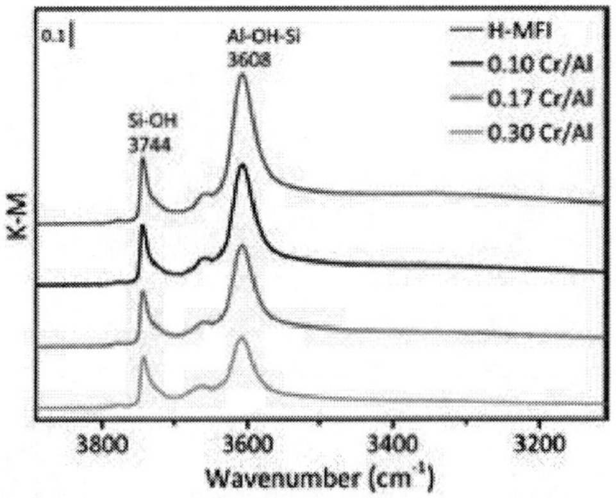

Fig. 8 DRIFTS spectra in the hydroxyl region of H-ZSM-5 and Cr/H-ZSM-5 samples with Cr loading 0.10, 0.17, and 0.30 Cr/Al. Measurements were taken at 100 °C under N_2 after treatment under N_2 at 650 °C for 1 h. Spectra were normalized by the intensity of the zeolite Si–O–Si overtone peak at 1987 cm^{-1}.

under N_2 flow at 650 °C are shown in Fig. 8. The two main peaks are due to bridging hydroxyls located at zeolite framework Al sites (3608 cm^{-1}) and external silanols (3744 cm^{-1}).[21,25] The two minor peaks at 3658 cm^{-1} and 3777 cm^{-1} are due to hydroxyls of extra-framework Al.[21,24] With increasing Cr loading, the intensity of the Al bridging hydroxyl peak decreased substantially, suggesting that Cr binds at Al sites. The silanol band decreased as well, showing that a small fraction of Cr resides at silanol sites at all loadings. However, since the bridging hydroxyl peak loses considerably more intensity with increased Cr loading, we conclude that even up to loading of 0.30 Cr/Al, most of the Cr is located at Al sites after the heat treatment in inert. Very little, if any, change in the extra-framework Al hydroxyl peaks was observed, implying that these sites do not play a significant role in Cr speciation.

NO adsorption on 0.10, 0.17, and 0.30 Cr/Al samples after treatment in N_2 at 650 °C allowed comparison of the distribution of Cr species present among samples with low, intermediate, and high dehydrogenation activity, respectively (Fig. 9). Intensity of all nitrosyl peaks increased with Cr loading, but trends in relative peak area were observed. First, the dinitrosyl characterized by 1767 cm^{-1}/1904 cm^{-1} bands grew with respect to the dinitrosyl bands at 1780 cm^{-1}/1911 cm^{-1}. Most strikingly, absorbance at low frequencies grew substantially with increasing Cr/Al. A peak at 1732 cm^{-1} with a shoulder at 1720 cm^{-1} grew with Cr/Al, and at 0.30 Cr/Al, a shoulder around 1695 cm^{-1} became clearly visible. Additionally, at 0.30 Cr/Al, increases in intensity around 1750 cm^{-1} and 1890 cm^{-1} suggest that a higher fraction of Cr resides at silanol sites. However, it is unlikely that peaks at 1732 cm^{-1} and 1720 cm^{-1} are due to Cr at silanol sites since resolved peaks in this range were not observed in Cr/Na-ZSM-5, and they were not reported previously for Cr/SiO_2 or Cr/silicalite.[23,27,31]

The presence of Cr at silanol sites at high Cr/Al is consistent with the quick deactivation observed in these samples, but as Cr/Na-ZSM-5 had the lowest activity among all samples tested, it is unlikely that Cr at silanol sites is

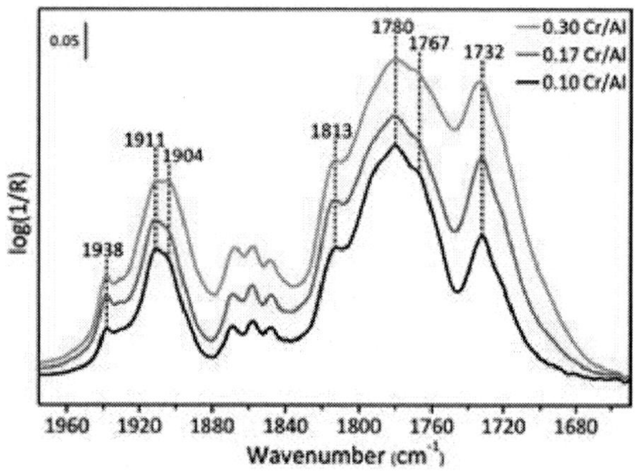

Fig. 9 FTIR-DRIFTS spectra of NO adsorbed at 30 °C on Cr/H-ZSM-5 samples with Cr loading 0.10, 0.17, and 0.30 Cr/Al after treatment under N_2 at 650 °C for 1 h.

responsible for the high activity observed at high Cr/Al. Instead, Cr species giving rise to nitrosyl peaks at 1732 cm^{-1} and 1720 cm^{-1} may be responsible for higher dehydrogenation activity with increasing Cr/Al. Previously reported Cr^{2+} dinitrosyls[23,27,31] do not display antisymmetric N–O stretching frequencies below 1742 cm^{-1}. In a prior study, pyridine co-adsorbed with NO on Cr/SiO_2 caused red-shifting of nitrosyl frequencies by 20–60 cm^{-1}. Electron donation from pyridine to Cr allowed stronger interaction between Cr 3d and NO π* orbitals, weakening the NO bond.[31] Similarly, the low frequency of the nitrosyl peaks that dominate at high Cr/Al in our samples suggests that these Cr species are more electron-rich than Cr species that are most abundant at low Cr/Al. We expect these to be Cr^{2+} species, as Cr^{+} is not generally observed in supported Cr catalysts. In one study, red-shifting of dinitrosyl stretching bands to 1736 cm^{-1} and 1868 cm^{-1} led earlier researchers to propose that a dichromate-like species in Cr/SiO_2–TiO_2 may display higher electron density than a monochromate.[47] The similarity of these reported frequencies to peaks observed in the present samples at 1732 cm^{-1} and 1868 cm^{-1} lead us to the tentative assignment of these peaks to stretching bands of Cr^{2+} dinitrosyls exhibiting higher electron density, possibly in the form of multinuclear Cr species (Cr_xO_y) which become more abundant as Cr loading increases.

It was previously concluded that multinuclear Cr in Cr/Al_2O_3 are more active for propane dehydrogenation than mononuclear Cr.[15] However, when multinuclear Cr aggregates become too large, dehydrogenation activity drops. This is consistent with the present result that highly active samples with high Cr/Al also deactivate more quickly. The higher surface density of Cr may encourage formation of multinuclear active sites, but excessive aggregation can result in lowered activity and blocking of micropore volume. Insufficient quantity of stable binding sites for Cr may also result in migration of Cr to external silanols, another potential source of deactivation.[21] While a detailed understanding of deactivation mechanisms was not the focus of this study, it must be kept in mind that the NO adsorption experiments were performed after treatment at 650 °C for 1 h. If catalysts deactivate over this short time scale at that temperature, the NO adsorption spectrum may not fully reflect the active state of the sample. Future work will be aimed at identifying unique active structures of Cr stabilized on the zeolite surface.

3.4 Impact of paired Al sites as binding sites for Cr

In an attempt to gain insight into the binding of Cr at Al sites, a sample was prepared using ZSM-5 that was exchanged with Ca^{2+} prior to Cr deposition. The divalent Ca^{2+} cation was expected to exchange preferentially at paired Al sites,[3] selectively blocking Cr from occupying these sites. The resulting catalyst, 0.5 wt% Cr/Ca-ZSM-5 (0.10 Cr/Al), displayed more than triple the initial activity of Cr/H-ZSM-5 of equal Cr loading. It was more active per mole of Cr than 0.30 Cr/Al Cr/

H-ZSM-5 and showed less deactivation (Fig. S17†). While the 0.30 Cr/Al Cr/H-ZSM-5 sample lost over 60% of its initial activity over 3 h, 0.10 Cr/Al Cr/Ca-ZSM-5 lost less than 35% of initial activity over the same amount of time. Thus, a highly active catalyst with enhanced stability resulted when the quantity of paired Al sites was reduced and the Cr loading kept reasonably low.

The DRIFTS spectrum of the hydroxyl region of Cr/Ca-ZSM-5 provides information about the location of Cr in this sample (Fig. 10a). While the silanol bands at 3744 cm^{-1} changed very little in 0.10 Cr/Al Cr/H-ZSM-5 and Cr/Ca-ZSM-5 compared with H-ZSM-5, clear changes were observed in the intensities of the bridging hydroxyl peaks at 3608 cm^{-1}. Comparison of peak intensities provides a rough estimate of concentration of Al bridging hydroxyls in the samples. The bridging hydroxyl peak height for 0.10 Cr/Al Cr/H-ZSM-5 was 81% of the peak height of H-ZSM-5, which yields a ratio of two Al sites occupied per Cr. If Ca^{2+} exchange removed hydroxyls at all paired Al sites, this would account for a loss of 39% of the bridging hydroxyls compared with H-ZSM-5. Assuming also a ratio of two Al sites consumed per Cr cation for 0.10 Cr/Al Cr/Ca-ZSM-5, it would be expected that 41% of Al sites would remain after addition of Cr. In fact, the bridging hydroxyl peak height of the Cr/Ca-ZSM-5 sample was found to be 43% of the peak height in H-ZSM-5. This suggests that a considerable amount of Al sites were titrated by Ca^{2+} exchange, forcing a greater fraction of Cr to bind at isolated Al sites.

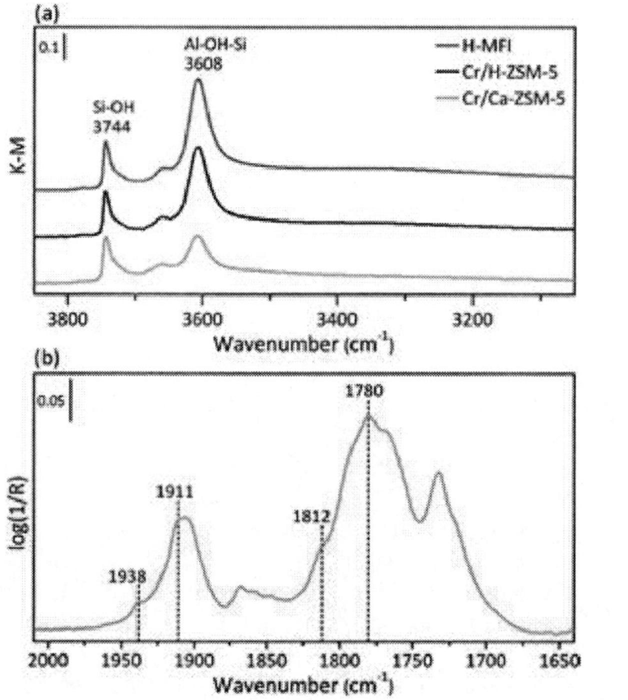

Fig. 10 (a) FTIR-DRIFTS spectra of the hydroxyl region of H-ZSM-5, Cr/H-ZSM-5 (0.10 Cr/Al), and Cr/Ca-ZSM-5 (0.10 Cr/Al) under N_2 at 100 °C after treatment at 650 °C for 30 min. H-ZSM-5 and Cr/H-ZSM-5 were treated under N_2 and Cr/Ca-ZSM-5 was treated under H_2. Spectra were normalized by the intensity of the zeolite Si–O–Si overtone peak at 1987 cm^{-1}. (b) FTIR-DRIFTS spectrum of NO adsorbed at 30 °C on Cr/Ca-ZSM-5 (0.10 Cr/Al) after treatment in H_2 at 650 °C for 30 min.

The NO adsorption DRIFTS spectrum (Fig. 10b) of Cr/Ca-ZSM-5 reduced in H_2 at 650 °C showed a drop in relative intensity of the pairs of peaks at 1780 cm^{-1}/1911 cm^{-1} and 1812 cm^{-1}/1938 cm^{-1} compared with 0.10 Cr/Al Cr/H-ZSM-5 (Fig. 3). Peaks at 1767 cm^{-1}/1904 cm^{-1}, 1732 cm^{-1}, and 1720 cm^{-1} became more prominent, similar to the trend observed with increasing Cr/Al for Cr/H-ZSM-5 (Fig. 9). The broad absorbance between 1732 cm^{-1} and 1767 cm^{-1} which was observed at high Cr/Al was also observed for Cr/Ca-ZSM-5. However, based on the spectrum of the hydroxyl region of Cr/Ca-ZSM-5, the broad absorbance in the NO adsorption spectrum seems not to result from Cr at silanol sites. Instead, this absorbance may be due to nitrosyls of multinuclear Cr species at Al sites. Furthermore, the sharp carbonyl peaks at 2195 cm^{-1} and 2208 cm^{-1} were less pronounced for Cr/Ca-ZSM-5 (Fig. S14†), suggesting that these are carbonyls of Cr located at paired Al sites. It is also possible that the pairs of peaks at 1780 cm^{-1}/1911 cm^{-1} and 1812 cm^{-1}/1938 cm^{-1}, which were lower in intensity when there were fewer available paired Al sites, are due to dinitrosyls of Cr located at α and β paired Al sites.

In Cr/Ca-ZSM-5, Cr was directed away from paired Al sites, binding instead at isolated Al sites and possibly forming multinuclear species. This catalyst, as a result, displayed much higher activity than when Cr was allowed to bind at paired Al sites. 0.10 Cr/Al Cr/Ca-ZSM-5 also displayed more stable activity on stream than 0.30 Cr/Al Cr/H-ZSM-5, which had similar initial activity. The lower surface density of Cr in Cr/Ca-ZSM-5 may be the reason for higher stability, as excessive aggregation was reduced with respect to 0.30 Cr/Al Cr/H-ZSM-5.

Conclusions

Speciation of Cr at Al sites in zeolite ZSM-5 meaningfully impacts ethane dehydrogenation activity and stability. Reduced Cr was found to anchor mainly at zeolite Al sites in all Cr/H-ZSM-5 samples, and clear trends in catalysis were observed with increasing Cr loading up to 1.6 wt% (0.30 Cr/Al). Specifically, at low Cr/Al, catalysts were associated with stable dehydrogenation activity, while at increased Cr/Al, catalysts were more active but less stable on stream. A variety of Cr structures present in Cr/H-ZSM-5 samples were differentiated by NO and CO probe molecule adsorption and co-adsorption DRIFTS measurements. Several Cr species that possessed high degrees of coordinative unsaturation (unsaturated to the extent that they could coordinate at least three adsorbate molecules at 30 °C) were observed. NO adsorption DRIFTS indicated that the most active samples contained greater amounts of more electron-rich, possibly multinuclear, Cr^{2+} species. The presence of Ca^{2+} on ZSM-5 before Cr introduction modified the distribution of reduced Cr species and resulted in a threefold increase in dehydrogenation activity.

This evidence suggests that highly active and stable Cr dehydrogenation catalysts may be designed by controlling the distribution of Al binding sites on zeolite supports, and that probe molecule spectroscopy is an effective tool for differentiating reduced Cr sites in zeolites. Although the present study does not definitively identify the structures and binding sites of Cr in ZSM-5, the results educate future investigation and indicate the value of understanding this catalyst system in greater detail. Additional research combining experiment and theory towards identification and further evaluation of the variety of dispersed Cr species stabilized by zeolites will be useful for development of catalysts containing uniform Cr active sites.

Conflicts of interest

The authors have no conflicts to declare.

Acknowledgements

The authors acknowledge Dr. Simon Bare and Dr. Adam Hoffman for their help in collecting XAS data. Use of the Stanford Synchrotron Radiation Lightsource, SLAC National Accelerator Laboratory, is supported by the U.S. Department of Energy, Office of Science, Office of Basic Energy Sciences under Contract No. DE-AC02-76SF00515. Co-ACCESS funded by the Chemical Sciences Program. Chemical Sciences, Geosciences, and Biosciences Division, Basic Energy Sciences of the U.S. DOE Office of Science is acknowledged for the X-ray absorption spectroscopy.

Notes and references

1 L. Liu, M. Lopez-Haro, C. W. Lopes, C. Li, P. Concepcion, L. Simonelli, J. J. Calvino and A. Corma, *Nat. Mater.*, 2019, **18**, 866.
2 C. Paolucci, I. Khurana, A. A. Parekh, S. Li, A. J. Shih, H. Li, J. R. Di Iorio, J. D. Albarracin-Caballero, A. Yezerets, J. T. Miller, W. N. Delgass, F. H. Ribeiro, W. F. Schneider and R. Gounder, *Science*, 2017, **357**, 898.
3 O. Bortnovsky, Z. Sobalik and B. Wichterlova, *Microporous Mesoporous Mater.*, 2001, **46**, 265.
4 P. Sazama, E. Tabor, P. Klein, B. Wichterlova, S. Sklenak, L. Mokrzycki, V. Pashkkova, M. Ogura and J. Dedecek, *J. Catal.*, 2016, **333**, 102.
5 D. Kaucky, A. Vondrova, J. Dedecek and B. Wichterlova, *J. Catal.*, 2000, **194**, 318.
6 A. R. Kulkarni, Z. J. Zhao, S. Siahrostami, J. K. Norskov and F. Studt, *Catal. Sci. Technol.*, 2018, **8**, 114.
7 M. A. Newton, A. J. Knorpp, V. L. Sushkevich, D. Palagin and J. A. Van Bokhoven, *Chem. Soc. Rev.*, 2020, **49**, 1449.
8 S. Moussa, P. Concepcion, M. A. Arribas and A. Martinez, *ACS Catal.*, 2018, **8**, 3903.
9 Y. Zhang, S. Wu, X. Xu and H. Jiang, *Catal. Sci. Technol.*, 2020, **10**, 835.
10 S. S. Arzumanov, A. A. Gabrienko, A. V. Toktarev, Z. N. Lashchinskaya, D. Freude, J. Haase and A. G. Stepanov, *J. Phys. Chem. C*, 2019, **123**, 30473.
11 A. Mehdad, N. S. Gould, B. Xu and R. F. Lobo, *Catal. Sci. Technol.*, 2018, **8**, 358.
12 N. M. Phadke, J. Van Der Mynsbrugge, E. Mansoor, A. B. Getsoian, M. Head-Gordon and A. T. Bell, *ACS Catal.*, 2018, **8**, 6106.
13 X. Niu, X. Nie, C. Yang and J. G. Chen, *Catal. Sci. Technol.*, 2020, **10**, 1881.
14 C. A. Gartner, A. C. vanVeen and J. A. Lercher, *ChemCatChem*, 2013, **5**, 3196.
15 R. L. Puurunen and B. M. Weckhuysen, *J. Catal.*, 2002, **210**, 418.
16 B. M. Weckhuysen, R. A. Schoonheydt, J. M. Jehng, I. E. Wachs, S. J. Cho, R. Ryoo, S. Kijlstra and E. Poels, *J. Chem. Soc., Faraday Trans.*, 1995, **91**, 3245.
17 G. A. Martino, C. Barzan, A. Piovano, A. Budnyk and E. Groppo, *J. Catal.*, 2018, **357**, 206.
18 B. M. Weckhuysen, H. J. Spooren, R. A. Schoonheydt and K. U. Leuven, *Zeolites*, 1994, **14**, 450.
19 B. M. Weckhuysen, D. Wang, M. P. Rosynek and J. H. Lunsford, *J. Catal.*, 1998, **346**, 338.
20 P. Michorczyk, K. Zenczak-Tomera, B. Michorczyk, A. Wegrzyniak, M. Basta, Y. Millot, L. Valentin and S. Dzwigaj, *J. CO_2 Util.*, 2020, **36**, 54.
21 J. Gao, Y. Zheng, Y. Tang, J. M. Jehng, R. Grybos, J. Handzlik, I. E. Wachs and S. G. Podkolzin, *ACS Catal.*, 2015, **5**, 3078.
22 J. Janas, J. Gurgul, R. P. Socha, J. Kowalska, K. Nowinska, T. Shishido, M. Che and S. Dzwigaj, *J. Phys. Chem. C*, 2009, **113**, 13273.
23 A. Zecchina, G. Spoto, G. Ghiotti and E. Garrone, *J. Mol. Catal.*, 1994, **86**, 423.
24 K. Hadjiivanov, A. Penkova, R. Kefirov, S. Dzwigaj and M. Che, *Microporous Mesoporous Mater.*, 2009, **124**, 59.
25 M. Mihaylov, A. Penkova, K. Hadjiivanov and M. Daturi, *J. Mol. Catal. A: Chem.*, 2006, **249**, 40.
26 A. Penkova and K. Hadjiivanov, *Catal. Commun.*, 2003, **4**, 485.
27 E. Groppo, C. Lamberti, S. Bordiga, G. Spoto and A. Zecchina, *Chem. Rev.*, 2005, **105**, 115.
28 M. Botavina, C. Barzan, A. Piovano, L. Braglia, G. Agostini, G. Martra and E. Groppo, *Catal. Sci. Technol.*, 2017, **7**, 1690.
29 A. Zecchina, E. Garrone, C. Morterra and S. Coluccia, *J. Phys. Chem.*, 1975, **79**, 978.
30 G. Ghiotti, E. Garrone, G. Della Gatta, B. Fubini and E. Giamello, *J. Catal.*, 1983, **80**, 249.
31 E. Garrone, G. Ghiotti, C. Morterra and A. Zecchina, *Z. Naturforsch., B: J. Chem. Sci.*, 1987, **42**, 728.
32 B. M. Weckhuysen, I. E. Wachs and R. A. Schoonheydt, *Chem. Rev.*, 1996, **96**, 3327.
33 I. V. Babich, Y. V. Plyuto, P. Van Der Voort and E. F. Vansant, *J. Colloid Interface Sci.*, 1997, **189**, 144.
34 B. M. Weckhuysen, R. Ramachandra Rao, J. Pelgrims, R. A. Schoonheydt, P. Bodart, G. Debras, O. Collart, P. Van Der Voort and E. F. Vansant, *Chem. - Eur. J.*, 2000, **6**, 2960.
35 A. Hakuli and A. Kytokivi, *Phys. Chem. Chem. Phys.*, 1999, **1**, 1607.
36 Z. Sobalik, J. Dedecek, D. Kaucky, B. Wichterlova, L. Drozdova and R. Prins, *J. Catal.*, 2000, **194**, 330.
37 J. Dedecek, D. Kaucky and B. Wichterlova, *Microporous Mesoporous Mater.*, 2000, **35–36**, 483.

38 J. Dedecek, Z. Sobalik and B. Wichterlova, *Catal. Rev.: Sci. Eng.*, 2012, **54**, 135.
39 J. J. Dedecek, S. Sklenak, C. Li, B. Wichterlova, V. Gabova, J. Brus, M. Sierka and J. Sauer, *J. Phys. Chem. C*, 2009, **113**, 1447.
40 S. Sklenak, P. C. Andrikopoulos, S. R. Whittleton, H. Jirglova, P. Sazama, L. Benco, T. Bucko, J. Hafner and Z. Sobalik, *J. Phys. Chem. C*, 2013, **117**, 3958.
41 A. Janda and A. T. Bell, *J. Am. Chem. Soc.*, 2013, **135**, 19193.
42 J. B. Peri, *J. Phys. Chem.*, 1974, **78**, 588.
43 A. Zecchina, C. Otero Arean and E. Groppo, *ChemCatChem*, 2010, **2**, 259.
44 A. S. Hoffman, C. Y. Fang and B. C. Gates, *J. Phys. Chem. Lett.*, 2016, **7**, 3854.
45 E. Groppo, A. Damin, C. Otero Arean and A. Zecchina, *Chem. – Eur. J.*, 2011, **17**, 11110.
46 M. P. McDaniel, *Adv. Catal.*, 1985, **33**, 47.
47 S. J. Conway, J. W. Falconer and C. H. Rochester, *J. Chem. Soc., Faraday Trans.*, 1989, **85**, 79.

611729 Roles of Interaction between Components in CuO-ZnO-ZrO_2-Al_2O_3/HZSM-5 Bifunctional Catalysts for Direct Synthesis of Dimethyl Ether Via CO_2 Hydrogenation

Abstract:

Catalytic hydrogenation of CO_2 to dimethyl ether (DME) has received lots of interests, since DME not only can be used as a promising intermediate for high value-added chemicals, but also an alternative fuel owing to its high cetane number (55-60), low auto-ignition temperature, and considerable reduction of CO and NO_x in the exhaust gases. Recently, a lot of research focused on bifunctional catalyst for one-step DME synthesis via CO_2 hydrogenation. In this study, the roles of interaction in the CuO-ZnO-ZrO_2-Al_2O_3 (CZZA)/HZSM-5 bifunctional catalysts between two catalyst components on the catalytic activity and deactivation behavior for DME synthesis via CO_2 hydrogenation were systematically investigated. We found that CZZA catalyst showed excellent stability during methanol synthesis, however, there was severe deactivation in the CZZA/HZSM-5 bifunctional catalysts. The changes of catalyst properties were evaluated by characterizing the fresh and used samples via XRD, TGA, BET and N_2O-chemisorption. Our studies show that it is vital to optimize the level of contact intimacy between two components in the development of bifunctional catalyst for DME synthesis, in order to realize high activity and good stability.

613078 Synthesis, Characterization, and Catalytic Testing of Iron(III)-Containing SSZ-70

Abstract:

Real-feed hydrocracking tests using Al-SSZ-70 (International Zeolite Association three-letter framework code *-SVY) show evidence of overcracking of jet fuel to naphtha, which suggests that the acid sites are too strong. In contrast, the acid sites in B-SSZ-70 are too weak to be useful. Previous studies have shown that Brønsted acid site strength can be modified through isomorphic substitution into the tetrahedral positions according to Si < B < Fe < Ga < Al. As a result, Fe-SSZ-70 was prepared through a direct synthesis method. Characterization of the coordination environment of the iron(III) showed that it was tetrahedrally coordinated in the SSZ-70 framework in the as-made materials, but some iron(III) became octahedrally coordinated upon calcination. However, enough iron(III) remained in the framework to be catalytically active. Catalytic testing using n-decane as a model compound resulted in higher isomerization activity and less cracking for Fe-SSZ-70 over Al-SSZ-70 at comparable conversions. These results support previous work that showed that Fe does indeed have an intermediate acid site strength to that of Al and B.

599247 Catalytic Performance of Iron Encapsulated in ZSM-5 for CO_2 and CO Hydrogenation

Abstract:

Catalytic conversion of CO_2 into chemicals and fuels, such as lower olefins, is one of the most viable ways of reducing atmospheric CO_2 to alleviate the negative effects of climate change. One option for reducing CO_2 to chemicals and fuels is through reverse water-gas shift (RWGS) combined with Fischer-Tropsch Synthesis (FTS) [1]. Supported metal catalysts have been widely researched for CO_2 hydrogenation, but they have limited applicability due to low activity, selectivity, and deactivation caused by coke deposition and sintering of metal particles [2].

In this work, ZSM-5 zeolites are used to encapsulate Fe (Fe@ZSM-5) via a seed-directed growth technique. Fe@ZSM-5 improves the catalytic performance by suppressing metal nanoparticle agglomeration, preserving the number of active sites and tuning the selectivity toward light olefins via secondary reactions between the primary products and ZSM-5.

Initial XRD characterization results indicate the crystalline structure of both the Fe oxide and ZSM-5 is preserved within the Fe@ZSM-5. TEM images of Fe@ZSM-5 contain clear voids that likely result in increased CO_2 and CO uptake, with TPD profiles that are broadened and shifted toward higher temperatures when compared with Fe/ZSM-5 control catalysts prepared via wet impregnation. Together, the characterization data suggest possible trapping of CO_2 and CO within the Fe@ZSM-5 and successful encapsulation of Fe. Packed bed reactor studies for RWGS and FTS show that Fe@ZSM-5 catalysts are active for both CO_2 and CO hydrogenation, with increased selectivity toward C_2-C_4 olefins versus the Fe/ZSM-5 control. However, Fe@ZSM-5 exhibits rapid deactivation, which can possibly be improved with a mesoporous ZSM-5 shell. The results of this study demonstrate that the seed-directed growth technique is a viable strategy to control catalytic selectivity during CO_2 and CO hydrogenation reactions.

1. Marc D. Porosoff, et al. *Energy & Environmental Science*9.1 (2016): 62-73.

2. Xie, Jingxiu, et al. *ACS catal.* 6.6 (2016): 4017-4024.

613033 Quantification of Intraporous Hydrophilic Binding Sites in Lewis Acid Zeolites and Consequences for Sugar Isomerization Catalysis

Abstract:

Lewis acid sites catalyze aqueous-phase sugar isomerization at higher turnover rates when confined within low-defect hydrophobic zeolite micropores, because intraporous silanol groups serve as hydrophilic binding sites that stabilize extended water networks that entropically destabilize isomerization transition states. Silanol defect density varies widely among Sn-Beta zeolites prepared by hydrothermal and post-synthetic routes, leading to unintended or unknown consequences for reactivity. We developed a suite of methods to characterize and quantify intraporous silanol groups in Sn-Beta zeolites, and link the number of such groups to glucose isomerization rate constants. Sn-Beta zeolites with varying silanol density were prepared by grafting Sn to different extents into various dealuminated Beta supports, synthesized from parent Al-Beta zeolites with varying initial Al content (0.4–3.2 Al (unit cell)$^{-1}$) and residual defects by the mineralizing agent used (fluoride, hydroxide). Intrapore packing densities of methanol were quantified from single-component adsorption isotherms by comparing the relative uptakes of methanol (293 K) and dinitrogen (77 K) at the point of micropore filling, and decreased systematically from 0.98 to 0.07 among Sn-Beta zeolites with increasing SiOH density. IR spectra show that methanol ($P/P_0<0.2$, 303 K) arranges in isolated clusters within micropores of Sn-Beta with low silanol densities (<1 silanol (unit cell)$^{-1}$), but form extended hydrogen-bonded networks within micropores with high silanol densities (>1 silanol (unit cell)$^{-1}$). Total SiOH groups were quantified by H/D isotopic exchange with D_2 TPSR (500-873 K), while intraporous SiOH groups were quantified from strongly H-bound CD_3CN in IR spectra (2275 cm^{-1}). Aqueous-phase first-order glucose isomerization rate constants (per defect-open Sn, 373 K) were 4× higher on Sn-Beta prepared post-synthetically from dealuminated Beta supports that were initially low-defect (parent <0.6 Al (unit cell)$^{-1}$) than high-defect (parent >0.6 Al (unit cell)$^{-1}$). These findings suggest minimizing intraporous defect density in Lewis acid zeolites allow increasing sugar isomerization turnover rates.

613003 HY Zeolite Supported Molecular Iridium Catalysts: A Comparison between Single-Site and Pair-Site Catalysts on Zeolites

Abstract:

Highly dispersed metal catalysts on zeolites are economically preferred and offer appealing catalytic properties towards many industrial applications. Although these catalysts have drawn broad attentions, the structure-performance relationship is difficult to conclude, because many of such catalysts are lack of uniformity and undergo structural changes during catalysis.

In this study, we precisely synthesized iridium single-site and pair-site catalysts supported on dealuminated HY zeolites using organometallic precursors. The initial structures of the catalysts after synthesis is characterized by infrared (IR) and extended X-ray absorption fine structure spectra (EXAFS), showing highly uniform single-site and pair-site structures. In-situ spectroscopies were used for monitoring structural changes of the two supported iridium catalysts under various gas environment, including CO or H_2 at temperatures ranging from 298 K to 353 K. We used ethylene hydrogenation as a model reaction and compared the catalytic performance of supported iridium pair-sites with supported single-site Ir species and supported tetrairidium clusters on HY zeolites. The results showed that the pair-site iridium catalysts have a magnitude higher initial activity in ethylene hydrogenation than that of their single-site analogs. On the other hand, the highly unsaturated pair-site structure shows quicker deactivation than tetranuclear clusters. The study offers fundamental knowledge and new chemistry associated with zeolite supported molecular catalysts with different nuclearities.

612688 Conversion of Methane to Methanol in Cu-Exchanged SSZ-13

Abstract:

Selective conversion of methane to higher valued methanol has remained a challenge, which has gained significant attention recently due to the increased availability of natural gas. The major challenge for this reaction is to only activate the first C-H bond without converting methane to the thermodynamically stable, overoxidized products CO and CO_2. One promising method to accomplish high selectivity to methanol is in the stepwise conversion of methane to methanol over Cu exchanged zeolites [1,2]. Despite significant efforts, the nature of the active sites during this process is still under debate.

We have developed a thermodynamic model for Cu-SSZ-13 used to evaluate the stability of many unique Cu active sites and terminations including Cu monomers, as well as Cu-oxo and –hydroxy dimers and trimers bound to a series of local Al configurations. Using phase diagrams, we then studied the thermodynamically favored sites under different reaction conditions, and we found that the most stable Cu active site depends on the activation process used. Based on these predictions, we designed experiments to stabilize the desired active sites and evaluated the impact of the active site structure on the performance in the stepwise conversion of methane to methanol. We expect that the insights obtained here can be used to further improve and optimize the stepwise conversion of methane to methanol.

References

[1] Ravi, M.; Sushkevich, V. L.; Knorpp, A. J.; Newton, M. A.; Palagin, D.; Pinar, A. B.; Ranocchiari, M.; Bokhoven, J. A. V. Nature Catalysis **2019**, 2 (6), 485–494.

[2] Ipek, B.; Wulfers, M. J.; Kim, H.; Göltl, F.; ACS Catalysis **2017**, 7 (7), 4291–4303.

612492 Olefin Methylation and Chain Initiators in the Methanol to Hydrocarbon Reaction over Iron Zeolites

Abstract:

The methanol to hydrocarbons (MTH) reaction is a pathway for taking methanol or dimethyl ether (DME) and converting it into fuels and chemicals. The reaction is catalyzed by various acid zeolites and using the large-pore zeolite H-[Al]Beta, the products contain predominantly triptane and isobutane. The alkanes are formed via hydrogen transfer reactions of the isostructural olefins formed through the successive methylation of small olefins in the olefin cycle of the MTH reaction. The hydride donors (other olefins) then undergo further reactions to give aromatic products such as n-methylbenzenes; these side products are not valuable as a fuel and lead to catalyst deactivation. Zeolite catalysts containing framework iron (instead of aluminum) reduce the rate of hydrogen transfer reactions due to their weaker Brønsted acid strength but also reduce methylation rates: on Fe-zeolites thus, reaction temperature must be relatively high also leading to rapid deactivation. To reduce reaction temperatures, a chain initiator can be added to the feed to jumpstart chain growth and the olefin cycle. We have found, for example, that pure DME is essentially unreactive at 240 °C over H-[Fe]Beta, but after the addition of isobutene, the consumption rate of DME increased significantly. After stopping the isobutene flow, DME consumption rate decreased back to its initial low value. Co-feeding isobutene thus overcomes low catalytic rates by increasing the DME consumption rate at otherwise low reaction temperatures where the precursors to stimulate the formation of the carbon pool, do not form. These results suggest that, over Fe-zeolites, isobutene can initiate the methylation of olefins and the olefin cycle. This allows for the carbon from shale gas to be incorporated into larger molecules more suited for uses as fuels and chemicals then traditional MTH reaction products. The methylation of other molecules, including pentenes and cyclic alkenes, over Fe-zeolites is also reported.

612170 Effects of MFI Crystallite Properties on Propene Oligomerization Rate and Selectivity

Abstract:

Brønsted acid zeolites catalyze alkene oligomerization, a reaction relevant for upgrading light hydrocarbons to fuels.[1] First-order propene dimerization rate constants increase with decreasing zeolite void size, because transition states are larger than their alkoxide precursors and thus benefit preferentially from van der Waals stabilization.[2] First-order propene dimerization rate constants are independent of Al content on H-form zeolites (MFI, TON, Beta),[2] although some studies report rates that are sensitive to the bulk density and spatial distribution (i.e., proximity) of framework Al in MFI.[3,4] Here, we report propene dimerization rates (503 K) on commercial Al-MFI (Si/Al = 13-140) and B-Al-MFI (Si/Al = 60-150, Si/B = 26-32) synthesized in the presence of ethylenediamine (EDA) and tetrapropylammonium (TPA^+) structure-directing agents.[5] This suite of MFI samples contain different Al distributions (void environments and proximity). Propene dimerization rates (503 K, 169 kPa C_3H_6) on B-Al-MFI (>1.5 μm) are independent of Al content, but systematically decrease with increasing diffusion parameter, suggesting intracrystalline propene concentration gradients influence measured rates on these samples. Propene dimerization rates (503 K, 169 kPa C_3H_6) on commercial Al-MFI samples vary by 4× but are independent of diffusion parameter, suggesting these differences are kinetic in origin. These results demonstrate the need to account for intracrystalline diffusion effects on measured rates, and motivate the development of zeolite synthesis approaches to independently vary crystallite and active site propertoes.

References

(1) Degnan, T. F. *Catal.* **2000**, *13*, 349

(2) Sarazen, M. L., Doskocil, E. & Iglesia, E. *Catal.* **2016,** *344,* 553

(3) Mlinar A. N. et al. *Catal.* **2012**, *288*, 65

(4) Bernauer, M. et al. *Catal.* **2016,** *344,* 157

(5) Hur, Y. G. et al., *Ind. Eng. Chem. Res.*, **2019**, *58*, 11849

612152 Effects of Ni Site Density on Deactivation Rates and Mechanisms during Ethene Oligomerization on Ni-Beta Zeolites

Abstract:

Alkene oligomerization catalyzed by Ni^{2+} cations on zeolites proceeds via the coordination-insertion mechanism,[1] but rates decrease with time-on-stream due to deactivation[1,2] according to different mechanistic proposals[2,3]. Deactivation phenomena are complicated by the presence of activation periods at initial times, and the behavior of residual Brønsted acid sites present. We synthesized Ni-Li-Beta materials with varying Ni density (Si/Al = 12, Ni/Al = 0.06-0.25), wherein Li^+ replaces H^+ sites. Ethene oligomerization rates (453 K, 0.01-1 kPa C_2H_4) were measured in a differential reactor, and combined with *in situ* X-ray absorption (XAS) to investigate Ni structure during deactivation. *In situ* XAS revealed that Ni is present as isolated Ni^{2+} (without observing Ni^+, NiO, or Ni^0) during deactivation, reflecting poisoning of Ni sites by heavier molecular weight intermediates. The molar selectivity toward oligomer products at fixed ethene conversion did not change as Ni-Li-Beta deactivated to different extents, indicating non-selective deactivation of each Ni site. Deactivation transients were modeled using an apparent n^{th} order deactivation model, where the order in Ni site concentration (n), the apparent deactivation constant (k_d), and the product formation rate are fitted parameters. Second-order deactivation rates are measured at high Ni density, suggesting two Ni sites are involved in the deactivation mechanism, as proposed via formation of Ni-alkyl-Ni intermediates[2]. In contrast, first-order deactivation rates are measured at low Ni density, implicating a single-site deactivation mechanism. This quantitative approach allows modeling the loss in number of active Ni sites during deactivation of catalysts of different composition and at different reaction conditions.

References:

(1) Joshi, R.; Gounder, R. *ACS Catal.* **2018,** *8,*

(2) Mlinar, A. N.; Baur, G. B.; Getsoian, A. B.; Bell, A. T. *Catal.* **2012,** *296,* 156.

(3) Mlinar, A. N.; Shylesh, S.; Ho, O. C.; Bell, A. T. *ACS Catal.* **2014,** *4,*

601010 A Multidirectional Approach to Understanding and Controlling Zeolite Crystallization

Abstract:

Enhancing fundamental understanding of zeolite crystallization processes facilitates the design of microporous materials with tailored properties. Here, we will discuss collaborative efforts to finely control zeolite crystallization through unique approaches that are motivated by opposing applications that seek to either inhibit or optimize zeolite formation. Efforts to inhibit zeolite formation focus on the crystallization of structures (notably the GIS-type framework) that are known to accelerate the degradation of nuclear waste glasses [1]. We have identified several ways by which zeolite formation rates can be suppressed, thereby slowing the rate of degradation. These approaches include the addition of species, such as metals, that dramatically suppresses zeolite crystallization kinetics. Through the mechanistic insights gleaned from these studies, we have additionally developed a range of facile techniques for controlling both phase purity and the physicochemical properties of zeolites. We have demonstrated the applicability of these techniques for industrially relevant zeolite frameworks. For example, we have shown that the introduction of metals affords the ability to inhibit the transformation of FAU (a highly metastable zeolite) to GIS, while altering the chemical composition of the former. These findings have further implications for the development of fundamental understandings of zeolite synthesis, most notably in the preparation of transition metal-substituted zeolites, which are materials of growing importance for commercial applications. Finally, we will discuss new insights into the growth of zeolite GIS, which has a twofold benefit. First, we strengthened the link between intentional synthesis of GIS and waste glass stability tests (which unintentionally yield zeolite crystals), thus facilitating future stability testing. Second, we have been able to bypass intermediate phases, such as metastable FAU, and examine the direct crystallization of GIS where our studies reveal the ability to significantly alter the properties of GIS materials, most notably its silicon-to-aluminum ratio.

601092 DFT Analysis of Ethene Oligomerization on Metal Exchanged Zeolites

Abstract:

Ethylene oligomerization is gaining importance due to the abundance of ethene produced from steam cracking of ethane. The metal supported zeolitic materials, specifically the frameworks with larger pores, have emerged as promising catalysts for oligomerization reactions, due to their ease of recyclability as compared to homogeneous catalysts. Recent work from Joshi et al.[1] shows that on Ni-exchanged BEA catalyst, the Coordination Insertion (CI) mechanism is the major reaction pathway through which ethylene is converted to butenes. Further the authors show that isomerization of 1-butene to 2-butene can occur uniquely on the Ni sites in the absence of acid sites. Hence this study, focuses on using DFT in concurrence with microkinetic modelling on different metal exchanged zeolite catalysts to predict the isomer distribution of the oligomers, as a function of pressure & temperature.

Under ethylene oligomerization reaction conditions, it is proposed that the Ni-H coordinates with ethylene forming β-agostically bound ethyl complex, leaving a proton on the adjacent site. Hence the calculations were performed starting with Ni-ethyl complex as the active site. The isomerization pathway analysis shows that the there's a branching pathway where 1-butyl complex can undergo either ethylene assisted desorption to form 1-butene or can undergo isomerization to form 2-butene. Furthermore, the trends in activity and selectivity towards butene formation across different metal cations was identified by comparing the calculations on Ni^{2+} with other cations like Co^{2+}, Zn^{2+} and Ga^{3+}. Finally, the effect of pressure was studied by increasing the ethylene coordination on Ni^{2+}, which showed increased mobility of the Ni complex inside the framework. The mechanistic implications of increase in ethylene coordination, on isomerization and deactivation pathways will be discussed.

1. Joshi et al, *ACS Catal.*, 2018, **8**, 11407–11422.

607297 Seed-Assisted Synthesis of the Hierarchical Zeolites

Abstract:

Zeolites are nanoporous aluminosilicates that are utilized in various commercial applications. The small pores of zeolites impose diffusion limitations that can be mitigated via nanocrystals or hierarchical materials with reduced diffusion path lengths.[1] Synthesis of these materials often requires the use of an organic structure-directing agent (OSDA), which poses commercial limitations owing to high economic and environmental costs. One alternate approach to either decrease or eliminate organics in zeolite synthesis is the use of seed crystals.[2]

Here we describe our findings revealing the impact of synthesis parameters on the kinetics of nucleation and interzeolite transformations in seed-assisted zeolite crystallization. Our findings expose several limitations to the prevailing hypotheses,[3] and highlight the effect of seeds on crystal morphology and crystal growth kinetics.[4] Moreover, we will discuss an exciting discovery wherein the judicious selection of seeds can produce hierarchical self-pillared pentasil (SPP) zeolites[5] with intergrown nanosheets without the use of organics. Our findings reveal the nonclassical pathway of SPP formation involving the nucleation of crystals at the exterior surface of amorphous precursor particles (Figure 1A, arrow), leading to SPP crystals (Figure 1B). We also explore the effects of various parameters, such as seed structure, chemical composition, temperature, and time on the physicochemical properties of SPP zeolites and confirm that these materials exhibit improved performance in catalytic reactions compared to conventional ZSM-5 catalysts (Figure 1C and D).

Overall, in-depth analysis of zeolite crystallization enables the improved understanding of underlying mechanisms of formation as a basis to develop novel methods in the rational design of nanoporous materials.

1. Grand et al.; Cryst. Eng. Comm. 18 **(2016)** 650-664.
2. Li E., García-Martínez J., (ed.) Mesoporous Zeolites: Preparation, Characterization and Applications. WILEY-VCH (2015).
3. Itabashi et al.; J. Am. Chem. Soc. 134 **(2012)** 11542-11549.
4. Jain and Rimer; Micropor. Mesopor. Mater. 300 **(2020)**
5. Zhang et al.; Science **(2012)** 1684-1687.

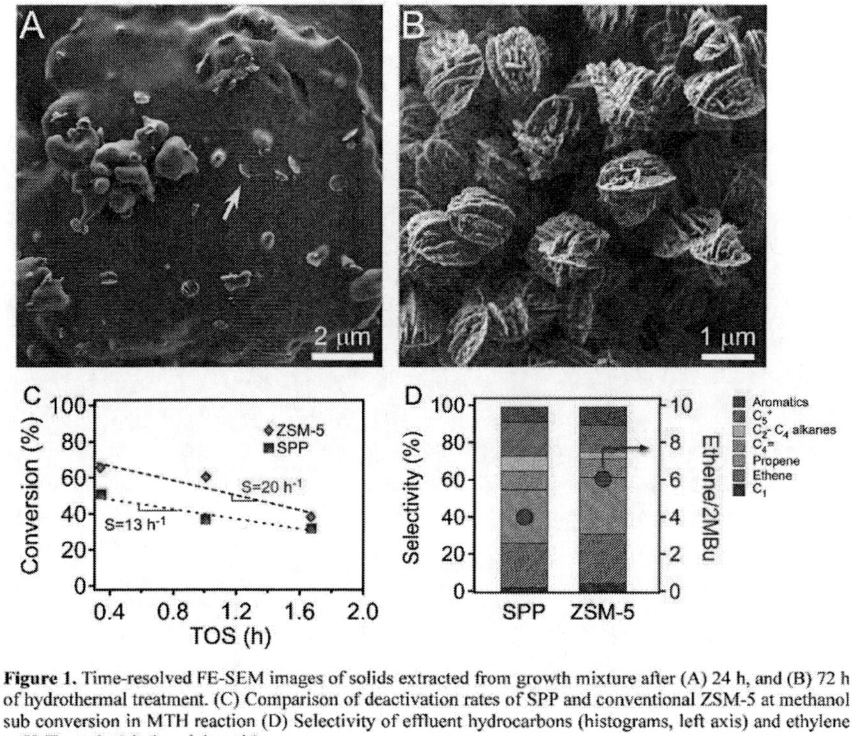

Figure 1. Time-resolved FE-SEM images of solids extracted from growth mixture after (A) 24 h, and (B) 72 h of hydrothermal treatment. (C) Comparison of deactivation rates of SPP and conventional ZSM-5 at methanol sub conversion in MTH reaction (D) Selectivity of effluent hydrocarbons (histograms, left axis) and ethylene to 2MBu ratio (circles, right axis).

608457 Bridging the Gap of Batch-Wise to Continuous Zeolite Synthesis through Microfluidic Reactor Design

Abstract:

Batch processes are difficult to scale because of dynamic internal environments and inherent variability between runs. Constantly changing conditions make collecting reliable kinetic data extremely difficult thus requiring highly precise reaction conditions. The conversion from batch-wise to continuous operation is key for implementing such key engineering controls. For challenging reactions like zeolite crystallization, segmented flow microfluidics provide a platform capable of forming, handling and characterizing solids in flow.

The flow-focusing device was constructed to achieve a high throughput, steady state generation of microdroplets with consistent size and composition. Wall interactions were eliminated by dispersing the sol-gel aqueous phase in an inert oil carrier phase. Hydrophobicity of the device walls ensured non-wettability as the gel formed uniform ellipsoid microdroplets. With each microdroplet modeled as an individual micro-batch reactor in flow, micro-scale design eases the modeling of heat and mass transfer as reduced length scales facilitate isothermal operation and tunable oscillatory mixing. Syntheses were carried out in parallel in traditional autoclave batch reactors and micro-batch droplets, and crystallinity was characterized *ex situ* using x-ray diffraction and Raman spectroscopy. Comparing the crystallinity curves, a clear reduction in induction time is observed in micro-batch systems as both characteristic LTA reflections and framework vibrations appear at shorter synthesis times. Engineering controls afforded by microfluidics have the potential to realize the goal of producing zeolites continuously.

610023 Understanding the Entropic Contributions of Irreversibly Bound Adsorbates on Zeolite Surfaces

Abstract:

The design of solid acid catalysts has greatly benefitted from understanding the energetics (enthalpy and entropy) of adsorbates, where the relative stability of surface species dictates the overall catalytic turnover frequency. While enthalpic contributions are relatively well understood, limited information exists for entropic contributions to the adsorption, particularly for strongly bound species representative of reaction intermediates and transition states. This is due to the fact that traditional methods used to measure an adsorbate's entropy, require the adsorption/desorption process to equilibrate. Strongly bound intermediate like alkylamines adsorbed on a Bronsted acid site however do not equilibrate on relevant timescales, given their negligible rate of desorption, which limits the applicability of existing methods. To overcome this issue, we apply the concept of adsorption assisted desorption by investigating the co-adsorption of multiple strongly bound adsorbates. While one strong adsorbate does not equilibrate due to limited desorption, two or more strongly bound adsorbate can continuously displace one another from an active site and attain adsorption/desorption equilibrium. Applying this concept to alkylamine adsorption on Brønsted acid sites in H-ZSM5 (Si/Al = 140), we demonstrate for the first time the experimental ability to measure the overall adsorption thermodynamics of strongly bound species. Despite the relatively strong adsorption, we find that the Langmuir adsorption isotherm can capture the adsorption of multiple alkylamines varying in size and basicity over a wide range of conditions. By comparing a homologous family of primary alkylamines, varying only in alkyl chain length, we find a fixed contribution to both adsorption enthalpy and entropy per methylene unit (Figure 1).

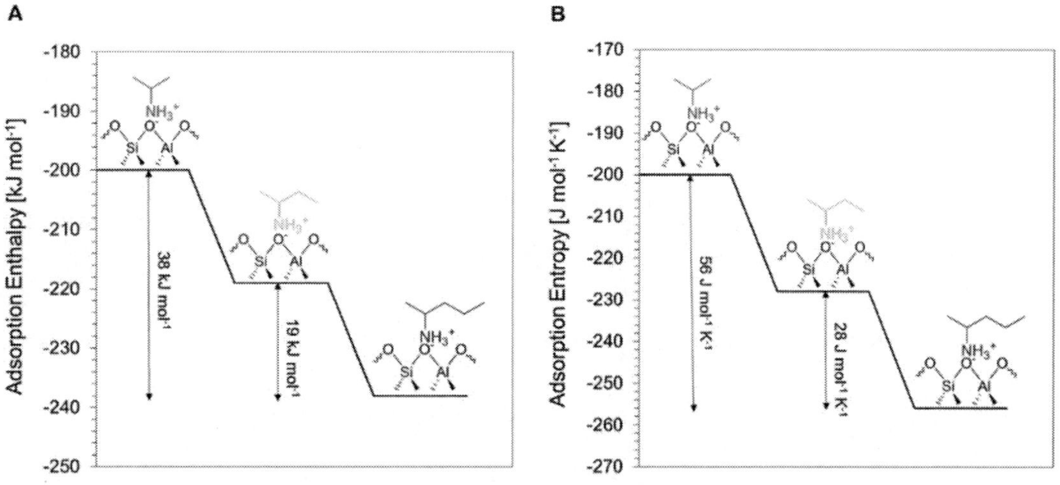

Figure 1: Adsorption Energetics of isopropylamine (red), sec-butylamine (green), sec-pentylamine (blue), over H-ZSM5 (Si/Al = 140). A. Adsorption Enthalpy difference for every methyl unit increase. B. Adsorption entropy difference for every methyl unit increase.

602339 Ag Ion-Exchanged ZSM-5 Zeolites for Hydrocarbons Trapping Applications

Abstract:

Significant attention has been paid to the reduction of vehicle emissions during cold-start due to the stringent emission regulations. A potential approach to regulate the cold-start hydrocarbon (HC) emissions is to utilize HC traps. Light (e.g. C_2H_4, C_3H_6) and heavy (e.g. C_7H_8) HCs need to be trapped simultaneously in practical applications. The goal of this study is to develop suitable HC traps for both light and heavy HCs. Herein, a series of Ag containing ZSM-5 (Si/Al = 11.5) zeolites with different Ag loadings (0.5 - 4.6 wt.%) were prepared via ion-exchange. Their trapping performance was evaluated using the U.S. DRIVE low temperature combustion of diesel (LTC-D) protocol. The results showed that Ag/ZSM-5 is able to adsorb C_2H_4 and C_7H_8 simultaneously and their adsorption capacities increased with increasing Ag loading (Fig. 1a). This indicates that Ag is able to act as an adsorption site for both C_2H_4 and C_7H_8 in the presence of H_2O via π-backbonding. The oxidation state of Ag can also influence HC adsorption (Figs.1b and 1c). Specifically, the initial C_2H_4 adsorption capacity (303 μmol/g) decreased to 202 and 125 μmol/g after reduction at 200°C and 600°C, respectively. Reduction at 200°C and 600°C resulted to a decrease in the number of Ag^+ sites due to Ag_n^+ and Ag^0 formation, respectively, and thus a decrease in C_2H_4 adsorption capacity. The decrease in C_2H_4 adsorption capacity can be recovered after re-oxidation at 600°C. Unlike C_2H_4 adsorption, a negligible decrease in C_7H_8 adsorption capacity was observed after reduction. This behavior can be attributed to the adsorption of C_7H_8 on both Ag and the zeolite framework, while C_2H_4 can be adsorbed only on Ag.

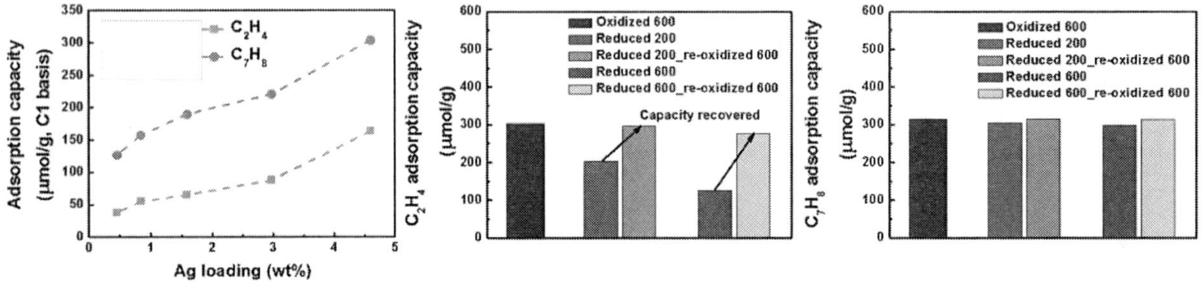

608581 Development of Mesopore-Containing CON-Type Zeolite with Unique Acidic and Catalytic Properties

Abstract:

The development of mesopore on the zeolite framework is an effective method for enhancing the catalytic performance of zeolites [1]. Recent work by M. Yoshioka *et al* showed that boroaluminosilicate CON-type zeolite ([Al,B]-CON) is an effective catalyst for the selective production of propylene in methanol-to-olefins (MTO) reaction [2]. In order to enhance the catalytic performance of [Al,B]-CON, we investigated the effect of the hydrothermal alkaline-treatment on the physicochemical properties and catalytic performance of [Al,B]-CON samples in MTO reaction.

[Al,B]-CON was prepared by following the composition and procedure reported by S. Park *et al.* [3]. For the alkaline-treatment, [Al,B]-CON (H-form) was added into the NH_4OH solution (1.1 or 2.2 mol L^{-1}) containing CTABr (0.03 mol L^{-1}). The obtained samples were designated as [Al,B]-CON_*x*, where *x* represents the NH_4OH concentration used for the alkaline treatment.

XRD peak patterns of samples indicate that the CON structure has retained after the alkaline-treatment. The number of acid sites were comparable before and after the alkaline-treatment but the mesopore volume increased at the expense of the micropore volume. We could confirm the incorporation of mesopores by TEM and N_2 adsorption analyses. *in-situ* FT-IR spectra after the 2,4,6-trimethylpyridine adsorption indicated the enhanced diffusion properties of the sample through the alkaline-treatment. In MTO reaction [Al,B]-CON_1.1 showed the enhanced catalytic life with the higher propylene selectivity (Fig. 1) which could be attributed to the positive effect of alkaline-treatment on both acidic and diffusion properties of the CON-type material.

[1] J.Y. Ying *et al,, US20050239634* (2005).
[2] M. Yoshioka *et al., ACS Catal.* (2015) 5, 4268–4275
[3] S. Park *et al.*, Microporous Mesoporous Mater. (2020) DOI: https://doi.org/10.1016/j.micromeso.2020.110213

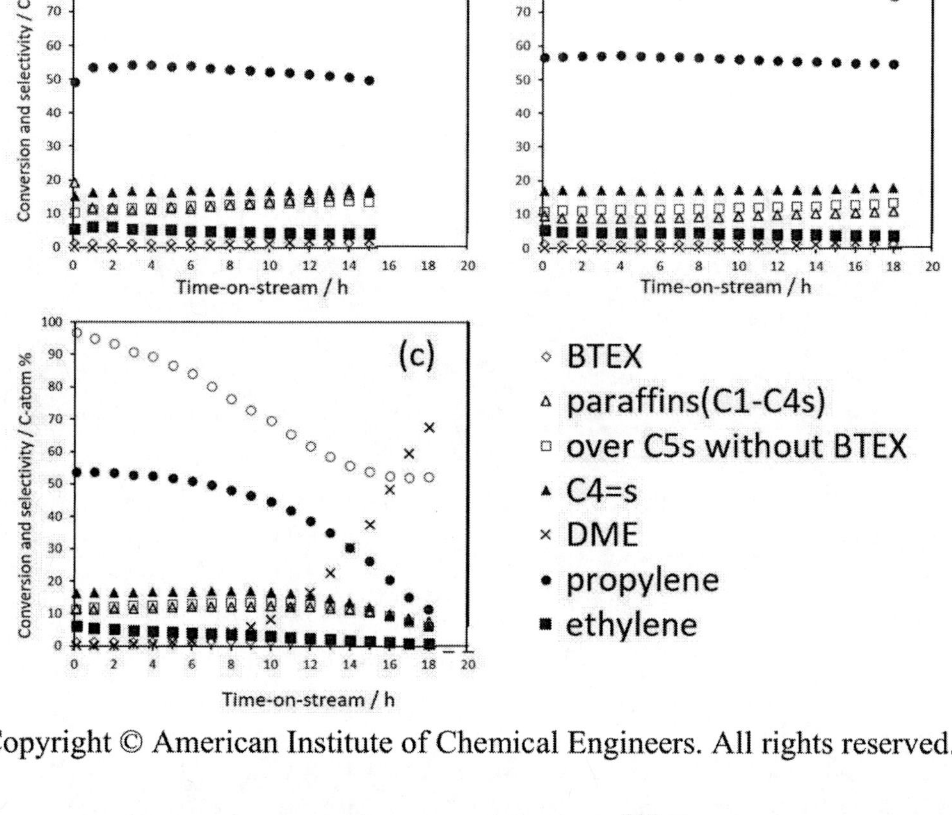

597287 Understanding Dealumination Mechanisms in Protonic and Cationic Zeolites

Abstract:

Zeolites can be partially hydrolyzed by steam via Al extraction which leads to the formation of a moiety with four hydroxyl groups. This irreversible dealumination process can decrease the crystallinity of zeolites and effectively modify their acidity, porosity, and catalytic activity. This work addresses the hydrolysis mechanism of the first Al–O bond in protonic (H^+) and cationic ($M = Na^+$, K^+, and Ca^{2+}) LTA zeolites. Periodic density functional theory calculations demonstrate that cationic LTA zeolites are more kinetically stable than H-LTA by exhibiting higher reaction barriers. Analysis revealed a H_2O dissociation reaction mechanism involving proton transfers and rotations in both protonic and cationic zeolites, which further leads to the Al–O bond breaking. Our results also indicate that a lower Si/Al ratio tends to favor hydrolysis reactions, while the cation identity has a small effect on the reaction barriers. A correlation between the Al–O bond length in the initial adsorbed states and the reaction activation energies is identified.

601261 Enhancing Hydrophobicity and Catalytic Activity of Nano-Sn-Beta through Post-Synthetic Treatment with Fluoride

Abstract:

Elucidation of structure-function relationships can enhance the performance of catalytic materials used for a broad range of reactions relevant to biomass valorization and pharmaceutical production. After considerable focus had been on examining structure-function behavior of the catalytic site, it has become increasingly evident that molecular interactions beyond the catalytic site in the outer sphere, such as solvation, significantly impact catalytic activity. The key challenge is to control this catalytic reaction environment. In this work, a highly active catalyst is achieved by tuning the molecular interactions of the catalyst with the substrates through post-synthetically modifying the hydrophobicity of the material. First, nano-Sn-Beta is synthesized through a hydrothermal route and is used to overcome diffusion limitations that were observed for conventional Sn-Beta. While conventional Sn-Beta is synthesized in the presence of fluoride to create materials with few defects and large crystals, the synthesis of nano-Sn-Beta utilizes hydroxide as the mineralizing agent, leading to smaller crystals that are more hydrophilic because of the greater number of silanols. Building on the few examples that modify the hydrophobicity of MFI, we propose a method that utilizes a post-synthetic treatment of the calcined nano-Sn-Beta with tetraethyl ammonium fluoride (TEAF) to heal defects and reduce the silanol content throughout the crystal, thereby increasing its hydrophobicity. The treated catalyst exhibits a significant increase in catalytic performance (70% increase in TOF0) over the untreated material for the Lewis acid catalyzed epoxide ring opening with alcohols. Enhancing the hydrophobicity of nano-Sn-Beta can increase the stability of the catalyst as well as the scope for different chemical reactions. Overall, this work will demonstrate the post-synthetic modification of nano-Sn-Beta to produce a hydrophobic zeolite framework and the effect of hydrophobicity on catalytic activity and selectivity.

603929 Hydrothermal Ageing Mechanism of SAPO-34 Studied By Kinetic Analysis of Temperature Programmed Desorption Data

Abstract:

SAPO-34 is a zeotype suitable for use in the methanol to olefins (MTO) process. This catalyst is known to deactivate when exposed to high levels of water at elevated temperature during the MTO reaction. Gaining a fundamental understanding of this hydrothermal ageing will allow the development of improved catalyst formulations.

Samples of SAPO-34 were hydrothermally aged at high temperature for varying times. The aged samples were characterized using ammonia temperature programmed desorption (TPD). TPD is a commonly used thermal analysis technique for investigating adsorption sites. These adsorption sites are important features in the performance of a catalyst and can be influenced by formulation and chemical structure.

Multiple overlapped peaks are common in TPD results and deconvolution of these is important for data analysis. Historically, the shapes of these peaks would be estimated before mechanistic modelling was carried out; introducing assumptions which can be a source of error. In the approach used in this study, no *a priori* assumptions are made about the peak shapes, rather a modified Sestak-Berggren model is used to fit the data directly.

This model is a numerical fit and the estimated exponent parameters (n and m) indicate which mechanisms could be plausible for the system under investigation. These are analysed, and model discrimination carried out, allowing the fitting of kinetic parameters. These parameters allow comparison between samples which have been aged for different times.

The Sestak-Berggren analysis of these data demonstrated that the adsorption energy values for each thermal event did not change significantly through hydrothermal aging, inferring there is no change in the nature or strength of the acid, adsorption sites. The change is in the pre-exponential factor, indicating clearly that it is the number rather than strength of the acid sites that reduces.

611357 Contrasting Diene Formation Pathways with Methyls and Larger Alkyls during Methanol-to-Olefin Reactions in MFI and CHA Using DFT

Abstract:

Brønsted acid zeolites, such as CHA and MFI, catalyze methanol-to-olefin (MTO) processes, in which a pool of alkene and aromatic species co-catalyze the production of light olefins. Olefins that do not egress as products may undergo hydride transfers to dienes. Dienes can cyclize to form aromatics and eventually polyaromatics, which induce catalyst deactivation by obstructing acid sites. Prior studies have demonstrated the rate of catalyst deactivation decreases with the introduction of formaldehyde scavenging Y_2O_3, implicating that formaldehyde promotes polyaromatic formation. Previous studies suggest hydrogen transfer reactions govern diene formation; however, these reaction pathways have not been extensively studied with DFT. Here, we investigate two potential pathways for diene formation during MTO processes–alkene disproportionation and CH_2O-mediated– in CHA and MFI zeolite frameworks to provide insight into catalyst deactivation. Multiple routes are studied for each mechanisms; however, we find that the sequential route, which involves a hydride transfer between a surface-bound alkyl and donor species (C_4H_8 or CH_3OH for alkene disproportionation and CH_2O-mediated, respectively), is preferred for both mechanisms. Therefore, the ability of C_4H_8 and CH_3OH to participate in hydride transfers determines which pathway dominates. Barriers of both pathways are comparable for C_1–C_3 surface-bound alkyls, suggesting the preferred mechanism is largely governed by the ratio of CH_3OH to C_4H_8 pressures; because of higher CH_3OH pressures at MTO conditions, the CH_2O-mediated pathway is likely preferred. Due to their larger size and therefore enhanced carbocation stability, surface-bound C_4 species react readily with CH_3OH at rates $> 10^7$-times faster than C_1 alkyls. Steric interactions hinder hydride transfer between surface-bound C_4 and C_4H_8–making these reactions unfavorable. Overall, we find dienes are likely formed via the reaction of CH_3OH and surface-bound tert-butyl in MFI and CHA—imparting valuable insight into the mechanism of diene formation that can potentially be exploited to mitigate catalyst deactivation.

602201 A Paradigm Shift in Catalyst Stability for Toluene Alkylation with Methanol to Produce *P*-Xylene

Abstract:

One of the commercially most important aromatic compounds is *p*-xylene, which is used for the production of various fine chemicals. Due to its cheap feedstocks, toluene alkylation with methanol is one of the most commercially attractive routes to produce *p*-xylene; however, challenges such as low catalyst stability and single pass *p*-xylene yield need to be addressed. Here we present a study of the catalytic properties of MWW-type zeolite catalyst, for toluene alkylation with methanol under high pressure conditions (600 psia).

To understand the catalytic behavior of MWW, we deconvoluted structure-function relationships for different topological features (supercages, sinusoidal channels, and external surface pockets) having profound impact on catalyst performance. Deconvolution was achieved by selectively passivating the active sites in specific pore systems, and then comparing their activity with the parent MWW catalyst.

Under high pressure operating conditions, the catalyst life increases multi-fold at high toluene conversion with side reactions such as methanol-to-hydrocarbon reactions being suppressed in the absence of water or hydrogen co-feed. Interestingly, the selectivity of xylene isomers changes significantly during the course of reaction, from thermodynamic equilibrium mixture to 55% *p*-xylene selectivity.

The combination of structure-deconvoluting experiments and DFT calculations reveals that the active sites in surface pockets are unselective and fast isomerization on external surface leads the reaction towards the thermodynamically determined xylene product distribution, while microporous sinusoidal channels and/or supercages preferentially form *p*-xylene. Overall, this study introduces a commercially viable route for *p*-xylene production as well as new insight into the reaction mechanism over MWW zeolite catalysts to aid in the development of optimized processes for various alkylation reactions.

605703 Finned Zeolite Catalyst: An Emerging Class of Hierarchical Material

Abstract:

Confined channels and cages of zeolites have been widely used as shape-selective heterogeneous catalysts in the (petro)chemical industry. A common objective in the design of zeolite is to overcome the inherent mass transport limitations of micropores. The advent of two-dimensional[1] or self-pillared zeolites[2] exhibit superior catalytic performance to conventional zeolites. In this talk, we will describe an alternative approach to reduce the internal diffusion limitations of zeolites via the introduction of fin-like protrusions on zeolite surfaces by secondary growth. We will discuss the synthesis of multiple frameworks with nano-sized fins (size a) which exhibit an identical crystallographic registry with the interior crystal (size b) and show their superior catalytic performance relative to conventional analogues.

This new class of mass transport enhanced zeolites were synthesized using secondary growth mixtures prepared with a finely tuned composition that allows for the epitaxial growth of fins on the surface of seed crystals. Here we will discuss examples of several finned zeolites with disparate 3-dimensional pore networks to demonstrate the broader applicability of this approach. We also demonstrate a proof of concept using commercial zeolite samples where finned analogues improve their catalytic performance. Our studies of synthesis and MTH catalytic testing as benchmark reaction are coupled with state-of-the-art characterization using techniques such as high-resolution electron tomography, operando spectroscopy, novel acid titration methods, and molecular modeling to correlate structural features of finned zeolites and their diffusion properties with enhanced catalyst performance.

1. Choi, M., et al. *Nature* 461 **(2009)** 246-249.
2. Zhang, X., et al. *Science* 336 **(2012)** 1684-1687.

601570 The Effect of Confinement on Self-Diffusivities and Hydrogen Bonding Network of Organic Liquids inside a Nanoporous Silica Catalyst

Abstract:

Transport properties of molecules confined in nanoporous materials are substantially different from those present in the bulk liquid. This is attributed to altered liquid-liquid interactions and liquid-surface interactions within the porous media. Changes in transport properties and molecular interactions can also significantly alter the reaction kinetics within the nanopore. In the present work, a combination of molecular dynamics (MD) and grand canonical Monte Carlo (GCMC) was used to study the confinement effects of organic liquids within a porous catalyst. Self-diffusivities of propanol, 1,3-propanediol, 1,4-butanediol, glycerol, acetone, propionic acid, and heptane were analysed in the bulk and inside a silica nanopore. In this regard, GCMC was used to obtain the intrapore density of liquids, and MD was used to calculate the self-diffusivities and to analyse the hydrogen bonding (H-bond) structure. The ratio of self-diffusivities in bulk to self-diffusivities inside the pore was taken as an estimate for comparison between the organic compounds. This factor was used for comparison with the published experimental results obtained from pulse-field gradient nuclear magnetic resonance (PFG-NMR) experiments. In this study, alkanes were used as a benchmark, as their lack of functional groups results in minimal interactions with porous media, and intermolecular interactions. Interestingly, polyols (1,3-propanediol, 1,4-butanediol, and glycerol) showed an anomalous enhanced rate of self-diffusion within the nanopore. This anomaly is in accordance with the experimental observations, obtained using PFG-NMR. It is well known that a reduction in the number of H-bonds increases the mobility of molecules, leading to increased self-diffusivities. To attain mechanistic insights into the phenomena behind this anomaly, the H-bonding structure, its population and lifetime, and intermolecular interactions were analysed inside the nanopore. Based on these calculations, the anomalous behaviour of polyols, in contrast to alcohols, carbonyl compounds, and carboxylic acids, was explained.

596237 Value-Added Carbon Materials By Conversion of Natural Gas Using Microwave Plasma Technology

Abstract:

The advanced plasma technology practically and cost-effectively converts natural gas to value-added chemicals and premium carbon materials such as graphene and conductive carbon black analogues (CCBA) with no CO_2 emissions and low capital and infrastructure expenditures. A microwave driven plasma drives hydrocarbon decomposition – producing a variety of carbon nanostructures without the use of catalyst. With lower-energy requirements than conventional thermal plasmas, reactions in microwave plasmas are driven by electron kinetics rather than thermodynamics, and their non-equilibrium energy distribution opens reaction pathways that are unavailable with conventional chemical or thermal plasma processes. The form and purity of carbon material can be controlled by optimizing the several interrelated parameters that include methane to hydrogen ratio in the feed stream and reactor conditions such as input energy and formation temperature. Primary products include nanographene comprised of 2-6 sheets per stack with lateral dimensions between 100 and 500 nm, and graphitic carbon particles with structure analogous to conductive carbon blacks. Analytical techniques including high-resolution transmission electron microscopy (HRTEM), X-ray diffraction (XRD), thermogravimetric analysis (TGA), Raman spectroscopy and electrical conductivity measurements are utilized to study the form and quality of these valued carbon materials. Optical spectra are collected and analyzed to determine the formation temperature of these carbons using blackbody radiation and C_2^* Swan band emission. The electrical conductivity of the as-produced CCBA material is higher than that of commercially available conductive carbon blacks. These results highlight the importance of advanced plasma technology for the economic utilization of natural gas by producing premium carbon materials.

604693 Mechanistic Understanding of Methane to Methanol Conversion on Graphene-Stabilized Single-Atom Metal Centers

Abstract:

The functionalization of methane to value-added chemicals, such as methanol, remains as one of the "grand challenges" in the chemical industry due to the energy-demanding C-H activation step. This restricts the extensive utilization of methane despite its vast reserves. A single Fe atom stabilized on N-functionalized graphene has been reported as an efficient catalyst for direct conversion of methane to methanol[1]. In this work, we investigate the C-H activation of methane and successive methanol production mechanisms through reaction with H_2O_2, using Density Functional Theory calculations. By elucidating different reaction paths, it is shown that methanol is likely to be produced even at ambient temperatures through a radical formation mechanism. Importantly we demonstrate that the active site is a di-oxo iron center, which is responsible for generating methyl radicals (and the site being hydroxylated). These radicals can react with either the formed hydroxyl on the active site or directly with H_2O_2 from the liquid phase to form methanol. In both cases the active site is regenerated through a H_2O_2 reduction to H_2O step. We demonstrate that the high activity is attributed to a fine balance of keeping the catalyst at an activated oxidized state (oxo-species) rather than the thermodynamically preferred hydroxylated state. Taken together, this work reveals novel mechanisms and catalytic functionalities for converting shale gas (methane) to valued chemicals (methanol), while addressing solvent effects and catalyst stability on the overall catalytic behavior.

[1] X. Cui et al., Chem 4(8), 1902-1910 (2018).

596160 Catalytic Conversion of CO_2 into Vinyl Acetate with Activated Carbon Based Zn Acetate Catalyst

Abstract:

Conversion of CO_2 to high-value products is limited commercially due to the stability of the CO_2 molecule. Very few commercially significant reactions convert greenhouse gas like CO_2 to value-added products. The following is one such two-step process: the carboxylation of CH_4 to form acetic acid, which is reacted with acetylene to form vinyl acetate:

$CO_2 + CH_4 \rightarrow CH_3COOH$; (2) $CH_3COOH + C_2H_2 \rightarrow$ vinyl acetate

This formation of acetic acid is thermodynamically limited, but can be addressed by coupling this reaction with a thermodynamically favorable reaction, e.g., reacting acetic acid with acetylene, and forming vinyl acetate monomer (VAM), a high-value industrial intermediate. The overall coupled reaction with $CH_4 + CO_2 + C_2H_2$ to VAM is thermodynamically feasible ($\Delta G_r^{\,o}$= +6.4 kJ/mol). Previously, Wilcox et al. demonstrated the catalytic reaction of VAM from this reaction at temperature as high as 400^0C under atmospheric pressure [1], but VAM formation was not stable at these conditions (Figure 1). The formation of VAM from $CH_4 + CO_2 + C_2H_2$ involves one, single reaction at low temperature and high pressure. The goal of this work is to investigate and synthesize an active and stable catalyst at thermodynamically favorable reaction conditions. Keeping this in mind, activated carbon based zinc acetate (ZnA/C) was synthesized and tested at different reaction conditions.

After characterizing ZnA/C with SEM-EDS, Raman and XRD, VAM reaction was run from temperatures as low as 130^0C and at a pressure of 9 bar for ~15 hours. A stable VAM formation was observed. Acetaldehyde was observed as a side product of VAM. When the temperature was increased to 160^0C, VAM Selectivity increased as well. VAM selectivity also increased with lower space velocity. Activity decreased slightly with time. Catalytic deactivation can be observed with the loss of acetate from the catalytic surface at high temperature.

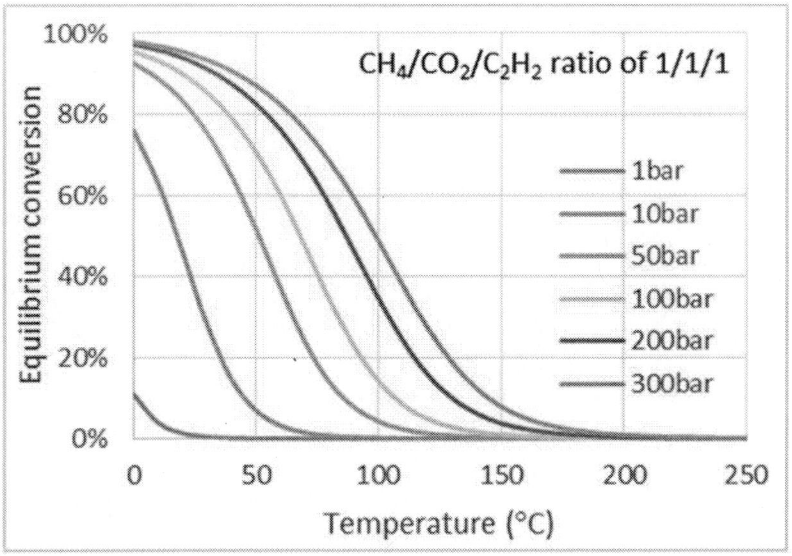

Figure 1. Equilibrium conversion of CH_4+CO_2+C_2H_2 at different temperatures and pressures. (calculated with HSC® 8.0)

601334 Towards Elucidating Critical Properties of Fe-ZSM-5 for Methane DHA

Abstract:

The abundance of domestic natural gas reserves has resulted in an increased interest in converting methane, the main component of natural gas, into higher-value chemicals. One promising route is the non-oxidative conversion of methane to aromatics via dehydroaromatization (DHA). However, at the high temperatures thermodynamically required for this reaction to proceed, coke formation becomes inevitable and results in rapid catalyst deactivation. An alternative energy source, microwave irradiation, can be applied to aid the catalytic reaction of methane DHA. Compared to conventional heating methods, microwave irradiation allows for selective heating, i.e. the active metal sites can be heated selectively while keeping the bulk catalyst and the gas phase at a lower temperature and thus avoid or reduce undesired side reactions, such as coke formation. However, microwave-assisted heterogeneous reactions are poorly understood to-date due to their complexity. In this project, we aim to deconvolute the complexity into two aspects: Investigating the properties that control microwave sensitivity of the catalyst on one hand, and properties that control the (thermo-)catalytic reactivity on the other hand. The latter will be the focus of the current presentation.

The present study is focused on Fe-ZSM5 as a model catalyst for DHA and identification of its key properties that control reactivity for DHA. By carefully controlling the properties of Fe-ZSM5 catalysts, we found a strong correlation both for the Si/Al ratio and Fe loading with catalyst activity and selectivity. Interestingly, this is in good agreement with parallel work in which we had identified that Si/Al ratio and Fe loading are also strongly correlated with the microwave sensitivity of these catalysts. Specifically, we find that catalyst reactivity can be correlated with Fe species distribution and acid site concentration, and that both vary depending on the extent of both Fe and Al incorporation into ZSM-5. Further characterizations of the Fe speciation and acid sites of the samples and correlation with the catalyst performance are currently on-going.

Through this study, we aim to improve our understanding of how the catalytic reactivity of ZSM-5 can be affected by the Al and Fe incorporation and lay the groundwork for the rational catalyst design for thermocatalytic and microwave-assisted methane DHA.

598656 Mechanistic Study of Structured Gallium and Platinum Promoted ZSM-5 in Ethane Dehydroaromatization Including Characterization By Pyridine Drifts

Abstract:

Ethane is the second-largest component of shale gas, its existence in the shale gas pipeline causes condensation, and this separated ethane is often known as stranded gas. Therefore, the utilization of stranded ethane has drawn significant interest from researchers due to its economic value. Previously in our group, it was theorized that the presence of Pt more readily promotes the dehydrogenation of ethane to ethylene, which is considered to be the primary step of dehydroaromatization reaction of ethane. The secondary step is when ethylene oligomerizes to the moderate acid sites (Bronsted acid sites). Additionally, according to the ammonia-temperature-programmed desorption (NH_3-TPD) preformed, the presence of the Pt has increased the strength and number of the moderate acid sites on the bimetallic catalyst, as compared to just the Ga/HZSM-5. The combination of Pt promoted Ga catalyst had bimetallic metal interactions that resulted in a favorable performance in ethane DHA.

In the current study three catalysts: GaPt/ HZSM-5, Pt/ HZSM-5, and Ga/ HZSM-5 were investigated for ethane dehydroaromatization (DHA) in a conventional fixed bed reactor at 615°C for 2 hours. Although, the Pt/HZSM-5 catalyst had a higher conversion, GaPt/HZSM-5 showed the best performance out of the three catalysts tested, having better aromatic selectivity. This follows the theorized roles of the promoters in the previous study. Pt/HZSM-5 had a higher selectivity towards ethylene compared to the GaPt/HZSM-5, proving it has an important role in the first step of ethane DHA.

The three catalysts in the current study were synthesized using the dry impregnation method with specific metal loading. In order to prove the roles of Ga and Pt promoters in each of these catalysts, an acid site characterization was done using pyridine DRIFTS at multiple temperatures of fresh and spent catalyst. The spent catalysts coke was removed in a muffle furnace set to 550°C for 6-hrs, TGA analysis confirmed the complete removal of coke deposits from the spent catalyst. The effects of the metal promoters on the concentrations of the weak and strong acid sites were expressed by ammonia-TPD techniques. Surface area analysis, XRD, and TEM were used to further discuss the catalyst structure before and after reaction.

612755 Evolution and Control of Active Sites in Methane DHA: Insights from Ab Initio Simulations Guiding the Choice of Catalyst Precursor and Pre-Treatment Protocols

Abstract:

Direct conversion of methane to aromatics in non-oxidative conditions generally utilizes a bifunctional catalyst having active molybdenum species supported on zeolite (Mo/HZSM-5). In this reaction, the nature of the active site for methane activation on Mo is suggested to be a carbide (or an oxycarbide) and the catalyst is known to deactivate from carbon deposition over time due to the formation of polyaromatic hydrocarbons. Herein, density functional theory (DFT) calculations are utilized to study the catalytically active carbide (Mo_xCy) clusters in C-H bond activation and C-C coupling reaction to form C_2 intermediates. A massively parallel cascade genetic algorithm (cGA) is applied to scan the potential energy surface (PES) for all possible low-energy structures (including the global minimum) of Mo_xCy. Catalytic activity of the global minimum and metastable clusters are accessed and compared for methane activation, highlighting the role of metastable clusters influencing the overall reaction rates. The evolution of the molybdenum carbide species from the precursor oxide and oxycarbide forms, dictating the anchoring of the carbide cluster on the zeolite support, is considered the key aspect in deciding the extent of coke formation on the catalyst. The mechanistic insights obtained from DFT calculations are thus providing a molecular level engineering approach, wherein the catalyst is rationally synthesized to obtain desired active sites anchored on the zeolite support, which are stable in providing high aromatic selectivity and overall reactivity. This is implemented in experiments at three stages; a) by altering the Mo precursors used in the synthesis of Mo/HZSM-5 catalyst, b) treating the zeolite support with the boric acid to alter support acidity and on c) changing the carburizing condition from pure methane to a mixture of hydrocarbon gases (mimicking the natural gas). In all three cases, the evolution and control of the active molybdenum carbide species, anchored on the zeolite support, is highlighted as a molecular level toolbox for obtaining desired reactivity and product selectivity, with a potential to achieve higher catalyst stability.

References:

1. 1) S. Balyan, M.A. Haider, T. S. Khan, K.K. Pant, Boric acid treated HZSM-5 for improved catalyst activity in non-oxidative methane dehydroaromatization, Catal. Sci. Tech. (2020) DOI: 10.1039/d0cy00286k
2. 2) T.S. Khan, S. Balyan, S. Mishra, K.K. Pant, M.A. Haider, Mechanistic Insights into the Activity of Mo-Carbide Clusters for Methane Dehydrogenation and Carbon–Carbon Coupling Reactions To Form Ethylene in Methane Dehydroaromatization, J. Phys. Chem. C 122 (2018) 11754–11764.

603070 Planning and Optimization of a Regional Hydrogen Supply Chain: An ASEAN Case Study

Abstract:

The world is seeking scalable and affordable low carbon and renewable energy sources to reduce atmospheric CO_2 concentration levels while meeting rising global energy demand. Hydrogen is receiving significant attention worldwide as a versatile energy carrier and feedstock that can achieve deep decarbonization across energy and industrial sectors. Researchers have proposed various approaches to evaluate hydrogen supply chains. For example, Almansoori and Shah[1] developed a multi-period hydrogen supply chain model, based on the hydrogen demand of fuel cell vehicles in the Great Britain from 2005 to 2034. Reuß et al.[2] carried out a techno-economic comparison between different transport and storage infrastructure options for scenarios with 25%, 50% and 75% fleet penetration of fuel cell electrical vehicles in Germany. Seo et al.[3] developed a novel hydrogen supply chain model to compare centralized storage and decentralized storage modes. However, most previous work has focused on country-specific supply chains, such as for the UK, Germany, or South Korea, and considered the transport sector as the only hydrogen demand, neglecting that hydrogen can also be utilized in power and industrial sectors. There is no study on regional hydrogen supply chains with multiple H_2 demand sectors. A regional supply chain can capture the benefits of integrating energy-abundant areas with high energy-consuming areas.

In this research, a multi-period regional hydrogen supply chain model is developed for a better understanding of different pathways for a hydrogen economy in Singapore and the ASEAN countries through 2050. A comprehensive superstructure including hydrogen production, storage, delivery, and utilization is proposed. Both renewable and non-renewable hydrogen production technologies with applicable carbon capture and sequestration are included, reflecting the diversity of energy sources available in the region for hydrogen production. To compare different options for an ASEAN hydrogen supply chain through 2050, we carry out techno-economic analyses of hydrogen production taking into account location-dependent attributes, such as cost and carbon emissions. The variation of hydrogen demand over a long-term planning horizon is considered, as well as uncertainty of energy source prices and technological progress, to find the competitive production technologies in each year. Since the volumetric energy density of hydrogen is low, hydrogen must be transported and stored at high pressure or in liquid form, or via liquid organic hydrogen carriers or ammonia. Consequently, different transmission-storage-distribution pathways are also considered and evaluated to address the trade-off between the energy density, energy consumption, and facility cost. Based on the superstructure, a mixed-integer non-linear programming (MINLP) model is formulated to optimize both the total cost and carbon emissions of the hydrogen supply chain. The most effective pathways for producing, transporting, and storing hydrogen are being identified. Several what-if scenarios are explored to derive insight on relative emissions and costs associated with various hydrogen pathways for the ASEAN region.

1. Almansoori, A., Shah, N., 2009. Design and operation of a future hydrogen supply chain: Multi-period model. *International Journal of Hydrogen Energy*, 34(19): 7883-7897.

2. Reuß, M., Grube, T., Robinius, M., Stolten, D., 2019. A hydrogen supply chain with spatial resolution: Comparative analysis of infrastructure technologies in Germany. *Applied Energy*, 247: 438-453.

3. Seo, S.-K., Yun, D.-Y., Lee, C.-J., 2020. Design and optimization of a hydrogen supply chain using a centralized storage model. *Applied Energy*, 262.

611098 Predicting Reaction Mechanism of Electrochemical CO_2 Reduction Using Machine Learning

Abstract:

Understanding the mechanism of an electrochemical reaction could, in principle, be one of the important factors in the accelerated discovery of highly efficient electrocatalysts. Cyclic voltammetry (CV) is one of the techniques providing important mechanistic insight into an electrochemical reaction by the virtue of the change in current density when a reactant or an intermediate species is adsorbed/desorbed on the electrode surface. However, interpreting a CV can be influenced by subjectivity from the experience of the electrochemist. Moreover, quantifiable confidence in such interpretations is wanting. Machine Learning (ML) algorithms can be applied in such cases to circumvent the heuristics in CV interpretations and user subjectivity to present robust, quantifiable confidence in predicting electrochemical reaction mechanisms. One of the most attractive electrochemical reactions in recent years has been the electrochemical CO_2 reduction (CO2R) to value-added chemicals due to its potentially high impact on the industries, society, and the environment. This work shows the implementation of an ML model (Neural networks) to predict the reaction mechanism of CO2R by learning from different data sets of cyclic voltammograms. The simulated and experimental cyclic voltammograms are used to train the neural network to identify the dependence on the parameters like CV scan rates and uncompensated resistance. For each individual CV, the voltammogram serves as an input to the model. The expected output predicts the likelihood of the occurrence of either simple reversible electron transfer reactions or irreversible product formation after an electron transfer reaction. This ML model confirms that the predictions can be made even by simple models as a substitute for the conventional subjective approach.

610876 Revealing the Mechanism of Intriguing Predominant Urea Formation from Thermal Degradation of CO_2-Loaded Aqueous Ethylenediamine

Abstract:

Amine scrubbing using aqueous amine solutions remains to be the most efficient near-term solution for post-combustion CO_2 capture from flue gas at coal-fired power plants.[1,2] Monoethanolamine (MEA) is one of the most extensively tested alkanolamines, but its commercial application tends to be limited because of degradation and corrosion issues and enormous energy requirement for solvent regeneration.[3–6] About (~10 %) of total capital cost of CO_2 capture is attributed to makeup of amine solvent loss.[7–10] Moreover, some degradation products are considered to be dangerous if exposed to the environment.

Ethylenediamine (EDA) has been reported to have several advantages over MEA for its higher CO_2 capacity, faster CO_2 absorption rate, easier producible implementation from renewable resources, and slightly lower degradation rates at moderate CO_2 loading.[11–16] EDA and MEA have a similar molecular structure as a primary amine; the former contains two primary amine groups and the latter has one primary amine and one hydroxyl group, while the end groups are separated by two -CH_2-groups in both molecules. Despite their structural similarity, the major thermal degradation products of CO_2-loaded EDA and MEA solutions have been found to be distinctly different.[11,17]

Thermal degradation of amines is generally understood to occur primarily in the stripper due to high temperature (~ 100 - 120 °C) and is known to increase with CO_2 loading.[18] Previous experimental studies proposed that thermal degradation of primary and secondary amines would be initiated by polymerization of carbamates, which can ultimately degrade to urea and imidazolidinones.[19,20] The MEA degradation is thought to proceed via 2-oxazolidinone (OZD) formation that may undergo further reaction with amines to form more stable products.[21] Our previous study predicted isocyanate formation from dehydration of carbamic acid to be highly probable, and also showed that cyclization of intermediate isocyanate to OZD would be kinetically more facile than reaction with MEA to form urea in aqueous MEA solution. The formation of isocyanate, albeit very small quantity, has also been experimentally reported.[22,23] In contrast to the case of MEA, the major thermal degradation products of EDA have been identified to be 1,3-Bis(2-aminoethyl)urea and 2-imidazolidone from mass spectrometry and high performance liquid chromatography measurements, while urea formation appears to be preferred.[11,17]

In this talk, we will present molecular mechanisms underlying thermal degradation of EDA in CO_2-loaded solution, especially the underlying reason for the preferential formation of urea, rather than cyclic compounds, in contrast to the case of MEA. The thermodynamic favorability of previously proposed reaction pathways, including the relative stabilities of intermediates and products, were evaluated using quantum mechanical calculations with an implicit solvent model. The free-energy barriers for key elementary reactions involved in the EDA degradation were determined using *ab initio* molecular dynamics simulations combined with metadynamics. Besides the simulation results, further analysis of the interactions between short-lived intermediates and surrounding solvent molecules provide an explanation as to why EDA and MEA show different degradation behaviors. This work highlights the importance of kinetic effects associated with the local structure and dynamics of solvent molecules around the intermediates that may significantly influence the degradation process of amine solvents.

References

[1] G. T. Rochelle, *Science*, 2009, **325**, 1652–4.

[2] P. Luis, *Desalination*, 2016, **380**, 93–99.

[3] M. R. M. Abu-Zahra, L. H. J. Schneiders, J. P. M. Niederer, P. H. M. Feron and G. F. Versteeg, *Int. J. Greenh. Gas Control*, 2007, **1**, 37–46.

[4] M. R. M. Abu-Zahra, J. P. M. Niederer, P. H. M. Feron and G. F. Versteeg, *Int. J. Greenh. Gas Control*, 2007, **1**, 135–142.

[5] D. M. D'Alessandro, B. Smit and J. R. Long, *Angew. Chem. Int. Ed. Engl.*, 2010, **49**, 6058–82.

[6] H. M. Stowe and G. S. Hwang, *Ind. Eng. Chem. Res.*, 2017, **56**, 6887–6899.

[7] A. B. Rao and E. S. Rubin, *Environ. Sci. Technol.*, 2002, **36**, 4467–4475.

[8] A. J. Reynolds, T. V. Verheyen, S. B. Adeloju, A. L. Chaffee and E. Meuleman, *Int. J. Greenh. Gas Control*, 2015.

[9] G. S. Goff and G. T. Rochelle, *Ind. Eng. Chem. Res.*, 2006, **45**, 2513–2521.

[10] S. A. Mazari, B. Si Ali, B. M. Jan, I. M. Saeed and S. Nizamuddin, *Int. J. Greenh. Gas Control*, 2015.

[11] S. Zhou, X. Chen, T. Nguyen, A. K. Voice and G. T. Rochelle, *ChemSusChem*, 2010, **3**, 913–918.

[12] M. Rabensteiner, G. Kinger, M. Koller, G. Gronald and C. Hochenauer, *Int. J. Greenh. Gas Control*, 2014, **27**, 1–14.

[13] S. Kadiwala, A. V. Rayer and A. Henni, *Chem. Eng. J.*, 2012, **179**, 262–271.

[14] J. Li, A. Henni and P. Tontiwachwuthikul, *Ind. Eng. Chem. Res.*, 2007, **46**, 4426–4434.

[15] Z. Wang, M. Fang, Y. Pan, S. Yan and Z. Luo, *Chem. Eng. Sci.*, 2013.

[16] A. P. Salvi, P. D. Vaidya and E. Y. Kenig, *Can. J. Chem. Eng.*, 2014, **92**, 2021–2028.

[17] J. Thompson, H. Richburg and K. Liu, in *Energy Procedia*, 2017.

[18] D. Hatchell, O. Namjoshi, K. Fischer and G. T. Rochelle, *Energy Procedia*, 2014, **63**, 1558–1568.

[19] H. Lepaumier, D. Picq and P. L. Carrette, in *Energy Procedia*, 2009.

[20] H. Lepaumier, E. F. Da Silva, A. Einbu, A. Grimstvedt, J. N. Knudsen, K. Zahlsen and H. F. Svendsen, in *Energy Procedia*, 2011.

[21] J. Davis and G. Rochelle, *Energy Procedia*, 2009, **1**, 327–333.

[22] T. Sakakura, J. C. Choi and H. Yasuda, *Chem. Rev.*, 2007, **107**, 2365–2387.

[23] T. E. Waldman and W. D. McGhee, *J. Chem. Soc. Chem. Commun.*, 1994, 957.

611409 Techno-Economic and Sustainability Analysis of the Production of Acetic and from Methane and Carbon Dioxide

Abstract:

Methane and carbon dioxide together represent 90% of greenhouse gases, and one way of diminishing their emissions is by converting them into added value chemicals. Acetic acid is one of such chemicals, and efforts at the laboratory scale have been carried out to prove it can be directly and indirectly synthesized from CH4 and CO2. However, there are very few studies that evaluate the possibility of implementing any of these processes at a larger scale. In this thesis, techno-economic and sustainability analyses were used to assess the feasibility of upscaling two pathways to produce acetic acid: (1) by direct conversion of methane and carbon dioxide using a plasma-assisted reactor and (2) by indirect catalytic conversion, using syngas (obtained from dry methane reforming) as an intermediate. ASPEN Plus was used for simulation, and with literature values and plant cost estimation techniques, the capital and operational expenses were estimated and compared to those of the commercial route.

The indirect and direct pathways attained return on investment (ROI) values of 1.5% and 5.5%, respectively, and unless credit is given to the reduction in the carbon footprint, neither process can economically compete with the commercial route, which attained an ROI of 19%. In terms of sustainability, the indirect and direct processes captured 0.7 and 2.5 times more carbon dioxide compared to the emitted amount by the commercial pathway. For the direct pathway it was found through a sensitivity analysis the carbon credit value can be reduce down to a value of $50/tonne to render the process economically viable.

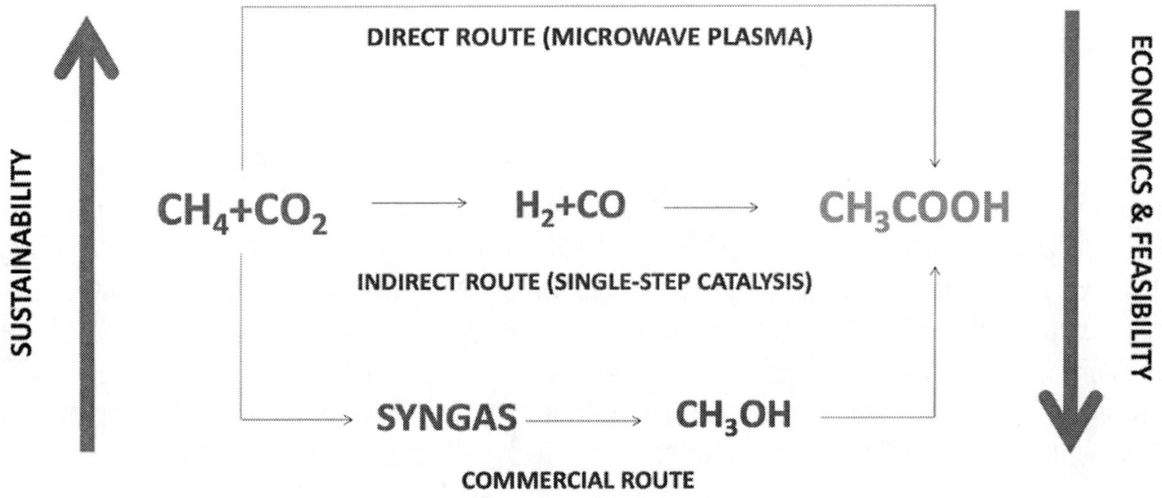

THE CARBONSAFE INITIATIVE: LAUNCHING THE CONVERGENCE OF CARBON CAPTURE AND STORAGE PROJECTS BY 2025

Mary Sullivan[1], Kanwal Mahajan[2], Darin Damiani[3], and Tomas Mora[4]

[1]National Energy Technology Laboratory, Pittsburgh, PA 15236-0940, U.S.A

[2]National Energy Technology Laboratory, Morgantown, WV 26507-0880, U.S.A

[3]U.S. Department of Energy, Washington, DC 20585, U.S.A.

[4]KeyLogic Systems, Inc., Morgantown, WV 26505, U.S.A.

Abstract

The Carbon Storage Program implemented by the U.S. Department of Energy (DOE) Office of Fossil Energy and managed by the National Energy Technology Laboratory (NETL) is helping to develop technologies that safely and permanently store carbon dioxide (CO_2) without impacting human health and the environment or hindering economic growth. Building upon almost two decades of knowledge and experience gained from the field lab testing projects and Regional Carbon Sequestration Partnership (RCSP) Initiative efforts, the Program initiated the Carbon Storage Assurance Facility Enterprise (CarbonSAFE) Initiative. The CarbonSAFE Initiative addresses key gaps on the critical path toward carbon capture, utilization and storage (CCUS) deployment by reducing technical risk, uncertainty, and cost of a geologic storage complex with capacity to store 50+ million metric tons of CO_2 from industrial sources over a 30-year period. The CarbonSAFE Initiative is taking a phased approach: (I) Integrated CCS Prefeasibility, (II) Storage Complex Feasibility, and (III) Site Characterization, Permitting and Carbon Capture Assessment. Subject to the availability of funding, a fourth phase would include construction and completion of wells and infrastructure leading to authorization to begin injection operations. A total of 13 projects were completed under Phase I, which provided high-level evaluations of potential CCS scenarios in regions throughout the United States. Six projects were awarded under Phase II to confirm the adequacy of storage complexes through initial site characterization. Phase III projects commenced in 2020 and will involve the entire integrated process, including a capture assessment, detailed site characterization, and preparing an application for a Class VI Underground Injection Control (UIC) permit to construct. These efforts are designed to help launch the industry by 2025 and contribute to increasing confidence that geologic storage of captured CO_2 can be a safe enterprise on a broad scale.

Keywords: U.S. Carbon Storage DOE Program; National Energy Technology Laboratory (NETL); carbon capture, utilization and storage (CCUS) research; geologic storage technologies; Regional Carbon Sequestration Partnerships (RCSPs); field projects; accomplishments; CarbonSAFE

1. United States Leadership in Commercial Deployment of CCUS

One of the great challenges facing the world today is the need to reduce global greenhouse gas (GHG) emissions and thereby mitigate the potential impacts of climate change without provoking undue detriment to global economies. In the United States (U.S.), energy-related CO_2 emissions have declined over the last decade in part due to a trend in fuel-switching from coal-based to natural gas-based sources in power generation, as well as increased production of renewable electricity, and improved energy efficiencies of vehicles. However, the U.S. Energy Information Administration (EIA) projects that U.S. energy-related emissions will remain essentially unchanged for the next 30 years unless carbon reduction programs and policies are adopted [1, 2]. Numerous international organizations, from the Intergovernmental Panel on Climate Change to the International Energy Agency (IEA), recognize that CCUS is an essential clean energy technology if a low-carbon economy is to be achieved at the lowest possible cost. Geologic Carbon storage includes the capturing of carbon dioxide (CO_2) from a stationary source; compressing it to a liquid-like state; transporting it via pipeline, ship, truck, or other means; and injecting it underground into deep reservoirs safely and securely. Although not part of CarbonSAFE, CO_2 recycling to produce economically valuable products and services such as improved materials, processes, fuels, and chemicals can be added to the CCUS value chain.

The United States is ideally situated for commercial CCUS projects because of its abundant carbon storage resources in proximity to many industrial CO_2 sources such as coal- and gas-fired power plants, gas-processing plants, ethanol plants, cement manufacturers, hydrogen production facilities and a host of others.

The United States also has over 50 years of CO_2-enhanced oil recovery experience and more than 5,000 miles of CO_2 pipeline infrastructure. Moreover, enhancements to federal and state policies are providing the needed incentives for advancing commercial CCUS projects. According to the IEA, 45Q tax credits could trigger new capital investments of up to $1 billion (B) for CCUS technologies over the next 6 years, resulting in an additional 10 to 30 million (M) metric tons (MT) of CO_2 capture capacity [3]. In the U.S power sector alone, 45Q credits could incentivize retrofit technologies in coal and natural gas power plants and capture of 49 MMT of CO_2/year (yr) by 2030 [4]. To date, there are over 30 carbon capture projects in various stages of project development in the U.S. [5, 6, 7]. With respect to regulations, recent U.S. Environmental Protection Agency (EPA) approval of primacy for North Dakota and Wyoming for Underground Injection Control (UIC) Class VI wells (used for geologic saline storage of CO_2) has provided developers an opportunity to work more closely with state jurisdictions, which might help facilitate the process and expedite permitting in these two states.

Over the past two decades, the U.S. Department of Energy (DOE) Office of Fossil Energy (FE) and its partners have reduced the cost of carbon capture from power and industrial sources and advanced the integration of CCUS systems in storage complexes with different geologic settings. One example of such an integrated system was NRG Energy's, Inc. Petra Nova project, the world's largest coal-fired post-combustion CO_2 capture system. The captured CO_2 was sent for use in enhanced oil recovery (EOR) in a partially depleted oil field in Jackson County, Texas. Lessons learned from Petra Nova and other industrial projects, such as the Air Products and Chemicals, Inc. hydrogen production facility in Port Arthur, TX and the Archer Daniels Midland Company ethanol facility in Decatur, IL, will result in lower costs for future projects. The next large-scale coal-fired post combustion CO_2 capture and compression project is expected to reduce costs by nearly 30% compared to Petra Nova's cost [8].

The knowledge gaps impeding growth in commercial CCUS deployment are both technical and non-technical. FE's Carbon Storage Program is working to close these remaining gaps by supporting a portfolio of research and development (R&D) projects and initiatives that investigate the technical gaps and aid in addressing the non-technical aspects of CCUS, such as the regulatory process, finance and markets, and public perception. Perhaps one of the best-known and most successful endeavours is the Regional Carbon Sequestration Partnerships (RCSPs). Over the last 17 years the RCSPs have characterized the CO_2 storage resource potential in their respective regions, completed pilot and large-scale CO_2 injection projects that cumulatively stored more than 11 MMT of CO_2, demonstrated the effectiveness of various technologies to verify storage containment in multiple reservoir types, and established a broad national network among companies and professionals supporting the deployment of commercial-scale CO_2 storage projects throughout the United States. Perhaps most importantly, the RCSPs have helped to validate storage security by virtue of its injection projects showing no evidence of impacts to human health or the environment. Through these successes the RCSP Initiative has laid the foundation for the CCUS industry, with key findings and recommendations captured in a series of Best Practice Manuals [9] made available to the public.

2. Convergence of DOE-FE's Carbon Capture and Carbon Storage programs

At present, the current state of CCUS technology is sufficient to develop and execute a CCUS project. However, cost reduction and performance optimization are among the biggest challenges to an industry launch by 2025. Increasing the number of CCUS projects at commercial scale enables development of the following:

- Cost-effective and accurate approaches for appraising site-specific storage resources and supporting investment-level decisions.
- Fully optimized and cost-effective long-term monitoring technologies and tools for commercial-scale onshore storage sites.
- Information and approaches for more targeted data acquisition and understanding of risk(s) in support of the permitting process.

DOE-FE's Carbon Storage Program is addressing these knowledge gaps primarily through its Carbon Storage Assurance Facility Enterprise (CarbonSAFE) Initiative, as described in more detail below. In parallel, DOE-FE's Carbon Capture Program is advancing R&D post-combustion carbon capture technologies that have the potential to reduce capture cost to $30/MT (at 95% purity). Both DOE-FE programs are planned to converge in 2025 with the intent to deploy integrated CCUS projects that would link commercial-scale storage complexes with state-of-the-art capture facilities.

3. CarbonSAFE Initiative

The Carbon Storage Program envisioned CarbonSAFE as the natural next step following the success of the Regional Carbon Sequestration Partnership (RCSP) efforts [10, 11], which provided the technical and non-technical foundations for scaling up CCUS operations to commercial scale. RCSP successes have already been demonstrated by contributions to major demonstration projects (Archer-Daniels-Midland [ADM], Air Products, Petra Nova), which cumulatively have stored more than 11 MMT of captured CO_2. The CarbonSAFE Initiative was created to ensure carbon storage complexes with capacities of 50+ MMT of CO_2 will be ready for integrated CCUS system deployment starting in 2025. The vision of the CarbonSAFE Initiative is to provide lessons learned and examples of the development of a CCUS storage complex from the feasibility study up until the point of injection. The progress made by this effort will instil greater confidence that commercial-scale CCUS projects are technically feasible and can be integrated in an economically feasible manner. The CarbonSAFE Initiative was first envisioned in 2016 as consisting of four phases:

- Phase I: Integrated CCS Prefeasibility – 13 projects awarded in 2016
- Phase II: Storage Complex Feasibility – 6 projects awarded: 3 in 2016, 3 in 2018
- Phase III: Site Characterization and CO_2 Capture Assessment – 5 projects awarded in 2020
- Phase IV: Permitting and Construction – potential future phase, dependent on appropriations

Thirteen prefeasibility studies were conducted in ten states and two offshore locations in the lower 48 states. Prefeasibility projects conducted economic feasibility analyses for targeted sites under varying scenarios. In addition, extensive regional data was collected, analyzed, and modeled; multiple storage sites and infrastructure were evaluated; and business plans for a variety of CCUS scenarios were developed. Currently, the program is implementing the Storage Complex Feasibility and Site Characterization and CO_2 Capture Assessment efforts. Research conducted within the six Storage Complex Feasibility projects focused on one or more specific reservoirs within the defined storage complex, encompassing (a) data collection; (b) geologic analysis; (c) identification of contractual and regulatory requirements; (d) subsurface modeling to support geologic characterization, risk assessment, and monitoring; and (e) public outreach.

The Site Characterization, Permitting and CO_2 Capture Assessment phase (Phase III) began in 2020. These projects will converge with Front-End Engineering Design (FEED) studies awarded by DOE's National Energy Technology Laboratory (NETL) Carbon Capture Program for commercial-scale post-combustion capture technologies implemented in the power sector.* The key objectives of this phase are to (a) complete a detailed site characterization of a commercial-scale CO_2 storage site (50+ MMT of captured CO_2 within a 30 yr period); (b) apply for and obtain UIC Class VI permit(s) to construct injection well(s); (c) complete a CO_2 capture assessment; (d) perform all work required to support a National Environmental Policy Act determination for each proposed project; and (e) address all other permitting processes and requirements, liability relief, and finance agreements in support of the business model for eventual commercial operations. Once the UIC Class VI permit applications to construct are submitted, additional activities will include working with the regulators to satisfy their requirements until construction permits are granted. Figure 1 shows locations of the CarbonSAFE Phase III projects with FEED studies across the U.S. The characterization projects in Phase III will converge with five of nine FEED studies awarded by DOE's NETL Carbon Capture Program for commercial-scale carbon capture systems on coal and natural gas power plants.

Subject to availability of funding, the final phase of CarbonSAFE, Construction of the Storage Complex, will involve the drilling and completion of the injection and monitoring wells and development of risk and mitigation plans to obtain authorization by the regulator to begin injection operations. Issued final reports for all project phases are located at the DOE's Office of Scientific and Technical Information site (https://www.osti.gov/search). Project-specific data produced by the CarbonSAFE effort are available on the Energy Data Exchange (edx.netl.doe.gov).

---

* The Funding Opportunity Announcements (FOAs) for the CarbonSAFE Phase III projects and Carbon Capture Program FEED studies were run independently with no dependence between them in terms of project selection. CO_2 capture assessments are being completed under CarbonSAFE for sources that did not receive FEED funding from the Carbon Capture Program.

Figure 1. Map of the contiguous U.S. showing locations of CarbonSAFE Phase III projects and Carbon Capture Program FEED studies

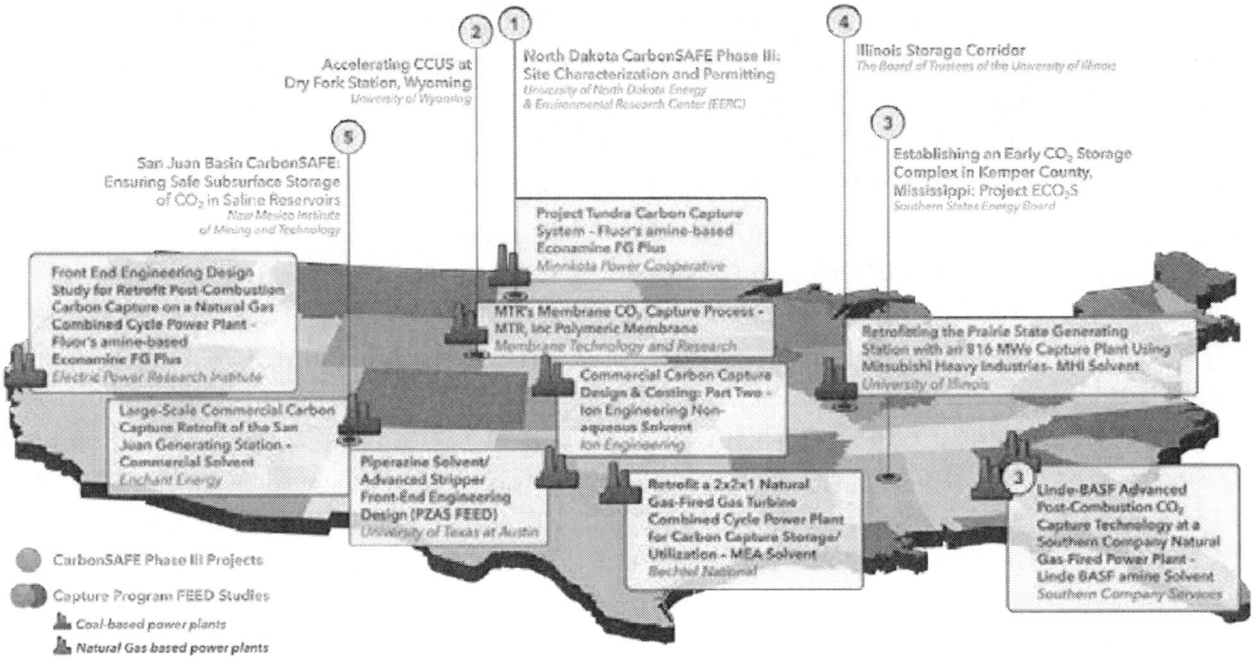

3.1 Integrated CCS prefeasibility projects

While not all prefeasibility projects were able to receive funding to continue their investigations into the next phase, a significant amount of useful and actionable research was produced. Extensive regional data was collected, analyzed, and modeled; multiple storage sites and infrastructure were evaluated; and business plans for a variety of CCUS scenarios were developed. A summary of each project's sources and storage estimates were provided in a previous publication [12]. Further characterization of the subsurface is needed at each site to verify the general initial findings. Common themes identified for making an integrated CCUS project successful from the prefeasibility projects include the following:

- Collaboration with oil and gas companies and strong host site support is essential.
- Uncertainty in federal policy results in lack of commitment from industry and the public for CCUS projects.
- Capture costs will need to decrease to make integrated CCUS projects economical.

- Reservoir heterogeneity requires improved models and data to increase plume behavior accuracy.
- Pressure management may be necessary to meet target rates and volumes in applicable geologic settings.
- Delineating the area-of-review for a geologic storage project is a critical component of permitting and the technical risk assessment.
- Regional state-of-stress and geomechanical characterization are important for site selection.

3.2 Financial and economic analysis

The prefeasibility projects performed a regional storage resource assessment to explore economic strategies for capturing, transporting, and storing large-scale anthropogenic sources in potential 50 MMT storage sites. Revenue from 45Q tax credit incentives for dedicated and associated storage scenarios and the potential sale of CO_2 to EOR fields were considered to offset the costs of these projects. Associated storage revenues were calculated at \$20-33/MT of CO_2 sold for EOR at an oil market price range of \$55-60/barrel (bbl). Table 1 includes the total cost of several fully integrated systems for CO_2 capture, compression, and transportation from sources to injection sites in the Onshore U.S. Prefeasibility projects with primary storage sites in the U.S. Offshore and Coastal regions in the Gulf of Mexico, Coastal Louisiana, and Offshore Washington are not summarized in this section.

Table 1. Total cost and revenue (shown in parentheses) for prefeasibility projects in onshore U.S.

Key	Project Name	Source(s) by Location	Econ. Cases	Revenue Type	Total Cost ($/MT)
1	Integrated Mid-Continent Stacked Carbon Storage Hub	Coal-fired power plant in Lincoln County, NE Ethanol plant corridor in eastern NE	3	Dedicated storage	31 to 67
			3	Oil+Net CO_2 stored for EOR	(1.72) to 57
2	Nebraska Integrated Carbon Capture and Storage Prefeasibility	Coal-fired power plant in Lincoln County, NE	1	Dedicated storage	20
			1	CO_2 sold for EOR	32
3	Integrated Carbon Capture and Storage for Kansas (ICKan)	Two coal-fired power plants in Finney and Pottawatomie County, KS	6	Dedicated storage	2 to 27
			6	CO_2 sold for EOR	14 to 39

		Ethanol plant corridor in northeast KS			
4	Integrated CCS Prefeasibility Study at Dry Fork Station	Coal-fired power plant in Campbell County,WY	1	CO_2 sold for EOR and dedicated storage	(0.14)
5	CarbonSAFE Illinois East Basin[1]	Chemical plant in Vigo County IL	1	CO_2 sold for EOR and dedicated storage	-
6	CAB-CS: Central Appalachian Basin CarbonSAFE Integrated Prefeasibility Project	Coal-fired power plant, a Gas-fired power plant, and Chemical plant in eastern Ohio. Potential sources: steel mill and gas-fired power plant in eastern and southern Ohio	6	dedicated storage	9 to 66
			6	CO_2 sold for EOR and dedicated storage	9 to 68
			6	CO_2 sold for EOR	1 to 67
7	Northern Michigan Basin CarbonSAFE Integrated Prefeasibility Project	Coal-fired power plant in Otsego Co., MI. Potential sources: steel mill and gas-fired power plant in northern and central MI	4	Dedicated storage	21 to 116
			4	CO_2 sold for EOR and dedicated storage	17 to 114
			4	CO_2 sold for EOR	12 to 108
8	California CO_2 Storage Assurance Facility Enterprise	Three integrated energy corridors in Southern San Joaquin Valley, CA	1	CO_2 sold for EOR and dedicated storage	(40)
9	CarbonSAFE Rocky Mountain Phase I: Ensuring Safe Subsurface Storage in the Intermountain West	Two coal-fired power plants in Emery County, Utah.	4	Dedicated storage [NA]	50 to 51
			9	CO_2 sold for EOR [NA]	17 to 27
10	Integrated Commercial CCS Prefeasibility Study at Rock Springs Uplifts, Wyoming[2]	Coal-fired power plant in Sweetwater Co., WY.	1	CO_2 sold for EOR and Dedicated storage	-

[1]Capital investment was estimated at $58.7 M. Annual operating costs were estimated at $12.7 M to supply 1.7 MMT/yr of CO_2.
[2]Capital investment was estimated at $787 M. Annual operating costs were estimated at $54 M to supply 2 MMT/yr of CO_2.
[NA] Economic analysis did not include 45Q tax credit incentives.

The economic analysis for each prefeasibility study can be categorized into three types of scenarios: coal-fired power plant point-to-point transport (Scenario 1), multi-source point-to-point transport (Scenario 2), and long-range energy corridor transport (Scenario 3). The scenarios are meant to offer potential commercial synergies for reducing the unit cost of CO_2 storage and the risk of investment. One representative prefeasibility study for each scenario is summarized in the following sections.

3.2.1 Scenario 1: Coal-fired power plant point-to-point transport

The University of Utah performed a prefeasibility study in central Utah (Table 1, Key 9). The PacifiCorp's Hunter Power Plant and Huntington Power Plant are located in Emery county near a 7,000-foot deep brine-bearing sedimentary rock formation. A commercial amine-based capture technology (Mitsubishi Heavy Industries (MHI) KM-CR Process® with KS-1™ solvent) was evaluated to estimate the capture and compression costs. Approximately 1.84 MMT and 7.2 MMT of CO_2 per yr can be captured from each plant at $50 and $61/MT, respectively. The proposed locations of two storage sites named Buzzards Bench and Drunkards Wash have estimated storage capacities of 201 MMT and 380 MMT of CO_2, respectively.

Table 2 lists the point-to-point economic scenarios for each power plant. Cost analysis supports the primary core scenario consisting of 65% capture (equivalent to 1.84 MMT of CO_2 per year) from each coal-fired power plant and transporting and injecting it into a sandstone formation at the Buzzards Bench or Drunkards Wash area. Simulation studies suggest that one well is sufficient to store 50 MMT of CO_2 over 30 years in each storage complex. The second scenario involves selling CO_2 to an EOR field located in the Uinta Basin in Utah and a natural source of CO_2 in the McElmo Dome in southwestern Colorado. Findings suggest that larger-scale infrastructure deployments to oil fields in the Uinta Basin for EOR applications could result in a cost savings of up for 67%/MT of CO_2. None of the scenarios considered splitting tax credit incentives from storing CO_2 in both dedicated and associated storage sites. Detailed discussion of the results of this study are provided by McPherson [13].

Table 1. Economic scenarios without 45Q tax credit incentives

Case	Source(s) and Sink(s)	Pipeline (Miles)	Total Cost ($M/yr)	Capture ($M/yr)	Trans. ($M/yr)	Storage ($M/yr)	Total Unit[1] ($/MT)

1	Hunter – Buzzards	22	86	78	0.8	6.6	50.40
2	Hunter – Drunkards	11	87	78	2.4	6.7	51.37
3	Huntington – Buzzards	9	86	78	1.3	6.6	50.67
4	Huntington – Drunkards	7	87	78	1.8	6.7	50.98
5	Hunter – Uinta EOR	100 >	29	78	1.5	(51)	16.87
6	Hunter – McElmo Dome	300 >	42	78	14.5	(51)	24.57

Note: Numbers in parentheses indicate negative costs or profit derived from 45Q tax credit. [1] Total Unit cost includes storage and monitoring costs for 30 years.

3.2.2 Scenario 2: Multi-source point-to-point transport

Battelle performed a prefeasibility study in northern Michigan (Table 1, Key 7). Multiple CO_2 sources, rather than a single source, were considered in the analysis. These include: 1) the existing Wolverine Power Cooperative-owned natural gas-fired power station in Otsego County that will be retrofitted to a combined cycle facility with CO_2 capture supplemented by CO_2 from the DCP Midstream-owned natural-gas separation facility, also located in Otsego County; 2) a potential new source (PNS) assumed to be located close to the saline storage site(s) and the EOR fields that is a conventional natural gas combined cycle (NGCC) facility with capture; 3) a second variation of the PNS that incorporates the emerging natural gas-fired technology being developed and built by NET Power, LLC, based on the Allam Cycle; and 4) Project TIM, a proposed steel manufacturing facility by New Steel, Inc. in Shiawassee County. The saline storage sites, Saint Peter Sandstone (SPSS) Site 2 and SPSS Site 7, were down selected for detailed economic analysis. Site 2 is located in Montmorency County and Site 7 is located in Kalkaska and Missaukee counties in Michigan. Simulation studies suggest that Site 2 requires six injection wells to inject 50 MMT of CO_2 and has a projected 50 MMT CO_2 plume area of 18,200 acres. Site 7 requires only three injection wells and has a projected plume area of 14,000 acres.

Table 4 lists the multi-source point-to-point economic scenarios evaluated. Interest costs during construction and escalation were included for each of the storage, pipeline, and capture facilities during the construction period prior to the commencement of operations in 2025. Economic analysis of Net Power, an emerging gas-power technology, was not included. The three types of scenarios include: 50 MMT of CO_2 stored at Site 2 for each of the identified sources, 50 MMT of CO_2 supplied to the EOR fields (i.e., carbonate reefs), and a hybrid case in which the CO_2 is divided between saline storage (50%) and CO_2-EOR (50%). In the hybrid case, only one group of EOR reefs is needed to store 25 MMT CO_2. The availability of EOR (ongoing and expanded) as an option for stacked storage with saline storage is an attractive feature for the Northern Michigan Basin area, as this can help defray some of the capital and operating costs for saline storage, increasing the feasibility of the project. Results show that either a 20% reduction in the cost of capture or a $30/MT of CO_2 sale price is necessary for a new natural gas-fired combined cycle with 100% CO_2-EOR storage to achieve a breakeven net revenue requirement. Core Energy, a key member of the Battelle Carbon Storage – Northern Michigan Basin project team, operates in EOR fields and could benefit from this project. Detailed discussion of the results of this study are provided by Kelly [14] in Table 4.

Table 2. Economic scenarios with 45Q tax credit incentives

Case	Source(s) and Sink(s)	Pipeline (Miles)	Total Costs ($M)	Capture ($M)	Trans. ($M)	Storage ($M)	Total Unit Cost[1] ($/MT)
1	Wolverine – SPSS-2 site	21	1,056	691	40	325	54-74
2	Wolverine – EOR and SPSS-2 site	74	1,017	691	131	195	55-73
3	Wolverine – Niagaran Reefs	105	881	691	190	-	47-60
4	New NGCC – SPSS-2 site	17	1,016	658	33	325	44-58
5	New NGCC – EOR and SPSS-2 site	46	937	658	84	195	43-56
6	New NGCC – Niagaran Reefs	83	808	658	150	-	38-48
7	Project Tim – SPSS-2 site	154	1,450	848	277	325	89-116
8	Project Tim – EOR and SPSS-2 site	183	1,370	848	327	195	89-114
9	Project Tim – Niagaran Reefs	225	1,250	848	402	-	85-118

[1]Cost components and anticipated amount of revenue for a scenario on a 30-year levelized basis, expressed in 2018 $/tonne (the project-start date) on a low or high cost of capital basis.

3.2.3 Scenario 3: Large-scale energy corridor transport

The University of Kansas, along with the Kansas Geological Survey, conducted a preliminary feasibility study of two coal-fired power plants in Kansas connecting to large storage corridors in Nebraska (Table 1, Key 3). The Westar Energy's Jeffrey Energy Center (JEC) and S.E. Holcomb station were selected to evaluate multiple capture levels using a commercially available Linde-BASF system as the basis for the capture technology. The estimated capture costs for JEC (2.5 MMT/yr) and S.E. Holcomb Station (1.7 MMT/yr) were \$45–67/MT and \$46–\$72/MT, respectively. In addition, about 34 ethanol plants within the region could supply CO_2 to EOR and storage sites located in southwest Kansas and the Permian basin in Texas. Capture and compression costs from these ethanol plants were estimated at \$19/MT and represent a yearly production 10.9 MMT of CO_2 from 3.6 B gallons of ethanol.

Saline aquifers that are overlain by oil reservoirs were studied in the state of Kansas. As a primary site option, the Patterson site, located in southwest Kansas, has an estimated storage capacity of 60 MMT. Previous studies in the region estimated that four of these fields in the NHSC have an estimated associated storage potential for EOR of 1.8-2.3 MMT/yr [15]. To date, 5.5 M barrels (bbls) of oil have been produced from the combined Patterson and Heinitz fields before and after unitization, and an additional 1.2 Mbbls of oil are expected from the waterflood (6.7 Mbbls ultimate recovery). As a secondary site option, the Davis Ranch and John Creek fields, located in the Forest City Basin in Wabaunsee county, have an estimated storage capacity of 24 MMT.

Table 5 lists the scenarios presented by the University of Kansas. All cost estimates include both capital costs and operating expenses over a 2-yr construction period followed by 20 years of operation, and a 10% cost of capital. The most feasible option, as outlined in Case 4, consists of delivering 4.3 MMT/yr of CO_2 from ethanol plants in Kansas and Nebraska to several EOR fields located in southwest Kansas. The total project cost can be further reduced by supplying 2-2.5 MMT to dedicated storage and selling 1.8-2.3 MMT of CO_2 to EOR in the nearby EOR field. Profits from the sale of CO_2 for EOR could offset transportation. Detailed discussion of the results of this study are provided by Holubnyak [16].

Table 3. Economic scenarios with 45Q tax incentives

Case	Source(s) and	Pipeline	45 Q	Pipeline	Capture	Total

	Sink(s)	(miles)	Revenue Type	Costs ($/MT)	Costs[1] ($/MT)	Cost ($/MT)
1	Holcomb – Patterson Site	25	EOR	(23.3)	46-72	23
			Saline	(35.1)	46-72	11
2	JEC – Davis Ranch/John Creek	50	EOR	(22.9)	45-67	22
			Saline	(34.8)	45-67	10
3	34 ethanol plants – Permian Basin	1,546	EOR	9.0	10	19
			Saline	0.8	7	8
4	15 ethanol plants – SW Kansas	737	EOR	5.0	9	14
			Saline	(2.5)	5	2

Note: Numbers in parentheses indicate negative costs or profit derived from 45Q tax credit. [1]The derived costs for scenarios 1 and 2 used lowest capture costs from pre-FEED studies

3.3 CarbonSAFE storage complex feasibility projects

CarbonSAFE feasibility projects focused on the next phase of advancing carbon storage toward commercial scale. Work conducted under the feasibility projects included activities such as completing at least one characterization well, updating of static earth models, updating of dynamic and geomechanical modeling, 3D seismic analysis, core and log analysis updates and well testing updates. Initial characterization results demonstrated the likely viability of multiple 50+ MMT geologic storage complexes in deep saline formations. A risk assessment exercise was also conducted to identify and assess technical and nontechnical risks that could prevent potential candidate storage complexes within the study area from serving as commercial storage sites. The assessment identified and evaluated six technical risk categories: 1) CO_2 injectivity, 2) storage capacity, 3) lateral migration of CO_2, 4) lateral pressure propagation, 5) vertical migration of CO_2 or formation brine, and 6) induced seismicity. The risk assessment results can be used to complete a site characterization plan (e.g., future site characterization, modeling and simulation, monitoring activities) and management strategies for the project. Three feasibility projects that have not yet advanced to the characterization phase are briefly described below.

3.3.1 Wabash CarbonSAFE

The Wabash CarbonSAFE project, led by the University of Illinois at Urbana-Champaign, determined the feasibility of commercial-scale CO_2 storage in the Mt. Simon sandstone formation located beneath the future Wabash Valley Resources

(WVR) ammonia production plant. Targeted to begin operation in 2021, the ammonia plant is under construction to repurpose and expand the former integrated gasification combined cycle plant located near Terre Haute, Indiana. Approximately 1.65 MMT of CO_2/yr will be produced as a by-product, which can then be captured through a process using Rectisol.

The selected storage site, located on the eastern side of the Illinois Basin, is on the WVR plant site. The primary target reservoir is the Mt. Simon sandstone formation. This formation is estimated to have a storage capacity of 12-172 billion metric tons of CO_2 in the Illinois Basin [17]. Secondary storage targets include units in the Knox Group, and the St. Peter Sandstone. A stratigraphic well was drilled in the WVR facility for the acquisition of 265 (ft) feet of core and to obtain geologic data to update models and confirm reservoir properties. Core samples were collected from the Maquoketa Group (secondary seal), the Eau Claire Formation (primary seal), and the Upper Mt. Simon Sandstone. An extensive suite of geophysical logs was completed to evaluate CO_2 storage potential, hydrocarbon potential, and general lithological and petrophysical characteristics of the sedimentary column. In addition, 20 kilometers (km) of 2D seismic data is being processed to assist with subsurface geologic characterization. Future vertical seismic profiles (VSPs) can be performed to improve delineation of geologic structures. Final reservoir simulation, based on the stratigraphic well's core and log analyses and drill stem test data, would be performed to assess the CO_2 injection potential of the Mt. Simon sandstone.

3.3.2 CarbonSAFE Illinois Macon County

The CarbonSAFE Illinois Macon County project, led by the Illinois State Geological Survey, assessed the feasibility of developing a CO_2 storage complex that can store more than 50 MMT of CO_2 in the central Illinois Basin. The closest potential CO_2 sources are the Abbot Power Plant located in Champaign County, Illinois; City Water, Light, and Power located in Springfield County, Illinois; and ADM located in Decatur County, Illinois. These three sources emit a total of approximately 3.0 MMT/yr of CO_2. A fourth, more distant potential source is the 1,600-megawatt (MW) coal-fired Prairie States Energy Campus, which began operations in 2012 and emits greater than 10 MMT of CO_2/yr.

The original project site in Macon County was switched to the Podolsky Oil Company's field in Christian County, Illinois to accommodate the larger potential sources of CO_2. The target reservoir was the Lower Mt. Simon sandstone formation.

Initial characterization work included drilling the T.R. McMillen #2 (TRM) stratigraphic well for the acquisition of 148 ft of core and to obtain geologic data to update models and confirm reservoir properties. The well was drilled through the targeted storage interval into the Precambrian basement to a total depth of 6,490 ft. Preliminary results indicated that the Mt. Simon sandstone at the location of the TRM #2 well was suitable for commercial-scale CO_2 storage. Initial modeling showed that 30 years of injection in three wells with 80 acre-spacing projected a 190 MMT CO_2 plume extension of 2.7 miles with little interaction with the adjacent Illinois Basin–Decatur Project (IBDP). A 30 km 2D seismic survey and a VSP were performed to obtain data needed to interpret the subsurface structure of the site. Seismic data is being processed and analyzed. Additionally, stacked storage in oil-bearing carbonate oil reservoirs in the Mt. Auburn trend of the Sangamon Arch in the Illinois Basin are being studied for CO_2-EOR potential.

Previous characterization efforts at the IBDP site suggested that induced microseismicity migrated from sediment into Precambrian basement. At the site, the microseismicity was monitored using deep borehole sensors. About 1400 events were recorded, with moment magnitudes between 1.1 and -1.7. [18, 19]. Induced seismicity risk was assessed at the Macon site through the characterization of geomechanical properties of rocks and fault analysis, and the stress state field measurements in the basement. Well log cross sections displaying the geomechanical properties at the Macon and IBDP sites were prepared to illustrate the differences between the sites. Laboratory tests performed on the cores from the TRM well were also performed on the basement rock. In situ stress direction characterization at the Macon site was completed using indicators observed in a Formation Micro-Imager log and well tests (e.g., Diagnostic Fracture Injection Test). Future characterization work would include the integration of these data and well site studies with improved geomechanical models and dynamic reservoir models.

3.3.3 Integrated Midcontinent Stacked Carbon Storage Hub

The Integrated Midcontinent Stacked Carbon Storage Hub project, led by Battelle Memorial Institute, is a combination of three prefeasibility projects located in Kansas and Nebraska. The intent of the project is to develop a source corridor that can run southwest from eastern and central Nebraska toward Red Willow County, Nebraska. Detailed discussion of the Phase I results of this study are provided by Duguid [20]. The feasible pipeline routes connecting sources along the source corridor to potential saline sinks (i.e., dedicated storage) or CO_2-EOR sinks (i.e., associated storage) in the storage corridor account for environmentally and culturally sensitive areas. The team identified 17 technically and economically feasible CO_2-EOR oilfields and three saline reservoir fields in the storage corridor. The designated oilfields have estimated storage of 577.4 MMT of CO_2 with the potential to produce 181.9 Mbbls of oil via EOR, equivalent to an estimated gross revenue for stacked storage and EOR of $30.9 B.

Three saline sinks were evaluated: Madrid field in southwest Nebraska, Patterson-Heinitz-Hartland field (PHH) in southwestern Kansas, and the Sleepy Hollow Field (SHF) in southwest-central Nebraska. One stratigraphic well was drilled at the SHF site to obtain geologic data through well logging and 110 ft of core for analysis. Two additional stratigraphic wells were drilled at the PHH site to collect geologic data through well logging and obtaining 778 ft of core for analysis. The effort also included the acquisition of 26 square miles (mi^2) of new 3D seismic data to interpret the subsurface structure in the PHH site. Results from seismic data processing revealed two major reverse faults that offset the reservoir and seal intervals and constitute an uplifted block in the Patterson area. In addition, three- and four-way structural closures were identified that can assist with trapping CO_2 in the Arbuckle-Osage reservoirs. Updated models indicated that the original 50 MMT target injection amount was not viable in the SHF site. Numerical simulation for the PHH site indicated that three wells would be required to store 50 MMT of CO_2 in deep stacked saline Paleozoic formations over 30 years. Future recommended characterization work would include fault analysis at the PHH site and conducting a 3D seismic survey and drilling a characterization well at the Madrid site.

3.4 CarbonSAFE site characterization and CO_2 capture assessment projects

Three of the five Site Characterization and CO_2 Capture Assessment projects build upon previous phases that indicated the commercial viability of storage complexes located in North Dakota, Mississippi, and Wyoming. The two remaining projects, located in New Mexico and Illinois, demonstrated the viability of their respective sites through previous work conducted, unrelated to the CarbonSAFE effort. The site characterization efforts conducted by these five projects, and the association of these projects with capture assessment projects, are discussed below.

3.4.1 Illinois Storage Corridor (Phase III)

The Illinois Storage Corridor is a region with proven geologic storage performance and numerous industrial carbon sources. Numerous DOE-FE funded projects in the Illinois Basin are already addressing the challenges posed by CCUS deployment at scale through the field validation of storage complexes and new technologies. These projects include two pilot carbon capture technology projects [21] two commercial-scale CCUS projects led by RCSPs [22], and Wabash and Macon CarbonSAFE phase II projects. As part of the CarbonSAFE phase III work, the Board of Trustees of the University of Illinois (Champaign, IL) will determine the feasibility of developing a storage complex near two facilities: One Earth Energy ethanol facility in Ford County, Illinois and Prairie State Energy Campus (PSEC) facility, a 1,200-MW coal-fired electricity generation plant in St. Clair County, Illinois. Annually, these two sources have the potential to capture 0.45 and 6 MMT of CO_2, respectively. A Carbon Capture Program funded FEED study at PSEC will evaluate the use of the commercial MHI's KM CDR Process™ CO_2 capture technology, and a capture assessment at the One Earth Energy ethanol facility will be completed under CarbonSAFE. Future characterization work at each site will include the drilling and logging of one characterization well, coring new samples, and acquiring 26 miles of 2D seismic data and 4-6 mi^2 of 3D seismic data.

3.4.2 San Juan Basin CarbonSAFE Phase III: Ensuring safe subsurface storage of CO_2 in saline reservoirs

New Mexico Institute of Mining and Technology (Socorro, NM) is performing a comprehensive site characterization study of a mature oil and gas field in the San Juan Basin, northwest New Mexico. The field, operated by Hilcorp Energy, has great stacked storage potential in alternating sequences of deep saline formations and oil-bearing reservoirs. Saline storage targets include the Salt Wash Member, Bluff sandstone, and Slick Rock Member.

The selected site is approximately 20 miles from the San Juan Generating Station (SJGS), an 847-MW coal-fired electricity generation plant. Annually, this plant has the potential to capture 6-7 MMT of CO_2. The CCUS project will consist of supplying 2 MMT of CO_2/yr to the storage complex and transporting the remaining CO_2 via the Cortez pipeline network to EOR fields in the Permian Basin in Texas. A Carbon Capture Program funded FEED study at SJGS will be completed using MHI's commercial amine-based solvent CO_2 capture technology. Ongoing and upcoming efforts at the storage complex site include the acquisition and evaluation of existing field data (e.g., 3D seismic, logging and coring samples), the drilling and logging of one characterization well, and coring of new samples.

3.4.3 Establishing an early CO_2 storage complex in Kemper County, Mississippi: Project ECO_2S

Southern States Energy Board (Peachtree Corner, GA) will fully characterize the storage complex facility in Kemper County, Mississippi. Previous characterization efforts recognized Kemper County as a world class geologic setting capable of securely storing over 1,200 MMT of CO_2 over a 30,000-acre area. Detailed discussion of the Phase II results of this study are provided by Reistenberg [23]. Three high-quality sandstone reservoirs include the lower Cretaceous Paluxy Formation, Washita-Fredericksburg interval, and the Upper Cretaceous Lower Tuscaloosa group. The initial proposed CCUS scenario comprised capture from the Kemper County Energy Facility with minimal pipelines required to transport the CO_2 to the storage complex. Due to operational changes at the facility, the project now includes two additional sources totaling 22.5 MMT of CO_2/yr for the development of a regional hub. A Carbon Capture Program funded FEED study will be completed using Linde-BASF's OASE blue Advanced Amine solvent-based capture technology at Southern Company's Plant Daniel in Moss Point, Mississippi. Additional CO_2 capture assessments of Plant Ratcliffe in Kemper County, Mississippi and Plant Miller in Jefferson County, Alabama will be completed under this project. Ongoing and upcoming efforts at the storage complex site include the drilling and logging of three characterization wells, coring new samples, and acquisition of 4-6 mi^2 of 3D seismic data.

3.4.4 North Dakota CarbonSAFE Phase III: Site characterization and permitting

The University of North Dakota Energy & Environmental Research Center (Grand Forks, ND) will fully characterize a storage complex near Minnkota's Milton R. Young Station (MRYS) one of two storage sites considered in earlier feasibility efforts. Detailed discussion of the Phase II results of this study are provided by

Peck [24]. The site is capable of storing over 3 MMT of CO_2/yr from the plant's 455-MW Unit 2. The project will investigate three storage reservoirs for further study: Broom Creek, Inyan Kara, and Deadwood sand reservoirs. Previous work suggests that Broom Creek formation is capable of securely storing over 563 MMT of CO_2. Coordinating with this CarbonSAFE Initiative is the Project Tundra initiative - an integrated CO_2 capture and storage project led by Minnkota Power Cooperative Targeted to begin operation in 2022-2023, the MRYS Unit 2 will be using a commercial amine-based Econamine FG Plus^{SM} capture technology from Fluor Corporation. A FEED study of this technology is being studied under the Carbon Capture program. Ongoing effort at the MRYS site will facilitate the completion of all necessary regulatory requirements for two Class VI well permits. Future characterization work will include the drilling and logging of two characterization wells, coring new samples, and 11 mi^2 of 3D seismic acquisition.

3.4.5 Wyoming CarbonSAFE: Accelerating CCUS commercialization and deployment at Dry Fork Power Station and Wyoming Integrated Test Center

The University of Wyoming (Laramie, WY) is finalizing the characterization of the storage complex in Campbell County, Wyoming. Detailed discussion of the Phase I results of this study are provided by Quillinan [25]. The selected storage site is centrally located in the Powder River Basin, which produces more than 40% of the coal mined in the United States [26], and is accessible to existing CO_2 pipeline infrastructure and potential CO_2-EOR fields. The project will investigate four storage reservoirs for further study: the Minnelusa, Lower Sundance, Lakota/Fall River Group, and the Muddy Sandstone. Initial analysis suggests a P50 storage volume of 3.4 MMT/mi^2 for the combined reservoirs. The project would capture CO_2 from Basin Electric's Dry Fork Station (DFS), one of the newest 385-MW commercial-scale coal-fired power plants in the United States. The power station is also home to the Wyoming Integrated Test Center (https://www.wyomingitc.org/), which is used to study the capture, storage, and management of carbon emissions from DFS. A Carbon Capture Program funded FEED study at the DFS will be completed for the commercial Polaris^{TM} membrane CO_2 capture technology from Membrane Technology and Research, Inc. Future characterization work will include analysis of previous phase II acquisitions (legacy 2D seismic, newly acquired 3D seismic, and coring new samples), re-entering one well to gauge for monitoring, constructing a second well for data collection and injection tests, and a new VSP seismic data acquisition.

4. Summary

CCUS is becoming widely accepted as a viable and necessary option for fossil-based energy sources, such as coal- or gas-fired power plants and other industrial sources to lower their carbon dioxide (CO_2) emissions. Over the past two decades DOE research efforts have provided supporting evidence that CO_2 can be stored in geologic formations safely and securely across multiple regions of the U.S. Efforts now must shift focus on the justification of the business case for widespread CCUS deployment. Research will continue to focus on technologies and methods to reduce cost, reduce operational risk and increase efficiencies to make commercial-scale CCUS economical.

The CarbonSAFE Initiative is providing this understanding of how to safely, efficiently and cost-effectively characterize and permit project site(s) at the commercial-scale throughout the U.S. Phase I and Phase II projects have investigated potential integrated CCUS project scenarios and identified both technical and non-technical research gaps. Each Phase III CarbonSAFE project now works to fill those gaps. Projects will need to consider and negotiate various aspects and business arrangements as part of the process to evaluate and secure CO_2 source(s), develop a plan for pore space ownership, and complete a UIC Class VI permit application(s) to construct the injection wells. To achieve wide adoption of CCUS, it is imperative that we build upon these learnings to test, mature and prove CCUS technologies at commercial scale.

5. Disclaimer

This report was prepared as an account of work sponsored by an agency of the United States Government. Neither the United States Government nor any agency thereof, nor any of their employees, makes any warranty, express or implied, or assumes any legal liability or responsibility for the accuracy, completeness, or usefulness of any information, apparatus, product, or process disclosed, or represents that its use would not infringe privately owned rights. Reference herein to any specific commercial product, process, or service by trade name, trademark, manufacturer, or otherwise does not necessarily constitute or imply its endorsement, recommendation, or favoring by the United States Government or any agency thereof. The views and opinions of authors expressed herein do not necessarily state or reflect those of the United States Government or any agency thereof.

6. Acknowledgement

KeyLogic's contributions to this work were funded by the National Energy Technology Laboratory under the Mission Execution and Strategic Analysis contract (DE-FE0025912) for support services.

7. References

[1] National Petroleum Council, "Meeting the Dual Challenge: A Roadmap to At-Scale Deployment of Carbon Capture, Use, and Storage, Chapter One: The Role of CCUS in the Future Energy Mix," 2019. [Online]. Available: https://dualchallenge.npc.org/files/CCUS-Chap_1-042320.pdf. [Accessed November 2020].

[2] EIA, "U.S. Energy-Related CO2 Emissions in AEO2019 Reference Case, 1990 to 2050 (figure 1-10)," EIA, 2019.

[3] S. Bennett and T. Stanley, "US budget bill may help carbon capture get back on track," 12 March 2018. [Online]. Available: https://www.iea.org/commentaries/us-budget-bill-may-help-carbon-capture-get-back-on-track. [Accessed November 2020].

[4] D. Nagabhushan and J. Thompson, "Carbon capture and storage in the US power sector - The impact of 45Q Federal Tax Credits," Clean Air Task Force, The impact of 45Q Federal 2019. [Online]. Available: https://www.catf.us/wp-content/uploads/2019/02/CATF_CCS_United_States_Power_Sector.pdf. [Accessed November 2020].

[5] Carbon Capture Coalition, "Carbon Capture Jobs and Project," June 2020.

[Online]. Available: https://carboncapturecoalition.org/wp-content/uploads/2020/06/Carbon-Capture-Jobs-and-Projects-1.pdf. [Accessed November 2020].

[6] Clean Air Task Force, "Potential Carbon Capture Projects Database," undated. [Online]. Available: https://docs.google.com/spreadsheets/d/115hsADg3ymy3lKBy4PBQRXz_MBknptqlRtlfuv79XV8/edit#gid=1540463113. [Accessed November 2020].

[7] Clean Air Task Force, "Interactive Map of CCUS Projects in Development in the U.S.," 27 July 2020. [Online]. Available: https://www.catf.us/2020/07/ccus-interactive-map/. [Accessed November 2020].

[8] H. Tanakaa, T. Tsujiuchi, T. Kamijo, S. Kishimoto, Y.-J. Lin, S. Kawasaki, Y. Nakagami and T. Nojo, "Advanced KM CDR Process using New Solvent," October 2018. [Online]. Available: https://www.cfaenm.org/wp-content/uploads/2019/03/GHGT14_manuscript_20180913Clean-version.pdf. [Accessed November 2020].

[9] NETL, "BEST PRACTICES MANUALS," undated. [Online]. Available: https://netl.doe.gov/coal/carbon-storage/strategic-program-support/best-practices-manuals. [Accessed November 2020].

[10] T. Rodosta, M. Ackiewicz and A. Albenze, "Status Update and Results from the U.S. Department of Energy Regional Carbon Sequestration Partnership Initiative, Proceedings GHGT-12," *Energy Procedia,* vol. 63, pp. 6039-6052, 2014.

[11] T. Rodosta, W. Aljoe, G. Bromhal and D. Damiani, "U.S. DOE Regional Carbon Sequestration Partnership Initiative: New Insights and Lessons Learned, Proceedings GHGT-13," *Energy Procedia,* vol. 114, pp. 5580-5592, 2017.

[12] M. Sullivan, T. Rodosta, K. Mahajan and D. Damiani, "An overview of the Department of Energy's CarbonSAFE Initiative: Moving CCUS toward commercialization," *Separations: Materials, Devices and Processes,* 6 November 2019.

[13] B. McPherson, M. Cather, R. Middleton, T. Chidsey, J. Heath, M. Saunders and S.-Y. Lee, "CarbonSAFE Rocky Mountain Phase I: Ensuring Safe Subsurface Storage of Carbon Dioxide in the Intermountain West," 31 August 2018. [Online]. Available: https://www.osti.gov/biblio/1559990. [Accessed November 2020].

[14] M. Kelley, A. Haagsma, P. Champagne, J. Glier, N. Gupta, M. H. Yugulis, J. Hawkins, J. Main, A. Pasumarti, J. Sminchak, S. Weber, B. Mannes, R.

Pardini, K. Sanders and W. Harrison, "Integrated Pre-Feasibility Assessment for a Northern Michigan Basin CarbonSAFE CO 2 Storage Complex," 2 September 2018. [Online]. Available: https://doi.org/10.2172/1469190. [Accessed November 2020].

[15] M. K. Dubois, E. T. Williams, J. C. Youle and D. E. Hedke, "Potential for CO2 storage and enhanced oil recovery in four southwest Kansas oil fields, an extended abstract; in W. L. Watney, PI, Final Report for DOE Award Number: DE-FE0002056," DOE, 2015.

[16] Y. Holubnyak, M. Dubois, T. Bidgoli, D. Wreath, L. Watney, S. Stover, D. Newell, F. Fazelalavi, A. Hollenbach, J. Jennings, C. Steincamp, J. Schremmer, B. Jordan and B. Crabtree, "Integrated CCS for Kansas (ICKan) Final Technical Report," December 31, 2018.

[17] United States Department of Energy, "Carbon Storage Atlas: 5th edition," DOE, Washington, D.C., 2015.

[18] B. Goertz-Allmann, S. Gibbons, V. Oye, R. Bauer and R. Will, "Characterization of induced seismicity patterns derived from internal structure in event clusters," *Journal of Geophysical Research: Solid Earth,* vol. 122, p. 3875–3894, 2017.

[19] R. Bauer, R. Will, S. Greenberg and S. Whittaker, "Chapter 19: Illinois Basin-Decatur project," in *Geophysics and Geosequestration*, T. M. Davis, M. Landrø and M. Wilson, Eds., Cambridge University Press, 2019, pp. 339-369.

[20] D. H. Bacon, D. Blankenau, D. Divine, A. Duguid, I. Fukai, J. Glier, J. Hawkins, M. Jimenez, R. Joeckel, S. McDonald, R. Middleton, R. Peterson and M. Scharenberg, "Integrated mid-continent stacked carbon storage hub, Phase I Final Report," 12 October 2018. [Online]. Available: https://doi.org/10.2172/1478726. [Accessed November 2020].

[21] NETL, "Large Pilot Testing of Linde-BASF Advanced Post-Combustion Carbon Dioxide Capture Technology at a Coal-Fired Power Plant," 2018. [Online]. Available: https://www.netl.doe.gov/projects/project-information.aspx?k=FE0031581. [Accessed November 2020].

[22] S. McDonald, "Illinois Industrial Carbon Capture & Storage Project: Eliminating CO2 Emissions from the Production of Bio Fuels - A 'Green' Carbon Process," 11 July 2017. [Online]. Available: https://www.energy.gov/sites/prod/files/2017/10/f38/mcdonald_bioeconomy_2017.pdf. [Accessed November 2020].

[23] D. Riestenberg, R. Esposito and K. Gray, "Establishing an Early CO2 Storage Complex in Kemper County, Mississippi: Project ECO2S (FE0029465)," 28 August 2019. [Online]. Available: https://www.osti.gov/biblio/1617036.

[Accessed November 2020].

[24] W. Peck, K. Glazewski, D. Nakles, N. Azzolina, C. Crocker, R. Klapperich, B. Oster, D. Daly, A. Livers-Douglas, S. Butler, S. Smith, B. Botnen, I. Feole, J. He, N. Dotzenrod, A. Salazar, S. Patil and J. Crossland, "North Dakota Integrated Carbon Storage Complex Feasibility Study," 24 March 2020. [Online]. Available: https://doi.org/10.2172/1606011. [Accessed November 2020].

[25] S. Quillinan and K. Coddington, "Integrated Commercial Carbon Capture and Storage (CCS) Prefeasibility Study at Dry Fork Station, Wyoming," 30 May 2019. [Online]. Available: https://www.osti.gov/biblio/1557442/. [Accessed November 2020].

[26] U.S. Department of the Interior, U.S. Geological Survey, "Assessment of Coal Geology, Resources, and Reserve Base in the Powder River Basin, Wyoming and Montana," February 2013. [Online]. Available: https://pubs.usgs.gov/fs/2012/3143/fs-2012-3143. [Accessed November 2020].

607371 Direct Non-Oxidative Conversion of Shale Gas to Aromatics: Dynamic Data Reconciliation, Parameter Estimation, and Dynamic Modeling of a Fixed-Bed Reactor

Abstract:

Majority of the BTX is produced from either petroleum naphtha, pyrolysis gasoline by-product, and/or coal liquids from coke-oven [1]. However, the abundant availability of small and large shale gas reserves in the United States, at easily accessible and remote locations, has significantly increased the interest in development of new reaction pathways for BTX production. For example, the Cyclar process developed by BP and UOP uses C_3-C_4 feed on a H-ZSM-5 catalyst for production of aromatics [1–3]. Other existing routes involve a multi-step process in which synthesis gas (syngas) must be first produced from the shale gas through steam reforming, partial oxidation, or autothermal reforming reaction. In these multi-step processes, the syngas generation step accounts for half of the plant capital, and a large share of the plant operating costs [4]. The carbon utilization efficiency is low (<50%), and the valuable H_2 product is also lost because CO or H_2 needs to be added to remove the oxygen when producing aromatics. Oxygen is removed in form of CO_2 or H_2O, as a result the energy penalty and capital cost increase due to the requirement for separation and disposal of CO_2.

One way to overcome these drawbacks is to directly convert methane, the main component in the shale gas, to aromatics. One of the key technologies for direct conversion of methane to aromatics is the non-oxidative methane dehydroaromatization (DHA) [5]. The DHA technology can lead to high conversion of methane with reasonable selectivity. Due to coke formation, conversion and yield from the DHA reactor keep changing with time making it difficult to develop an accurate kinetic model and estimate its parameters. Existing works in this area have assumed that the reactor operates under a pseudo-steady state condition thus ignoring the catalyst deactivation. Li et al. [6] have performed optimization of catalyst and membrane reactors for non-oxidative methane conversion process at a single temperature without considering the catalyst deactivation effect. Kaidi et al. [7] have proposed a set of 3 global reactions where the parameters of their rate model were estimated using the experimental data under assumed pseudo steady state region. Iliuta et al. [8], Fayzullaev et al. [9], and Yao et al. [10] have performed parameter estimation for methane DHA reaction system assuming steady state operation thus ignoring the catalyst deactivation effect.

In this work, reaction rate models including a deactivation model are proposed and their parameters are estimated using the in-house experimental data from a novel Mo/HZSM-5 catalyst. As the transient change in the amount and composition of the coke is not available and the experimental data do not satisfy carbon and hydrogen balances, a two-stage dynamic data reconciliation and parameter estimation problem is solved while simultaneously estimating the unknown initial state of the reactor. The kinetic model is then used to develop a multi-tubular multiscale first-principles dynamic fixed bed reactor model considering heat and mass transfer resistances. This model is used to simulate the cyclic reactor operation and techno-economic optimization.

References

[1] Niziolek AM, Onel O, Floudas CA. Production of Benzene, Toluene, and Xylenes from Natural Gas via Methanol: Process Synthesis and Global Optimization. AIChE 2016;62:1531–56.

[2] Treese SA, Pujadó PR, Jones DSJ. Handbook of petroleum processing. 2015. https://doi.org/10.1007/978-3-319-14529-7.

[3] Sweeney WA, Bryan PF. BTX Processing. Kirk-Othmer Encycl. Chem. Technol., 2007. https://doi.org/10.1002/0471238961.02202419230505.a01.pub2.

[4] Wilhelm DJ, Simbeck DR, Karp AD, Dickenson RL. Syngas production for gas-to-liquids applications: Technologies, issues and outlook. Fuel Process Technol 2001. https://doi.org/10.1016/S0378-3820(01)00140-0.

[5] Alvarez-Galvan MC, Mota N, Ojeda M, Rojas S, Navarro RM, Fierro JLG, et al. Direct methane conversion routes to chemicals and fuels. Catal Today2 2011;171:15–23.

[6] Li L, Borry RW, Iglesia E. Design and optimization of catalysts and membrane reactors for the non-oxidative conversion of methane. Chem Eng Sci 2002. https://doi.org/10.1016/S0009-2509(02)00314-7.

[7] Gao K, Yang J, Seidel-Morgenstern A, Hamel C. Methane Dehydro-Aromatization: Potential of a Mo/MCM-22 Catalyst and Hydrogene-Selective Membranes. Chemie-Ingenieur-Technik 2016. https://doi.org/10.1002/cite.201500139.

[8] Iliuta MC, Iliuta I, Grandjean BPA, Larachi F. Kinetics of methane nonoxidative aromatization over Ru-Mo/HZSM-5 catalyst. Ind Eng Chem Res 2003. https://doi.org/10.1021/ie030044r.

[9] I. Fayzullaev N. Kinetics and Mechanism of the Reaction of Catalytic Dehydroaromatization of Methane. Int J Oil, Gas Coal Eng 2017. https://doi.org/10.11648/j.ogce.20170506.11.

[10] Yao B, Chen J, Liu D, Fang D. Intrinsic kinetics of methane aromatization under non-oxidative conditions over modified Mo/HZSM-5 catalysts. J Nat Gas Chem 2008. https://doi.org/10.1016/S1003-9953(08)60027-4.

608863 Multi-Functional Composite Catalyst Materials for the Generation of Propylene from 2-Butene

Abstract:

The petrochemical industry is facing numerous challenges in addressing the increasing global demand for important commodity chemicals, such as propylene and propylene derivatives. These changes in the market demand are driven in large part by a rise in population growth and an increase in consumer buying power, particularly in places like China, India, and Latin America. In order to meet future demand, industrial efforts are focused on identifying low value or underutilized refinery streams that could be modified and upgraded in order to produce propylene 'on-purpose'. Commercial routes for 'on-purpose' production of propylene exist currently, but they rely on the cross metathesis of 2-butene with ethene, which is also an essential commodity chemical. Recently, there has been growing industrial interest in the development of heterogeneous metathesis catalysts that can produce propylene utilizing 2-butene as the starting feedstock. The reaction relies on 2-butene isomerization to 1-butene followed by cross metathesis in order to produce propylene. However, this reaction typically generates undesired heavy by-products, such as pentenes and hexenes, which can be broken using a separate reaction with a cracking catalyst. Herein, we discuss the development of a dual-functional catalyst that incorporates metathesis and cracking functionalities within a single material, eliminating the need for multiple reactors and improving the process economics. The material combines group 6 metal oxides on silica supports and CHA or MFI-type zeolite crystals. We describe a variety of processing techniques, including a continuous aerosol flow reactor and a conventional spray dryer, in order to generate catalyst materials with tunable chemical composition and physical properties, which are characterized via conventional techniques. The materials are then tested in a fixed-bed reactor with a 2-butene feed stream and are optimized for maximum conversion and selectivity towards propylene.

609935 Mesoporous Silica Supported Metal Oxide As an Efficient Oxygen Carrier for Chemical Looping Partial Oxidation of Methane

Abstract:

Chemical looping partially oxidation of methane is one the most promising techniques to generate syngas. Developing high performance oxygen carriers are the key to this process. In this research, Fe_2O_3 nanoparticles embedded in mesoporous support SBA-15 (Fe_2O_3@SBA-15) was synthesized and applied to chemical looping partial oxidation to generate syngas. For comparison, bulk Fe_2O_3 was also synthesized and characterized. Techniques including XRD, BET, TEM, SEM, and TGA were used to characterize the material before and after reaction. Results showed that Fe_2O_3@SBA-15 have enhanced stability and reactivity, with a conversion rate 70% higher than bulk Fe_2O_3. This can be attributed to tube structure of the support SBA-15, which provides large space and surface area for Fe_2O_3 nanoparticles and prevent the agglomeration of nanoparticles during reaction. In addition, selectivity was also test in TGA. Results showed that very low concentration of CO_2 was observed during reaction, indicating high syngas selectivity >99%. This designed material has the potential to generate pure syngas at relatively low temperature. The atomistic thermodynamics methods and density functional theory calculations were carried out to investigate the selectivity enhancement mechanism. It was found activation energy barrier of CH_x radicals strongly depends on the particle size, and SBA-15 support has an influence on the surface termination geometry and oxygen vacancy formation, which significantly influences the correlation between the C–H bond activation and the CH_x radical adsorption. This study provides guidance on the design of novel oxygen carriers for chemical looping partial oxidation.

597076 Systematic Ligand Design for Methane to Methanol Transformation: A Computational Approach

Abstract:

The selective functionalization of inert hydrocarbons has attracted major interest over the past decade. Terminal metal-oxygen bonds (MO) have been a common unit for many heterogeneous and molecular catalysts that insert oxygen in inert CH bonds. Here we present a systematic optimization of metal-oxide molecular catalysts for the conversion of methane to methanol. Using high-level theoretical chemistry calculations, we show that strong-field ligands stabilize the oxo form of MO (oxidative addition mechanism), and that weak-field ligands the more reactive oxyl form (radical mechanism). For example, we show that ammonia ligands deactivate the FeO^{2+} unit more than water, and that halogens can activate the inert early transition metal oxide units, such as NbO and ZrO. Specifically, the high spin oxyl states of NbO and ZrO offer lower energy barriers for both the activation of C-H and the subsequent formation of methanol. These oxyl states are stabilized over the lower energy (in case of naked ZrO) oxo-states by the attachment of halogens to the metal. Our results suggest practical rules of how to adjust the balance of oxo/oxyl character and reactivity of metal-oxygen units. More metals will be employed in the near future and the connection of the oxo/oxyl character to selective C-H activation will be investigated.

AUTHOR INDEX

AIChE
120 Wall Street, FL 23
New York, NY 10005-4020

ISBN 978-1-7138-2310-0